Bob Brant

# Build Your Own
# Electric Vehicle

## TAB Books
Division of McGraw-Hill, Inc.
Blue Ridge Summit, PA 17294-0850

FIRST EDITION
FIRST PRINTING

**Library of Congress Cataloging-in-Publication Data**

Brant, Bob.
    Build your own electric vehicle / by Bob Brant.
        p.  cm.
    Includes index.
    ISBN 0-8306-4232-3        ISBN 0-8306-4231-5 pbk.
    1. Electric vehicles—Design and construction—Amateurs' manuals.
    I. Title.
    TL220.B68  1993
    629.25'02—dc20                                        93-29185
                                                          CIP

Acquisitions Editor: Kimberly Tabor
Editorial team: Susan Bonthron, Editor
                Lori Flaherty, Managing Editor
                Joanne Slike, Executive Editor
Production team: Katherine G. Brown, Director
                Wendy L. Small, Layout
                Susan E. Hansford, Typesetting
                Kelly S. Christman, Proofreading
Design team: Jaclyn J. Boone, Designer
                Brian Allison, Associate Designer
Cover design: Holberg Design, York, Pa.
Cover photograph: Courtesy of Ford Motor Company. The Ecostar Van    TAB1
    is a Ford vehicle currently in pilot production.                 4283

# Contents

# Acknowledgments

This book is dedicated to Darwin Gross—the engineer's engineer, the scientist's scientist, author, musician, wayshower to many, and the most humble, creative genius, anyone could ever have the good fortune to meet. I had the privilege of creating this book, but credit for any future-directed ideas belongs to Darwin, who made me feel like a newspaper person must have felt around Nikola Tesla. Special credit must be extended to Jim Harris of Zero Emissions Motorcar Company, master builder and extraordinary mechanic, without whose help and two electric vehicles, this book would not have been possible. Special credit must also be extended to Ken Koch of KTA Services Inc., whose enormous expertise and superb photographic collection contributed greatly.

Credit must also be extended to the many other electric vehicle enthusiasts who made this book possible. While space does not permit listing all their names here, know that I appreciate each of your contributions and have done my best to direct readers to you or your organization in the appropriate chapter.

Warm thanks and appreciation to my wife, Bonnie Brant, for challenging the ideas and proofreading, ensuring my written expression fell somewhat within the confines of the English language; and to my mother, Mary Brant, for first sparking my interest in books, writing, and life.

# Preface

This book owes its shape to Stacey Pomeroy, the How-To Acquisitions Editor at Tab/McGraw-Hill. Stacey and the marketing team reasoned that, beyond actually building an electric vehicle, it was also important to address the broader issues and expand on the past, present, and future of electric vehicle technology.

## MY INTENTIONS ARE HONORABLE

My intentions in writing this book were to create a useful guide to get you started; to encourage you to contact additional sources in your own "try-before-buy" quest, to point you in the direction of the people who have already done it (i.e., electric vehicle associations, consultants, builders, suppliers and integrators); to familiarize you with the electric vehicle components; and finally, to go through the process of actually building/converting your own. My intention was never to make a final statement, only to whet your appetite for electric vehicle possibilities. I hope you have as much fun reading about the issues, sources, parts, and building process as I have had.

## THE EV'S TIME HAS COME

They say that nothing is stronger than an idea whose time has come. If so, welcome the wave of renewed interest in electric vehicles that promises to wash over all previous fears, doubts, and myths, and restore EVs to their rightful place in the transportation hierarchy.

I hope this book leaves you with the firm conviction that EVs are not only sexy and fun but also highly useful. All automobile owners on the planet today can benefit both themselves and the environment by using electric vehicles in some form. And the benefits to yourself, the environment, and your children's children will only multiply as other people also wake up and catch on.

## WHO I AM

I've been a GM man all my life—with a few Chrysler products thrown in, plus a Volkswagen and an Audi—until I met my present 1992 plum-colored Ford Thunderbird at the January auto expo in Portland, Oregon, and became a believer. Along the way, I've converted a diesel to gas, the Volkswagen to a supercharged dune buggy, and done my share of tinkering with engines, trannies, exhaust systems, suspensions and paint jobs. But my first love has always been electrical.

I grew up in New York City, rode its electric-powered subway trains at an early age, and played with my own electric train sets. Later on, I got my BSEE and worked on NASA projects such as Apollo, Lunar Excursion Module, Earth Resources Technology Satellite, and others for a large aerospace firm. Still later, my frame of reference was further expanded by credits toward MSEE and MBA degrees, and by working with companies/agencies that did the Lunar Rover, maglev, telemetry, and fly-by-wire projects.

My involvement in electric vehicles stems from the time my graduating engineering class at the University of Denver built a self-propelled electric robot for an "Engineer's Day" function.Every electric vehicle breakthrough that's happened from that time through my present occupation as a computer consultant has only made me more fascinated with the concept, more convinced of its substantial personal and environmental benefits, and more curious why stronger steps have not been taken to make electric vehicles a reality. So I had to write this book.

# Introduction

**W**hy should you plunk down your hard-earned money for an electric vehicle? What's in it for you? Besides, if it was such a good idea, wouldn't the government and automobile industry have made millions of them by now? And isn't our government taking steps to make sure there will be no or minimal pollution from internal combustion engine-powered vehicles? Who needs electric vehicles?

On the other hand, what lies behind the renewed interest in electric vehicles? Increasingly people seem to be buying, driving, and owning them. You can hardly pick up a newspaper or magazine or watch television without hearing of them. Nor is the interest limited to the United States—it's international.

Why are environmental issues being raised by more groups, in more locations, than ever before? Why are air-quality issues of particular concern to elected representatives and government officials of virtually all urban areas throughout the world? And why are Zero Emission Vehicles (ZEV) inexorably linked to air-quality issues as at least part of the solution?

Finally, why is every major automobile manufacturer on the planet doing electric vehicle research, and why has the U.S. Department of Energy's electric vehicle research budget multiplied more than fivefold from fiscal year '89 to '93? *Build Your Own Electric Vehicle* answers these and other questions.

This book can help you in three ways. First, it's designed to give you the historical background and bring you up to speed on EVs, including why the new ones are better than those made 150 years ago (or even 10 years ago). Second, it will enable you to better participate in electric vehicle discussions by increasing your knowledge of their components and the issues involved in building or purchasing one. Third, it provides a fairly detailed plan for converting an EV yourself, and directs you to a number of suppliers for any parts you might need.

Although the title says *build* you are actually going to *convert* your own electric vehicle from an existing internal combustion vehicle chassis. The details of why this is the most cost-effective and easiest method are covered in the book. You won't have to build anything from scratch. Instructions, photographs, and illustrations assist you in converting your electric vehicle and getting it running in the shortest possible time.

## A UNIVERSAL APPROACH

You have probably heard many times: "Give a man a loaf of bread and he can feed his family for a day. Give him the grain and show him how to plant and harvest and he can feed his family for a lifetime." This book builds on the same principle.

My goal is to give you the tools and a foundation you can build upon and use over and over again. Think of it as if you were building a house. I am going to give you the tools and plans and show you the process. The exact house you build is up to you.

What you learn from this book will make it easier than ever for you to buy, convert, or build your own electric vehicle at the best possible price. But technology is changing rapidly, so consider the products and prices you read about here only as a starting frame of reference rather than the last word. You must do your own shopping to get the best products and prices at the time you buy, convert, or build.

Keep your eyes, ears, and mind open. New ideas are constantly being brought forth. If you look hard enough, chances are, someone has already done exactly what you wanted to do—maybe even better—and saved you all the hard work!

## CHAPTER OVERVIEW

*Build Your Own Electric Vehicle*'s 12 chapters are organized into three sections: philosophy, options, and process. By the time you finish it, you will be armed with information, ready to charge into buying, converting, or building any electric vehicle with confidence, enthusiasm, and zest. Chapters 1 through 5 cover electric vehicle philosophy issues; chapters 6 through 10 describe the options available to you, and chapters 11 and 12 cover the process of conversion and use. Here's the chapter lineup in more detail:

Chapter 1 describes what electric vehicles are, why they're efficient, and their tremendous advantages. It highlights the change in consciousness responsible for the upsurge in EV interest; describes why buying, converting, or building an EV should be of interest to you; dispels EV myths; and describes how future trends and technology will make EVs even better.

Chapter 2 describes how electric vehicles save the environment, reduce dependence on foreign oil, and produce no toxic byproducts, and how technology will improve them.

Chapter 3 covers past, present, and future electric vehicle history. EVs have been around for more than 100 years but improved technology makes today's models better than those of even a decade ago.

Chapter 4 describes how to pick the best electric vehicle for you, including purchase decisions (buy, build, or convert); conversion decisions (van, car, or pickup); conversion trade-offs (speed versus range; cost).

Chapter 5 lists additional electric vehicle sources, including clubs and associations; manufacturers and suppliers; converters and consultants; and books, articles and papers.

Chapter 6 discusses chassis for your electric vehicle conversion: Optimizing weight, aerodynamic drag, rolling resistance, and drivetrain trade-offs; designing for range versus speed; and load and component trade-offs.

Chapter 7 discusses motors for your electric vehicle conversion: How they work and definitions; dc versus ac characteristics; today's and tomorrow's best motor solutions.

Chapter 8 discusses controllers for your electric vehicle conversion: Solid-state vs. simple switch; SCRs, IC chips, and MOSFET and IGBT power devices; dc versus ac; today's and tomorrow's best controller solutions.

Chapter 9 discusses batteries for your electric vehicle conversion: How they work; their real world characteristics; how to buy, care for, and maintain them; today's and tomorrow's best battery solutions.

Chapter 10 discusses battery chargers and electrical systems for your electric vehicle conversion: The charging cycle and ideal and real world chargers; build-your-own, offboard, and onboard charger trade-offs; tomorrow's best charger solution; high-current, low-voltage, and charging systems; and system components and wiring.

Chapter 11 covers step-by-step conversion of a Ford Ranger pickup into an electric vehicle, including preconversion steps; conversion chassis, mechanical, electrical and battery steps; and postconversion testing and finishing steps.

Chapter 12 describes how to maximize your electric vehicle enjoyment, including licensing, insuring, driving, and maintenance steps.

# 1
# Electric vehicles are right for today

*What is desirable and right is never impossible.*

Henry Ford (inscribed on plaque in Ford Fairlane Mansion)

Why buy, convert, or build an electric vehicle today? Easy. Electric vehicles are fun to drive, economical to operate, and save you a bundle when you buy, convert, or build "smart." They require no periodic tuneups, filter replacements, or SMOG certification, and virtually no maintenance. They are zero-emission vehicles, odorless and virtually noiseless. With no volatile fuels aboard, they are inherently safer. An electric motor will outlive an internal combustion engine several times over, and consume far less energy to do the same amount of work.

In very practical terms, the 1993 Ford Ranger pickup electric vehicle conversion shown in FIG. 1-1 (which you'll learn how to convert in chapter 11) goes 75 mph, gets 60 miles (or better) on a charge, uses conventional lead-acid batteries and off-the-shelf components, and can be put together by almost anyone. Its batteries cost about $1200 and last about 3 years, its conversion parts cost about $5000, and it costs 2.2 cents per mile to operate (versus about 6.25 cents per mile for its internal combustion version). Its maintenance cost is negligible compared to the additional 2.5 cents per mile (conservative estimate) required for its internal combustion engine version.

As a result of your wise decision to own an electric vehicle, you help save planet Earth, conserve scarce nonrenewable energy sources, make the atmosphere cleaner for you and your neighbors today, and take steps toward assuring a legacy of clean air for your children and their heirs in the future. What more could you ask for?

In this chapter you'll learn what an electric vehicle is, and the change in consciousness responsible for the upsurge in interest surrounding it. You'll discover the truths and untruths behind electric vehicle myths. You'll also learn about the EV's advantages, and why its benefits—assisted by technological improvements—will continue to increase in the future.

**1-1** It's not just a Ford Ranger pickup truck—it's the converted electric vehicle you'll meet in chapter 11.

## ELECTRIC VEHICLE vs.
## INTERNAL COMBUSTION ENGINE VEHICLE

To really appreciate an electric vehicle, it's best to start with a look at the internal combustion engine vehicle. The difference between the two is a study in contrasts.

Mankind's continued fascination with the internal combustion engine vehicle is an enigma. The internal combustion engine is a device that inherently tries to destroy itself: numerous explosions drive its pistons to turn a shaft. A shaft rotating at 6,000 revolutions/minute produces 100 explosions every second. These explosions in turn require a massive vessel to contain them—typically a cast iron *cylinder block*. Additional systems are necessary for:

- Cooling—to keep the temperatures within a safe operating range.
- Exhaust—to remove the heated waste products safely.
- Ignition—to initiate the explosions at the right moment.
- Fueling—to introduce the proper mixture of air and gas for explosion.
- Lubricating—to reduce wear on high-temperature, rapidly moving parts.
- Starting—to get the whole cycle going.

It's complicated to keep all these working together. Complexity means more things can go wrong (more frequent repairs, and higher repair cost). Figure 1-2 summarizes the internal combustion engine vehicle systems.

Unfortunately, the internal combustion engine vehicle's legacy of destruction doesn't just stop with itself. The internal combustion engine is a variant of the generic combustion process. To light a match, you use oxygen (O) from the air to burn a carbon-based fuel (wood or cardboard matchstick), generate carbon dioxide ($CO_2$), emit toxic waste gases (you can see the smoke and perhaps smell the sulfur), and leave a solid waste (burnt matchstick). The volume of air around you is far greater than that consumed by the match; air currents soon dissipate the smoke and smell, and you toss the matchstick.

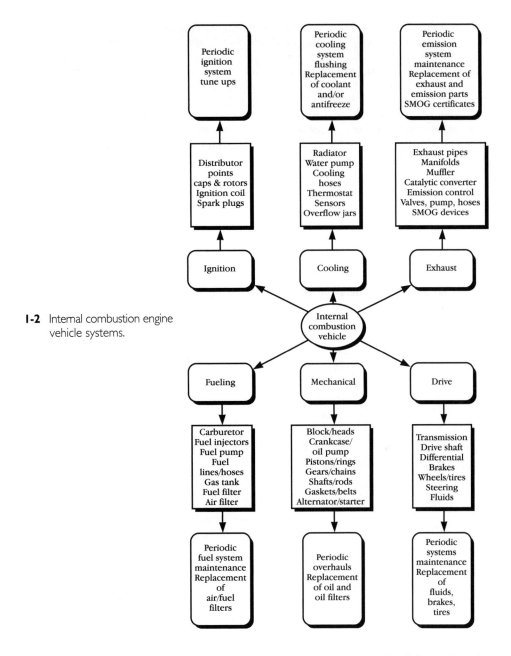

**I-2** Internal combustion engine vehicle systems.

Today's internal combustion engines represent its most evolved form, but the byproduct of its carbon-burning combustion process—like that of the match—always yields heat and pollution. Everything going into and coming out of the internal combustion engine is toxic, and it's still one of the least efficient mechanical devices on the planet. Unlike lighting a single match, the use of hundreds of millions of internal combustion engine vehicles threatens to destroy all life on planet Earth. You'll read about environmental problems caused by internal combustion engine vehicles in chapter 2.

In contrast to the hundreds of internal combustion engine moving parts, the electric motor has just one. That's why electric vehicles are so efficient. To make an electric vehicle out of the car, pickup, or van you are driving now, all you do is take out the internal combustion engine along with all related ignition, cooling, fueling, and exhaust system parts, and add an electric motor, batteries, and a controller. Hey, it doesn't get any simpler than this!

Figure 1-3 shows all there is to it: Batteries and a charger are your "fueling" system, an electric motor and controller are your "electrical" system, and the "drive" system was as before (although today's advanced electric vehicle designs don't even need the transmission and drive shaft).

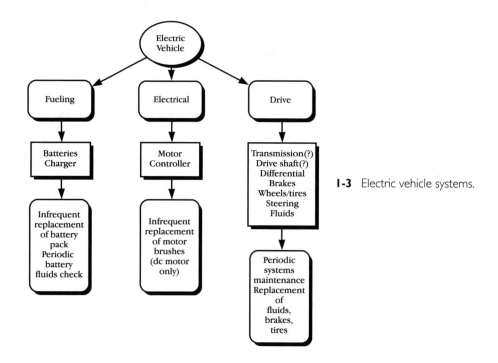

1-3 Electric vehicle systems.

A simple diagram of an electric vehicle looks like a simple diagram of a portable electric shaver: a battery, a motor, and a controller or switch that adjusts the flow of electricity to the motor to control its speed. That's it. Nothing comes out of your electric shaver and nothing comes out of your electric vehicle. EVs are simple (therefore highly reliable), emit zero pollutants, have lifetimes measured in millions of miles, need no periodic replaceables (filters, etc.) or tuneups, and cost significantly less per mile to operate. They are highly flexible as well, using electric energy readily available anywhere as input fuel.

In addition to all these benefits, if you buy, build, or convert your electric vehicle from an internal combustion engine vehicle chassis as suggested in this book, you perform a double service for the environment: You *remove* one polluting internal combustion vehicle from service and *add* one nonpolluting electric vehicle to service.

You've had a quick tour and side-by-side comparison of electric vehicles and internal combustion engine vehicles. Now let's take a closer look at electric vehicles.

## WHAT AN ELECTRIC VEHICLE IS

An electric vehicle consists of a *battery* that provides energy, an electric *motor* that drives the wheels, and a *controller* that regulates the energy flow to the motor per your instructions. Figure 1-4 shows all there is to it—but don't be fooled by its simplicity. Scientists, engineers, and inventors down through the ages have always said, "in simplicity there is elegance." Let's find out why the electric vehicle concept is elegant.

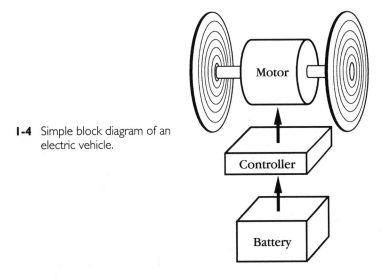

**1-4**  Simple block diagram of an electric vehicle.

### Electric motors are ubiquitous

Electric motors can be found in so many sizes, places, and uses that we tend to take them for granted. Universal in application, they can be as big as a house or smaller than your fingernail, and can be powered from any source of electricity. In fact, they are so reliable, quiet, and inexpensive that we tend to overlook just how all-pervasive and influential they are in virtually every civilized person's life.

Each of us encounters dozens, if not hundreds, of electric motors daily without even thinking about them: The *alarm clock* wakes you; you remove coffee beans from *refrigerator*, and put them in a *grinder*, in the bathroom you use electric *shaver, toothbrush,* or *hair dryer*; breakfast might be assisted by your electric *juicer, blender* or *food mixer*; you might clean your home with your *vacuum cleaner* or clean your clothes with your *washer* and *dryer*; next into your *automobile, subway, bus,* or *light rail transit* to ride to work, where you might go through an *automatic gate* or *door* or take an *elevator* or *escalator* to your floor; here you sit down at your *personal computer*, and use your *fax* or *copier* after you adjust the *fan, heater,* or *air conditioner*. Back at home in the evening, you might use an electric *garage door opener*, program your *VCR* or use an electric *power tool* on a project. On and on, you get the picture.

Why are electric motors ubiquitous? In one word—convenience. Electric motors do work so that you don't have to. Whether it's pulling, pushing, lifting, stirring, or oscillating, the electric motor converts electrical energy into motion, which is further adapted to do useful work.

What is the secret of the electric motor's widespread use? Reliability. This is because of its simplicity. Regardless of type, all electric motors have only two basic components: a rotor (the moving part) and a stator (the stationary part). Only one moving part. If you design, manufacture and use an electric motor correctly it is virtually impervious to failure and indestructible in use.

In internal combustion automobiles, in addition to your all-important electric starter motor, you typically find electric motors in the passenger compartment heating/cooling system, radiator fan, windshield wipers, electric seats, windows, door locks, trunk latch, outside rear view mirrors, outside radio antenna and more.

## Batteries are ubiquitous

No matter where you go, you cannot get away from batteries either. They're in your pocket tape recorder, portable radio, telephone, laptop computer, or portable power tool, appliance, game, flashlight, and many more devices. Batteries come in two distinct flavors: rechargeable and nonrechargeable. Like motors, they come in all sorts of sizes, shapes, weights, and capacities. Unlike motors, they have no moving parts. The nonrechargeable batteries you simply dispose of when they are out of juice; rechargeable batteries you connect to a recharger or source of electric power to build them up to capacity. Batteries of the rechargeable lead-acid type used in your car manage the recharging process invisibly via an under-the-hood generator or alternator that recharges the battery while you're driving.

Why are batteries ubiquitous? In a word—convenience. The battery, in conjunction with the starter motor, serves the all-important function of starting the automobile powered by the conventional internal combustion engine. In fact, it was the battery and electric starter motor combination, first introduced in the early 1920s and changed very little since then, that put the internal combustion engine car on the map—it made cars easy to start and easy to use for anyone, anywhere.

Rechargeable lead-acid automotive batteries perform their job very reliably over a wide range of temperature extremes and, if kept properly charged, will maintain their efficiency and deliver stable output characteristics over a relatively long period of time—several years. A lead-acid automotive battery is unlikely to fail unless you shock it, drop it, discharge it completely, or allow a cell to go dry. The only maintenance required in lead-acid batteries is checking each cell's electrolyte level and periodically refilling them with water. Newer sealed batteries require no maintenance at all.

Controllers have become more intelligent, too. The same technology that reduced computers from room-sized to desk-sized allows you to exercise precise control over a battery-powered motor. Regardless of the voltage source, current needs, or motor type, today's controllers—built with reliable solid state components—can be designed to meet virtually any need and can easily be made compact in size to fit conveniently under the hood of your car.

Why are electric vehicles elegant? When you join a reliable motor, battery, and controller together, you get an electric vehicle that is both reliable and convenient. Perhaps the best analogy is that you are now driving your entire car from an oversized electric starter motor, a more powerful set of rechargeable lead-acid batteries, and a very sophisticated starter switch. But it's only going to get better. Today's electric vehicles resemble your battery-operated electric shaver, portable power tool, or kitchen appliance. Tomorrow's electric vehicles will more closely resemble your portable laptop computer.

## HAVE YOU DRIVEN AN ELECTRIC VEHICLE LATELY?

Clean Air initiatives notwithstanding, nobody ever bought, built, or converted an electric vehicle because they wanted to save planet Earth. It might be reason #3, #4, or #5 but it certainly isn't reason #1. Let's look at the reasons why people do get into electric vehicles.

**People tell their friends about their electric vehicle experience**  Word of mouth and personal experience make a difference. The cumulative effect of numerous people attending EV symposiums, rallies, and Electric Automobile Association meetings all over the world—and experiencing first hand what it's like to ride or drive one—has gradually done the job. Almost universally, people enjoy their consciousness-raising electric vehicle experience, are impressed by it, and tell a friend. That's the real reason for the resurgence in interest in electric vehicles.

**Electric vehicles are fun to drive**  People are first and foremost practical, but also influenced by their emotions. Electric vehicles are also first and foremost practical—but also fun to own and drive. Owners say they become downright addictive. Tooling around in breezy electric vehicle silence gives you all the pleasure sailboat skippers enjoy while cruising down the highway. If you enjoy sailing, you'll probably enjoy driving an electric vehicle, because the sensation is similar. But you don't need any wind. Step down on the throttle and get pushed back into your seat whenever you like.

**Electric vehicles are fun to own**  Electric vehicles are a great way to meet people and make new friends in the 1990s. Whether you own a Solar Electric Destiny 2000 (seen in the movie *Naked Gun 2½*), converted Porsche 914, or Ford Ranger Pickup Truck, or built-from-scratch chassis with custom kitcar body, your electric vehicle is a sexy, quiet, technologically spiffy show-stopper.

Believe it, if the words *electric vehicle* appear prominently on the outside of your car, pickup, or van, you will not want for instant friends at any stoplight, shopping center, gas station (where you go only to put air into your tires!) or just playing stop and go in freeway traffic. Or you can park it in a conspicuous spot, lift the hood, and wait for the first passerby to ask, "where's the engine?" as FIG. 1-5 suggests.

There's a pride of ownership and an accumulation of knowledge that comes in handy at shows and demonstrations, but plan on having plenty of literature always available so you can keep your "electric vehicle discourses" to under five minutes in length. On the other hand, if you just want to meet people, make the letters on your sign real big and you will never want for company.

The first owners of anything new always have an aura of prestige and mystique about them. You will be instantly coronated in your own neighborhood. You are driving what others have only talked about. While everyone will own one in the future, you are driving an electric vehicle today. When TV sets were introduced in the 1950s, the whole neighborhood crowded into the house with the first tiny black & white TV screen. Expect the same with your electric vehicle project. But in the practical 1990s, as the first owner in your neighborhood, you can charge "consulting" fees to the neighbors that later drop by for building advice (nah, you wouldn't do that, would you?).

**Electric vehicle conversions are inexpensive to own and operate**  All this emotional stuff is nice, but let's talk out-of-pocket dollars. Ask any electric vehicle conversion owner, and they'll tell you it transports them where they want to go, is very

**1-5** "Hey, where's the engine?"

reliable, and saves them money. Let's examine separately the operating, purchase, and lifetime ownership costs and summarize the potential savings.

*Operating costs.* Electric vehicles only consume electricity. In between charge-ups, there are no other consumables to worry about except an occasional watering of the batteries. These figures are covered in more detail later, but the Ford Ranger electric vehicle pickup conversion of chapter 11 averages about 0.44 kWh (kilowatt-hours, a measure of energy consumption) per mile. At $0.05 per kWh for electricity (check your electric utility monthly statement for the prevailing rate in your area) that translates to:

$$0.44 \text{ kWh/mile} \times \$.05/\text{kWh} = .022 \text{ (2.2 cents) per mile}$$

Let's compare these costs with the EV's gasoline-powered internal combustion engine counterpart in a pickup chassis. The latter consumes gasoline; its ignition, cooling, fueling, and exhaust systems require filters, fluids, and periodic maintenance. The gasoline-powered pickup chassis (equivalent to the previous example) averages 20 miles per gallon or 0.05 gallons per mile. At $1.25 per gallon for gas that translates to:

$$0.05 \text{ gallons/mile} \times \$1.25/\text{gallon} = .0625 \text{ (6.25 cents) per mile}$$

Consumables and periodic maintenance must still be added. Assuming these cost $30 per month, and annual mileage is 12,000 miles per year, this translates to:

$$\$360/\text{year} \div 12,000 \text{ miles/year} = .03 \text{ (3.0 cents) per mile}$$

Adding the two figures together you're looking at 9.25 cents per mile operating cost for a gasoline-powered vehicle versus 2.2 cents per mile for its electric vehicle equivalent—more than a four-to-one cost difference favoring the electric vehicle.

While your average EV conversion—made with off-the-shelf components—might consume about 0.4 kWh per mile, General Motor's Impact electric vehicle is rated at 0.1 kWh per mile (0.07 kWh/km). This drops your electric vehicle operating costs to 0.5 cents per mile!

*Purchase costs.* Commercially manufactured electric vehicles are prohibitively expensive today—if you can find one at all. Tomorrow's electric vehicle costs will obviously drop to become equal to or less than internal combustion powered vehicles as more units are made (and manufacturing economies-of-scale come into play) because they have far fewer (and much simpler) parts.

But this book advocates the conversion alternative—you convert an existing internal combustion engine vehicle to an electric vehicle. You remove the internal combustion engine and all systems that go with it, and add an electric motor, controller, and batteries. If you start with a used internal combustion engine vehicle chassis you can save even more (with the advantage of having the drive train components already broken in, as later chapters point out). In round numbers, a typical electric vehicle conversion should cost you:

Electric motor, controller, and conversion parts cost   = $5000
Cost of a 120-volt battery pack (using 6-volt batteries) = $1200
Total electric vehicle conversion cost             = $6200

To this must be added the cost of your internal combustion engine vehicle chassis. If you start with a brand new vehicle, this could mean $10,000 or more (less any credit for removed internal combustion engine components). A good, used chassis might cost you $2000 to $3000 (or less if you take advantage of special situations as mentioned in chapter 6). So your total purchase costs are in the $8000 to $16,000 ballpark—a fraction of the "big-three" $100,000 electric vehicle costs and substantially less than already converted electric vehicles. Obviously, you can do better if you carefully buy and scrounge for parts. Equally obviously, you can also spend more if you elect to have someone else do the conversion labor, decide you must have a brand new Ferrari Testarosa chassis, or elect to build a Kevlar-bodied roadster with titanium frame from scratch.

This book only advocates an EV conversion as your second vehicle, and assumes you already own vehicle number one. As a second vehicle choice, logic (and Parkinson's law) dictates that the money spent for this decision will expand to fill the budget available—regardless of whether an internal combustion engine vehicle or EV is chosen. So second vehicle purchase costs for an internal combustion or electric vehicle are a wash—they are identical.

*Lifetime ownership costs.* Let's look first at the conversion components. A typical electric motor is rated at 100,000 hours. At 20 miles per hour, an electric motor-powered electric vehicle could go 2 million miles; at 50 miles per hour, it could go 5 million miles! No internal combustion engine even comes close to this.

A controller is by definition solid-state. There are no moving parts and nothing to wear out. If you don't cook, freeze, or drop it—the same factors that also spell death to lead-acid batteries—it's going to last a long time with no maintenance whatsoever. Other conversion components—contactors, relays, switches, meters, and wire—are similarly long-lived. In short, the electrical conversion components have enormously long lifetimes. Now let's look at the batteries.

Deep-cycle lead-acid batteries for electric vehicle use are rated at hundreds of deep discharge duty cycles (as you'll learn more about in the battery chapter). A leading West Coast electric vehicle and conversion shop dealer I talked to said that after three years of operation, they have yet to have a set of batteries come back for replacement.

The 120-volt operating voltage recommended by this book requires 20 of the 6-volt lead-acid batteries at about $60 each, or $1200 total. If a three-year lifetime is assumed at 12,000 miles per year (36,000 miles overall), battery replacement costs are:

$$\$1200/3 \text{ years} \div 36000 \text{ miles}/3 \text{ years} = .033 \text{ (3.3 cents) per mile}$$

So battery replacement costs every three years in an electric vehicle are roughly the same as monthly consumables and periodic maintenance costs in an internal combustion engine vehicle. When you factor in licensing costs (usually the same) and insuring costs (usually lower for electric vehicles—because of their higher safety rating), ownership costs for an internal combustion or electric vehicle are a wash—they are identical.

So what's the bottom line? The electric vehicle conversion costs $6200 (this might be reduced by zero-emission tax credits in your area), ownership costs are a wash, and you save on operating costs every time you use it—2.2 cents per mile (this might be reduced by lower electric utility charging rates in your area) for the electric vehicle versus 6.25 cents for the internal combustion engine vehicle.

What's the real bottom line? Your second vehicle purchase costs for an internal combustion or electric vehicle conversion are in reality identical, but lifetimes of all the electric vehicle's major components are measured in years, it has negligible maintenance costs, and your additional $1200 battery outlay only occurs every three years or so. All you do is plug it into your household 120/volt ac electrical system to recharge it. And it only costs 2.2 cents per mile to operate—a daily 60 mile round trip costs $1.32.

An electric vehicle is really more like your vacuum cleaner, washing machine, refrigerator or other highly reliable electrical appliance—you might get tired of it long before it wears out. But the economies of electric vehicles will only get better with time. And you can postpone your electric vehicle conversion's obsolescence indefinitely by doing modular upgrades to it using the latest technology components.

**Electric vehicles are safer for you**  EVs are a boon for safety-minded individuals. Electric vehicles are called ZEVs (zero emission vehicles) because they emit nothing—whether they are moving or stopped. In fact when stopped, electric vehicle motors are not running and use no energy at all.

This is in direct contrast to internal combustion engine powered vehicles that not only consume fuel but also do their best polluting when stopped and idling in traffic. EVs are obviously the ideal solution for minimizing pollution and energy waste on congested stop-and-go commuting highways all over the world, but this section is about save yourself: as an electric vehicle owner, you are not going to be choking on your own exhaust fumes.

It gets better: EVs carry no combustible fuels, 20,000-volt spark plug ignition circuits, hot exhaust manifolds, catalytic converters, or hot radiators on board. Those who enjoy gas tank explosions (caused by accidents—as seen in the movies), engine compartment fires (caused by ignition or hot manifolds—as seen by the side of the road), hot radiator coolant explosions (caused by improper radiator cap removal—many of us have experienced this firsthand) or starting forest fires (caused by hot catalytic converters parked over dry grass) simply have to go elsewhere for their entertainment.

Even better, you save wear and tear on yourself by not having to do needless chores associated with vehicle ownership. Contrast the numerous periodic internal combustion engine vehicle activities shown in FIG. 1-2 with the far simpler electric vehicle requirements shown in FIG. 1-3.

In the "yuk" but not really dangerous category—electric vehicle owners don't have to mess with oil (no dark slippery spots on your garage floor), antifreeze (no lighter slippery spots on your garage floor) or filters (the kind you hold in a rag far away from your body because they are filthy or gunky).

For critics who comment that lead-acid batteries emit potentially dangerous hydrogen gas when charging, and point out that electric vehicles have multiple batteries: When is the last time you heard of a death or injury resulting from charging a battery? It's possible but very unlikely.

How about the acid part in lead-acid batteries? You'll learn about battery details in chapter 9, but the acid is diluted sulfuric acid. It definitely hurts if it spills on you or anything else, but it doesn't explode or catch fire and can readily be counteracted by flushing with water.

In addition, this book advocates electric vehicle conversions built upon already safety-approved and tested internal combustion engine chassis. Anyone wincing at the thought of crashing into your internal combustion vehicle and its 400-pound cast-iron engine block has to think equally seriously about meeting your electric vehicle and its copper and steel 150-pound electric motor or its 1200-pound lead-acid battery pack.

**Electric vehicles are highly adaptable** EVs are found on mountain tops (railway trams, cable cars), at the bottom of the sea (submarines, Titanic explorer), on the moon (Lunar Rover), in tall buildings (elevators), in cities (subways, light rail, buses, delivery vehicles), hauling heavy rail freight or moving rail passengers fast (Pennsylvania Railroad Washington to New York corridor). Are they all electric vehicles? Yes. Do they run on rails or in shafts or on tethers or with nonrechargeable batteries? Yes. EVs were designed to do whatever was wanted in the past and can be designed and refined to do whatever is needed in the future. What do you need: big, small, powerful, fast, ultra-efficient? Design to meet that need. General Motor's Sunraycer and Impact electric vehicles are excellent examples of what can be done when starting with a clean sheet of paper. Closer to home and the subject of this book, do you want an EV car, pickup, or van? You decide.

Electric vehicles are easily and infinitely adaptable. Want more acceleration? Put in a bigger electric motor. Want greater range? Choose a better power-to-weight design. Want more speed? Pay attention to your design's aerodynamics, weight and power.

When you buy, convert, or build an EV today, all these choices and more are yours to make because there are no standards and few restrictions. The primary restrictions regard safety (you want to be covered in this area anyway), and are taken care of by using an existing internal combustion engine automotive chassis that has already been safety qualified. Other safety standards to be used when buying, mounting, using, and servicing your EV conversion components are discussed later in this book.

**Electric vehicles save the environment** EV ownership is visible proof of your contribution to help clean up the environment. Chapter 2 will cover in detail the environmental benefits of this choice. EVs produce no emissions of any kind to harm the air, and virtually everything in them is recyclable. Plus every electric vehicle conversion represents one less polluting internal combustion vehicle in circulation. Electric vehicles are not only the most modern and efficient transportation on earth—they help save the planet!

Until you've built, owned, driven, or ridden in an electric vehicle, you just don't get it. You don't understand what all the fuss is about. You really haven't a clue as to what I'm talking about here. But once you try, for most people there is no turning back. So before making any decision about electric vehicles, make a decision to experience one for yourself, firsthand. Chapter 5 tells you all about how to make the connection.

## THE BIG PLAYERS ARE NOW IN THE ACT

You've just learned why people want and are getting into electric vehicles. That's the demand side. The other side of the equation is the supply side. Let's look at the reasons why the big players—major automobile manufacturers, and city, state and national governments—are getting into electric vehicles.

### Timeout for a historical moment

Before getting into an area that will make EVs sound like something new, let's delve briefly into a subject that chapter 3 will cover in depth—electric vehicle history. Here are some facts for you. Because the electric motor came before the internal combustion engine, electric vehicles have been around since the mid-1800s, were manufactured in volume in the late 1800s and early 1900s, and declined only with the emergence and ready availability of cheap gasoline.

Even so, electric vehicle offshoots—tracked buses, trolleys, subways, and trains—have continued to serve in mass transit capacities right up until the present day because of their greater reliability and efficiency. All the diesel locomotives that eventually replaced every railroad's steam locomotives are actually diesel-electric—the electric motors drive the wheels (pure electric locomotives replaced any steam locomotives the diesel missed).

As mentioned earlier, the battery and electric motor combination borrowed from the electric vehicles of that era and applied as a starter motor for internal combustion engines was responsible for the great upsurge in the internal combustion engine vehicle's popularity. The starter motor systems employed in all of today's internal combustion engine vehicles are virtually unchanged from the original early 1920s concept.

### Los Angeles area environment forces action

You can go weeks without food and days without water, but only minutes without air. When inhabitants of the world's largest cities can see, smell, and taste their air but have trouble breathing it, it gets everyone's attention right away.

It is only logical that Los Angeles, with one of the world's largest internal combustion vehicle populations and biggest air pollution problems, take a leadership role. Yet even so threatened, the courage it took for individual members of the bureaucracy to move against the status quo was enormous. You'll read about the details of the Los Angeles Clean Air Initiative in chapter 2's environmental section. A place in environmental heaven has already been reserved for Los Angeles City Commissioner Marvin Braude, California Congressional Representatives George Brown and Howard Berman, and other "make it happen" individuals in the Los Angeles City Government, the California Air Resources Board (CARB), and the Los Angeles Department of Water and Power (LADWP).

In a nutshell, CARB is driving the solution to their pollution problem by restricting the types of new vehicles that can be sold in California, defining their characteristics, and allowing the manufacturers a well-defined time frame in which to comply.

Needless to say, EVs play a significant role. Under the CARB plan the exact percentage of how many vehicles sold in the state must be electric is mandated from 1998 on. The CARB plan makes so much sense, that other states from the equally polluted Northeast corridor of the U.S. (all the New England States, plus New York, New Jersey, Pennsylvania, Delaware, Maryland, DC and Virginia) have also adopted it. In addition, Florida, Texas, Oklahoma, and Arizona have adopted only slightly less restrictive versions of it.

Underscoring California's interest in capturing a leadership role on the road to electric vehicles was the creation of Calstart, a 41-member public/private research and development consortium whose membership comprises major utilities, large and small aerospace and high-tech companies, and educational and research institutions. Calstart's overall goal is to develop advanced technology electric vehicle subsystems that California manufacturers can build for the world's automakers—i.e., to turn aerospace technology into California jobs. Interim goals include developing propulsion systems, infrastructure and testing programs, prototype high-technology showcase electric vehicles, and funding mechanisms to jumpstart university research and involve small California businesses.[1]

## Large auto manufacturers jump on the bandwagon

Every manufacturer's dream is to be the sole supplier of a product for which there is unlimited demand—such as the hula hoop fad of several decades ago. Every manufacturer's nightmare is be stuck with a huge inventory that nobody wants—such as when the aforementioned hula hoop fad ended. That's why in the capital intensive automobile manufacturing business, they've elevated the science of monitoring customer perceptions and reactions into an art form. A curve in a fender or a new line in a trunk deck lid just will not be there if it meets with an adverse reaction in preproduction public opinion surveys.

While this approach guarantees that large automobile manufacturers will never be fad leaders, it also guarantees that they will never suffer disastrous downside losses, and that they will build Mustangs instead of Edsels. (GM's recent financial debacle is an organizational and capital equipment obsolescence problem rather than a design problem.)

That's why EVs have been a chicken-or-egg question for large automobile manufacturers. It's not their fault for bringing up the rear. As Ford's Roberta Nichols said in response to L.A.'s Clean Air Initiative, "We really try to do what the customer wants and, until recently, the customers haven't been asking for electric vehicles."

Large automobile manufacturers typically aren't known for their breakthroughs. New technology has to be introduced carefully so it doesn't cannibalize existing model sales. But, occasionally, they take part in real innovation despite themselves. Call it coincidence or a strange quirk of fate but General Motors has much to do with the recent upsurge in electric vehicle interest.

While you'll hear the complete story in chapter 3's history section, portions of it are worth repeating here.[2] AeroVironment, the company that built GM's "Sunraycer"—the winning entrant in the 2000-mile 1987 solar powered car race across Australia—had a unique window of opportunity to propose a designed-from-the-ground-up electric vehicle concept car to GM in the summer of 1988. GM's then chairman Roger Smith gave them the go-ahead and in January 1990 the GM "Impact" electric vehicle was unveiled at the Los Angeles Auto Show. The rest, as they say, is history.

The GM Impact was no ordinary internal combustion car chassis with an electric motor and batteries thrown in. It was designed-from-scratch to be an electric vehicle.

But AeroVironment's design approach was radical—rather than waiting for improved batteries to put into a standard vehicle, they would instead "improve the vehicle" and put more or less standard lead-acid batteries into it. The result was spectacular. Its sleek design and tremendous performance specs—it blew away a Mazda Miata and Nissan 300ZX in 0 to 60 mph standing start races—stole the hearts and fancy of everyone attending the January 1990 Los Angeles Auto Show.

General Motors was baffled. The prototype GM Impact electric vehicle, shown in FIG. 1-6, was considered only a show car by GM. It was never intended to be manufactured. But lo and behold it created such an interest and stir at the 1990 Los Angeles Auto Show that GM had to send out a survey team to find out why. What the survey team found out was that electric vehicles, suitably priced and configured, would be of interest to many, many people and the range and speed abilities of present electric vehicles with present technology was not a problem for consumers if the price was right.

**1-6** General Motors Impact electric vehicle—this is not your grandmother's EV!

The results set General Motors into action: facilities were designated, design and manufacturing teams were assembled and the production-model GM Impact electric vehicle was scheduled for 1995 model year introduction. General Motors' subsequent public relations blitz did more to promote the renaissance in EVs than any other event of the century. Unfortunately, the Impact electric vehicle program (and many others) fell victim to the restructuring and downsizing necessitated by General Motors' huge and unprecedented late 1992 losses. Only 50 Impact evaluation prototypes were funded for 1993 consumer and industry testing.

## Governments join the fray

Governments are even further removed from innovation and breakthroughs. They provide the framework for others to do the job—but very slowly, unless a well-funded, focused "man-on-the-moon-in-this-decade" Apollo-type program is involved. But the government has also done its job in helping to bring EVs into the mainstream. Chapter 3 has the details, but a few highlights are worth repeating here.

Public interest, starting in the 1960s, eventually resulted in the passage of the landmark Electric and Hybrid Vehicle Research, Development and Demonstration Act in 1976, and the establishment of the Department of Energy (DOE) in 1977. While little additional legislation has passed into law since then, an ever-increasing flow of House and Senate bills and a nearly constant stream of committee hearings has definitely contributed to raising public awareness about EVs.

While it would not be my choice to select a heavy, absolutely nonstreamlined, aerodynamically inefficient, four-ton commercial van as my electric vehicle test platform, the early Department of Energy programs pursued exactly that course. Not surprisingly, test data proved these big, heavy vehicles were slow and had limited range. But this message—broadcast apart from its context—was used to characterize all EVs.

For example, many electric vehicle utilities purchased "G-Vans" (production GM vans converted to electric), kept meticulous records, and performed additional studies on them. Unfortunately, when results were published or discussed, the G-Van's 53-mph top speed, 60-mile range and 0-to-30 mph in 12 seconds acceleration was never correctly associated with its 8120 pound weight, 36 batteries, and huge frontal area. Everyone formed the wrong conclusion about EVs, overlooking the true cause—the overweight, never-optimized-for-electric van platform. Contrast the G-Van's energy consumption of 0.654 kWh/mile versus the 0.110 kWh/mile of the 2200-pound General Motor's Impact EV, and the 0.301 kWh/mile of the 4948-pound "TEVan"—based on the Chrysler minivan body—over the same city driving cycle.[3]

Acceptance of these mistaken conclusions sidetracked official EV efforts for years while endless battery studies—the assumed culprit—were done and redone. The government stance absolutely infuriated EV consumer groups (Electric Automobile Association, et al.) and they worked harder than ever to prove the electric vehicle's true value. The interim result of all this was thousands of people driving all kinds of converted vehicles, using standard lead-acid batteries, getting speeds, ranges, and accelerations that the government said couldn't be done—including General Motor's high-profile Impact electric vehicle.

The good news is the Department of Energy's Electric and Hybrid research budget grew more than fivefold from $13.8 million in fiscal year 1989 to $75.3 million in 1993, and the government's research agenda will now be industry driven. Meanwhile, the United States Advanced Battery Consortium (USABC) was formed in January 1991 by General Motors, Ford, and Chrysler (with additional participation by government and industry) to focus on developing advanced battery technologies for use in electric vehicles. The results aren't in yet, but with both industry and the United States government on one's side, the future of EVs should be spectacular. This story—or variants of it—has been repeated in virtually every industrialized nation in the world.

## DISPEL THE ELECTRIC VEHICLE MYTHS

There have been four widely-circulated myths about electric vehicles that simply aren't true. Because the reality in each case is the 180-degree opposite of the myth, you should know about them.

### Speed myth

The myth is that electric vehicles can't go fast enough. Well, this is probably true if you are talking about a four-ton van carrying 36 batteries. The reality is that EVs can go as fast as you want—just choose the electric vehicle model (or design or build one) with the speed capability you want. Current technology EVs that use today's lead-acid batteries, such as the General Motor's Impact or AC Propulsion's converted Honda CRX (shown "smoking" its tires in the October 1992 issue of Road & Track[4]) definitely do not lack speed or acceleration.

The speed of an electric vehicle is directly related to its weight, body/chassis characteristics such as air and rolling resistance, electric motor size (capacity), and battery voltage. The more voltage, the more batteries you have, the faster any given electric motor size will be able to push the vehicle—but adding batteries adds also to the vehicle weight. All of these factors mean you can control how much speed you get out of your EV, and you're certainly not limited in any way. If speed is important, then optimize the electric vehicle you choose for it. It's as simple as that.

## Range myth

The myth is that electric vehicles have limited range. Nothing could be further from the truth but, unfortunately, this myth has been widely accepted. The reality is that electric vehicles can go as far as most people need. Remember, this book advocates an electric vehicle conversion only as your *second vehicle*—transcontinental trips in your EV are not its highest and best use at this time.

But what is its range? The average daily commute trip distance for all modes of vehicle travel (auto, truck, bus) is 10 miles, and this figure hasn't changed appreciably in 20 years of data-gathering.[5] An earlier study showed that 98 percent of all vehicle trips are under 50 miles per day; most people do all their driving locally, and only take a few long trips. One hundred-mile and longer trips are only 17 percent of total miles.[6] General Motor's own recent surveys (taken from a sampling of drivers in Boston, Los Angeles, and Houston) indicated:

- Most people don't drive very far.
- More than 40 percent of all trips were under 5 miles.
- Only 8 percent of all trips were more than 25 miles.
- Nearly 85 percent of the drivers drove less than 75 miles per day.[7]

Virtually any of today's 120-volt electric vehicle conversions will go 75 miles—using readily-available off-the-shelf components—if you keep the weight under 3000 pounds. This means an EV can meet more than 85 percent of the average needs. If you're commuting to work—a place which presumably has an electrical outlet available—you can nearly double your range by recharging during your working hours. Plus, if range is really important, optimize your electric vehicle for it. It's that simple.

## Convenience myth

The myth is that electric vehicles are not convenient. Detractors point out that the recharging infrastructure doesn't exist. A popular question is, "Suppose you're out in the woods and you run out of electricity, what do you do?" Well, my favorite answer is, "Suppose you're out in the woods and you run out of gas, what do you do?"

The reality is that electric vehicles are extremely convenient. Recharging is as convenient as your nearest electric outlet. You can get electricity anywhere you can get gas—there are no gas stations without electricity. Plus you can get electricity from many other places—there are few homes and virtually no business in the United States without electricity. All these are potential sources for you to recharge your electric vehicle. As far as the woods example goes, other than taking extended trips in U.S. Western deserts (and even these are filling up rapidly), there are only a few places you can drive 75 miles without seeing an electric outlet in the contiguous United States. Europe and Japan have no such places.

Plug-in-anywhere recharging capability is an overwhelming electric vehicle advantage. No question it's an advantage when your electric vehicle is parked in your home's garage, carport, or driveway. If you live in an apartment and can work out a charging arrangement, it's an even better idea: a very simple device can be rigged to signal you if anyone ever tries to steal your car.

How much more convenient could electric vehicles be? There are very few places you can drive in the civilized world where you can't recharge in a pinch, and your only other concern is to add water once in awhile. Electricity exists virtually everywhere, you just have to figure out how to tap into it. If your electric vehicle has an onboard charger, extension cord, and plug(s) available, it's no more difficult that going to your neighbor's house to borrow a cup of sugar. Except, of course, you probably want to leave a tip of cash money in this case.

While there are no electric outlets specifically designated for recharging electric vehicles conveniently located everywhere today, and though it's unquestionably easier and faster to recharge your electric vehicle from a 440-volt or 220-volt kiosk, the widely available 120-volt electric supply does the job quite nicely. When more infrastructure exists in the future, it will be even more convenient to charge your batteries. You will be able to recharge quicker from multiple voltage and current options, have "quick charge" capability via dumping one battery stack into another, and maybe even have uniform battery packs that you swap and strap on at a local "battery station" in no more time than it takes you to get a fillup at a gas station today. Just as it's used in your home today, electricity is clean, quiet, safe, and stays at the outlet until you need it.

### Cost myth

The myth is that electric vehicles are expensive. While true of electric vehicles manufactured in low volume today, and partially true of professionally done conversion units, it's not true of the do-it-yourself electric vehicle conversions this book advocates. The reality, as we saw earlier in this chapter, is that electric vehicles cost the same to buy (you're not going to spend any more for it than you would have budgeted anyway for your second internal combustion engine vehicle), the same to maintain, and far less per mile to operate. In the long term, future volume production and technology improvements will only make the cost benefits favor electric vehicles even more.

## DISADVANTAGES

There is a downside. If any one of the factors below is important to you, you might be better served by taking an alternate course of action.

**Extended trips**  As already mentioned, the electric vehicle is not your best choice for transcontinental travel at this time, or long trips in general. Not because you can't do it; alternate methods are just more convenient. As mentioned, this book advocates the use of the build/convert-it-yourself electric vehicle as a second vehicle. When you need to take longer trips, use your first vehicle, take an airplane, train, or bus, or rent a vehicle.

**Time to purchase/build**  Regardless of your decision to buy, build, or convert an electric vehicle, it is going to take you time to do it. No existing network of new and used electric vehicle dealers exists. So plan on taking a few weeks to a few months to arrive at the electric vehicle of your choice.

**Electric vehicle resale**  If you should decide to sell your EV, it will take longer—for the same reason. While a reasonably ready market exists via the Electric Auto Association chapter and national newsletters, it is still going to take you longer and be less convenient than going down to a local automobile dealer.

**Repairs**  Handy electric vehicle repair shops don't exist yet either. Although the build-it-yourself experience will enable your rapid diagnosis of any problems, replacement parts could be days or weeks away—even via expedited carriers. You could just stock-pile spare parts yourself, but the time to carefully think through this—or any other—repair alternative is *before* you make your electric vehicle decision.

**Heating, air conditioning, and power steering**  Regardless of whether you are sitting stopped in traffic or whipping down the interstate, heating or air conditioning will consume an additional 15 percent of your electric vehicle's energy on average. With the electric vehicle in its technological infancy and these two specialized areas lagging even further behind, the best heating and air conditioning solutions are still to come. Research this thoroughly before leaping into an electric vehicle if you live in a geographic area of the country where either one or both of these is important to you. Power steering is typically less of an issue since it represents only 5 percent additional energy drain and most lighter electric vehicles can manage quite easily without it. But ditto on research if it's important to you.

## THE FORCE IS WITH YOU

What you are seeing today in electric vehicles is just the tip of the iceberg. It is guaranteed that future improvements will make them faster, longer-ranged, and even more efficient. There are five prevailing reasons that guarantee electric vehicles will always be with us in the future. The only one not previously discussed is *technological change*. All the available technology has just about been squeezed out of internal combustion engine vehicles, and they are going to be even more environmentally squeezed in the future. This will hit each buyer right in the pocketbook—incremental gains will not come inexpensively. Internal combustion engines are nearly at the end of their technological lifetime. Almost all improvements in meeting today's higher Corporate Average Fuel Economy (CAFE) requirements have been achieved via improving electronics technology. If CAFE figures go much higher, little more assistance can be expected from electronics; major engine technology breakthroughs will be required.

On the other hand, electric vehicles are still in their technological infancy—a vast amount of basic improvement is still possible, in addition to any electronics technology gains. Unquestionably, the future looks bright for electric vehicles because the best is yet to come.

The four remaining reasons were covered in this chapter, and are summarized in FIG. 1-7:

- Fun to drive and own.
- Cost efficient.
- Performance efficient.
- Environmentally efficient.

Any one of the five reasons above is compelling by itself; the benefits of all five taken together are overwhelming. There are three more reasons that guarantee the do-it-yourself variation of the electric vehicle trend will always be with us as well:

*Adaptability.* EVs are easily and infinitely adaptable. When you do-it-yourself, any choice you wish to make for more speed, acceleration or range is readily accommodated—just do it.

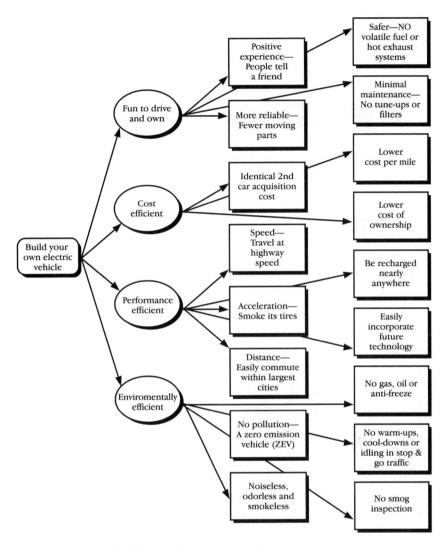

**1-7** Four reasons why EVs will always be with us in the future.

*Build now, upgrade later.* EVs are modularly upgradeable. See a better motor and controller? Bolt them on. Find some more efficient batteries? Strap them in. You don't have to get an entirely new vehicle, you can adapt new technology incrementally as it becomes available.

*Easily modify to meet special needs.* Even when EVs are manufactured in volume, the exact model needed by everyone will not be made because it would be prohibitively expensive to do so. But specialist shops that add heaters for those living in the north, air conditioners for those living in the south, and both for those living in the heartland will spring up. Chapter 5 will introduce you to the conversion specialists that exist today.

# 2

# Electric vehicles save the environment

*The needs of the many outweigh the needs of the few.*
Mr. Spock (from *Star Trek*, the movie)

Beyond your personal reasons for wanting to purchase, buy, or build an electric vehicle (the demand side), governments and large companies (the supply side) have become interested in the environmental benefits of EVs for reasons having to do with the planet, its resources and the air quality in its cities.

Why do electric vehicles save the environment? Because they are Zero Emission Vehicles—ZEVs—they add neither to the global warming problem of the planet nor the air quality problem of its cities. Since they derive their energy from electricity, they reduce our dependence on foreign oil and support our national energy security policy. In addition, they can dramatically alter the environmental cleanup equation because the amount of effort involved in the business of cleaning up the environment can be reduced by the widespread adoption of electric vehicles that don't dirty it in the first place.

In this chapter you'll learn about the positive role electric vehicles can play in each of these areas, and the tremendous near-term economic benefits they can provide to our electric utilities: a new market for electricity sales with no additional associated expense.

## SPACESHIP EARTH

The earth is a spaceship—much like the one that NASA is intending to launch in the late 1990s. It's a closed system. Other than the sunlight that falls on it, it has only finite resources than can be recycled but not replenished, and limited ability to cleanse itself.

Man's sophisticated scientific instruments have detected that the air we breathe and the water we use—the two essential ingredients for life—are becoming increasingly polluted, and this factor is far more noticeable in larger cities. Even more distressing, this pollution is so pervasive that it's being observed in locations far removed from any cities. In short, we're killing ourselves, and if we don't protect the legacy of what mother nature has bequeathed us there will be nothing left for our children and no life on earth. This is not a scary message, it's simply a statement of fact.

Ecologists call the earth a "self-contained ecosystem." This means that, except for the sunlight impinging on the earth's surface and the magnetic function,[1] there is nothing going in and nothing else coming out. We have to make do with what we have. Also, there is nothing you can do on the planet that doesn't affect the planet somewhere—if not everywhere—else. Destruction of the rain forest is a perfect example.

## WHO IS TO BLAME

No one and everyone is to blame. To quote from Vice President Al Gore's book:

"It is already clear that our information about the global environmental crisis does fall into a discernible pattern. For many, this pattern has become painfully obvious. But for others, it is still invisible. Why? The answer, in my opinion, is fear: too often we don't let ourselves see a pattern because we are afraid of its implications. Indeed, sometimes the implications suggest dramatic changes in our way of life. And, of course, those who have the heaviest investment in the status quo—whether it is economic, political, intellectual, or emotional—often organize ferocious resistance to the new pattern regardless of the evidence."[2]

Despite Mr. Spock, the needs of the many have seldom outweighed the needs of the few in post-industrial revolution society. It has always been a few controlling the extensive—yet finite—resources, regardless of the needs of any person, place, or thing. This is strictly business—not personal—on the part of individuals in the involved organizations. The process is driven by money, by profit. Of itself, the process is neither bad nor good.

When you are in a company, the standing orders of its marketing department are to promote your product at the expense of other competing products that might use similar or different technologies. For example, if you are selling plastic body panels to the big three automakers, you point out how "superior" your product is to your competitor's plastic panels and how, overall, plastic is superior to steel, aluminum, fiberglass, or composite fibers. Regardless of your company or the product or service you are selling, this positioning and marketing process is identical. At one extreme, it is genteel and upscale with well-defined rules—like a friendly game of bridge. At the other, it resembles brutal warfare in which the loser loses a company—like the aerospace or defense industry.

Think about it. While you are an employee of a company (or organization), from the president on down, you are committed to supporting its marketing and competitive policies. If you disagree with them, your choices are to argue for change, to shut up, or to get out. While "survival of the fittest" in nature implies constant innovation, adaptation, and change, "survival of the largest" rules in business. If you can profitably market more of your product than anyone else's, you simply drown out all competing voices. It matters not a whit that the competitive products were better; they simply cease to exist as competitive entities (merged, bought up, closed down, dismantled, etc.).

While today's rapidly changing high technology has somewhat changed the basic equation, the fact remains that when organizations become really big, they become more committed to maintaining the status quo than to true innovation. The reason is simple. They become locked-in victims of their own technology. When an organization has a substantial investment in older technology products, basic business acumen dictates they postpone or delay the introduction of newer technology products that would affect their revenue stream.

So who's to blame? No one and everyone. The line from Dr. Zhivago about, "one Russian ripping off wood from a fence to provide heat for his family in winter is pathetic, one million Russians doing the same is disaster" applies equally well to all of us and our internal combustion engine vehicles. Applied collectively, the legacy of the internal combustion engine is greenhouse effect, foreign oil dependence, and pollution. Let's succinctly define the problem and its solution.

## THE ENVIRONMENTAL PROBLEM

As you already read in chapter 1, heat, pollution, and greenhouse gases are still the byproducts of today's highly-evolved, carbon-burning internal combustion engines. Everything going into and coming out of an internal combustion engine is toxic, and it's still classified among the least efficient mechanical devices on the planet.

Far worse than its inefficient and self-destructive operating nature is the legacy of environmental problems (summarized in FIG. 2-1) created by the internal combustion engine vehicle when multiplied by hundreds of millions of vehicles.

- Dependence on foreign oil (environmental and security risk).
- Greenhouse effect (atmospheric heating).
- Toxic air pollution.
- Toxic solid waste pollution.
- Toxic input fluids pollution.
- Wasted heat generated by its inefficiency.

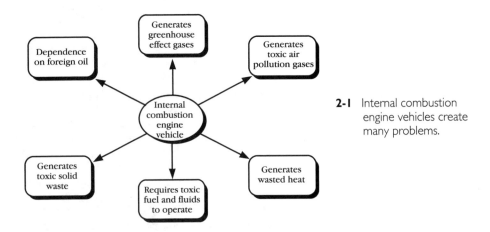

**2-1** Internal combustion engine vehicles create many problems.

Continuing our dependence on internal combustion engine vehicles commits us to die a slow death by asphyxiation, disease, and economic disintermediation. If it doesn't happen to us, it will certainly happen to our children and our children's children.

The problem is we're killing ourselves and our planet with petroleum-based vehicles. The people in the larger cities just happen to be doing it faster. This is why they have raised the flag first.

The solution is to find more efficient modes of transportation that allow us to continue life as we know it, and to find still more efficient methods of transportation that allow us to continue to exist for the next millennia rather than for just the next century or two. The internal combustion powered engine ain't it.

## THE ENVIRONMENTAL SOLUTION

Electric vehicles have existed for more than a hundred years (predating internal combustion engine vehicles). The aerospace-derived technology to improve them has existed for decades. Unquestionably, EVs will be the de facto transportation mode of choice for years to come, if life on our planet is to continue to exist in its modern form with the conveniences we count on today.

Figure 2-2 shows the reasons why. In direct contrast to the problems created by internal combustion engine vehicles, EVs:

- Use electricity (today, electricity comes mostly from coal).
- Are zero emission vehicles (ZEVs); they emit no pollutants.
- Generate no toxic waste (lead-acid batteries are 98 percent recyclable).
- Require no toxic input fluids (only occasional watering).
- Are highly efficient (motors and controllers 90 percent, batteries 75 percent).
- Benefit electric utilities (a new market for electricity sales).

Let's examine each of the internal combustion engine vehicle-caused environmental problems, and look at how the electric vehicle is a solution to each of them (and more).

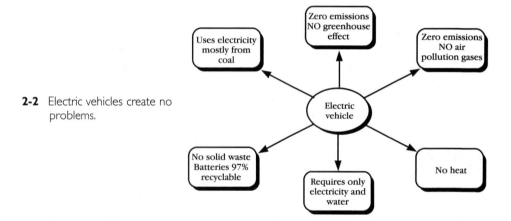

**2-2** Electric vehicles create no problems.

## DEPENDENCE ON FOREIGN OIL

Foreign oil dependency actually comprises four problems: United States dependence on oil; increase in foreign oil consumption; increase in environmental risk due to foreign oil consumption; and the eventual rise of long-term oil costs because oil is a finite resource. Let's look at each of these and how electric vehicles can help.

### United States transportation depends on oil

Although small amounts of natural gas and electricity are used, the United States transportation sector is almost entirely dependent on oil. A brief look at a few charts will demonstrate the facts. It doesn't take a rocket scientist to figure out this situation is both a strategic and economic problem for us.

Figure 2-3 shows the United States energy consumption by source from 1949 to 1991[3]—as you can see, petroleum accounted for 40 percent (33 quadrillion BTU) of the total energy consumption (82 quadrillion BTU) in 1991. Natural gas (20 quadrillion

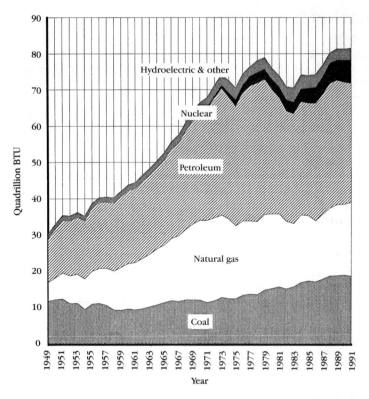

**2-3** United States energy consumption by source from 1949 to 1991.

BTU) came in second, coal (19 quadrillion BTU) was a close third, nuclear power (7 quadrillion BTU) was fourth and hydroelectric (3 quadrillion BTU) along with other sources brought up the rear. Our entire economy is obviously dependent on oil.

Figure 2-4 shows the petroleum products supplied by sector from 1949 to 1991 (where the petroleum shown in FIG. 2-3 was used). As you can see, 65 percent or 11 million barrels per day of the 17 million barrels per day total was consumed in the transportation sector in 1991. This is up from the 53 percent of the total used by transportation in 1949. Transportation is not only using more oil but taking a greater percentage of the total. In fact, the transportation sector derived more than 97 percent of its energy from petroleum in 1991.[4]

How did we get into this situation? We put ourselves into it. Figure 2-5 shows why at a glance.[5] The United States has the lowest price for gasoline in the world and the lowest tax rate on its gasoline. By subsidizing the widespread use of motor gasoline via these practices, we have "consumed" ourselves into a corner. Necessity is the mother of invention. Since none of us has the luxury of skipping going to work, many creative "work-around" solutions would come forward if gasoline suddenly cost $4 a gallon in the United States; for example, electric vehicles.

### Increasing foreign oil consumption

A few months before the Arab oil embargo of 1973, you could fill up your tank with "regular" for as low as 19.5 cents per gallon—thanks to the numerous gas station price

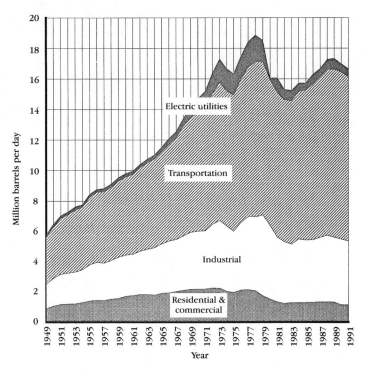

**2-4** United States petroleum products supplied by sector from 1949 to 1991.

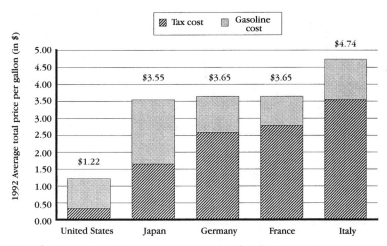

**2-5** Gasoline costs with tax component identified for selected countries.

wars going on in almost all areas of the country. A few months later, you waited in line to get gasoline at $2 a gallon and considered yourself lucky when you got a fillup.

While it's our own fault for letting it happen, the Organization of Petroleum Exporting Countries (OPEC) price hikes in crude oil have had a disastrous impact on our economy, our transportation system, and our standard of living. The Arab oil crisis of 1973 and subsequent ones were not pleasant experiences. After each crisis, the United States vowed to become less dependent on foreign oil producers—yet exactly the opposite has happened.

Figure 2-6 shows the widening gap between the United States' total oil production and oil consumption. The difference is made up by importing foreign oil. The cost is shown so that you can see the supply-demand-price relationship at a glance. In 1991, the 17.0 million barrels of oil per day demand was 42 percent supplied by net oil imports. And it's going to get worse—imported oil is projected to account for 53 percent to 69 percent of total oil use in 2010.

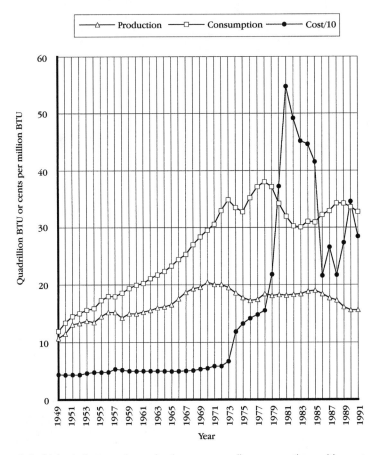

**2-6** United States oil production versus oil consumption with cost overlay from 1949 to 1991.

Figure 2-7 shows petroleum products supplied from 1949 to 1991—how the petroleum shown in FIG. 2-3 was used. As you can see, 43 percent or 7 million barrels per day of the 17 million barrels per day total was used as gasoline for motor vehicles in 1991. With motor gasoline as the largest component of petroleum, our basic motor vehicle needs are tied directly to foreign oil prices. No one can accurately predict what fuel prices

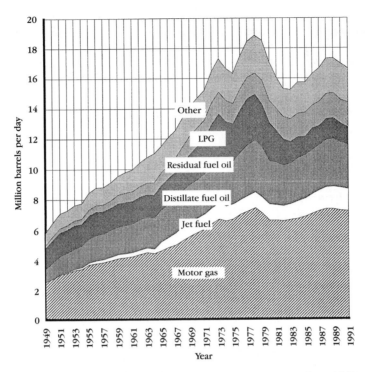

**2-7** United States petroleum products supplied by product from 1949 to 1991.

will be next year, and whether there will be a shortage or abundance of supplies. Everyone agrees that it's a bad situation, but nobody has taken steps to correct the problem.

## Increasing environmental risk

Almost everyone is familiar with the massive Exxon Valdez oil spill disaster, and the increasing frequency of smaller oil spills. Shipping oil around the planet in vast quantities places our environment at risk. But to put the matter in perspective, the Exxon Valdez disaster spilled 11 million gallons of oil. Each day we use 17 million barrels, multiplied by 42 gallons per barrel, or *714 million gallons per day.*

We use 65 times the amount that Exxon Valdez spilled every day. Multiply that figure by 40 percent to figure the amount of foreign oil imports probably enroute to us inside the hull of a supertanker like the Valdez—286 million gallons per day. By depending on foreign oil, statistics say that the huge quantities involved, and the long shipping distances required, mean that we are looking at a ticking environmental time bomb. No question that it will go off. The only questions are when and where.

## Increasing long-term oil costs

There is a fixed amount of oil/petroleum reserves in the ground around the world, and there isn't going to be any more. We're going to run out of oil at some point. Before that happens, it's going to get very expensive.

Figure 2-8 illustrates this aspect of the problem, caused by oil being a nonrenewable energy resource. The curve centered on the year 2000 resembles one drawn by M. K. Hubbert[6] many years ago but it's derived a different way. According to the *Annual Energy Review 1991*, there were 999.2 billion barrels of world crude oil reserves on January 1, 1991, and world petroleum consumption in 1990 was 65.92 million barrels of oil per day:

$$999.2 \times 10^9 \text{ barrels} \div (65.92 \times 10^6 \text{ barrels/day} \times 365 \text{ days/year}) = 41.53 \text{ years}$$

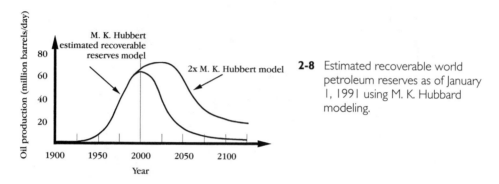

**2-8** Estimated recoverable world petroleum reserves as of January 1, 1991 using M. K. Hubbard modeling.

This reliable government-published sourcebook says that at 1990s rate of consumption there are 41.53 years of oil left in the world. But *proven* oil reserves used for the calculation can differ widely from *estimated recoverable reserves*, so to be conservative, the previous figure was doubled to 83 years to get the curve on the left in FIG. 2-8. This curve shows oil used up until 1990 and a smooth plot of remaining oil available to be used until 2073 (1990 + 83) with a little residual afterwards.

Another problem is that many of our oil wells are capped and just sitting there, waiting for the time when prices rise. Nobody really knows how much oil is left until you're ready to put your money on the table and pay for it. So the curve on the right in FIG. 2-8 doubles the area under the curve on the left—twice the estimated recoverable reserves. It might also be too conservative. But whether you double, triple, or quadruple the reserve figure, the amount of oil is fixed—at some point you run out. Before that point is reached, oil will become very expensive to find (new fields more remote, deeper, more difficult to get to), recover (new, more expensive processes to recover from capped or unproductive wells), and use (environment plus corporate average fuel economy place more restrictions on the type of fuel that can be used—driving its price up).

## How electric vehicles can help

The bad news is that the transportation sector is almost totally dependent on petroleum fuels that are more than 40 percent derived from foreign sources. Worse news is that by 2010, transportation sector fuel consumption could increase by one third and foreign sources could be providing two-thirds of our oil needs. This is an economically and strategically intolerable situation.

How can electric vehicles reduce dependence on foreign oil? Easy. Electric vehicles don't use oil at all. The primary source for electricity used in recharging electric vehicles is coal-fired electrical generating plants. Coal, although also a nonrenewable energy resource, is in far more plentiful supply. Again consulting the *Annual Energy Review 1991*, there were 1,788,944 million short tons of coal as of June 1991 and that world coal consumption in 1990 was 5171 million short tons:

$$1,788,944 \times 10^6 \text{ short tons} \div (5171 \times 10^6 \text{ short tons/year}) = 345.96 \text{ years}$$

In other words, at 1990s rate of consumption there are 345.96 years of coal left in the world. Even the most pessimistic projections for coal reserves show it will outlast petroleum by a factor of several times—many centuries—at projected use levels.

You are substituting one nonrenewable energy resource for another when you make the decision to switch from internal combustion engine powered vehicles to electric vehicles. But the United States has the second largest proven coal reserves (265,193 million short tons—behind China), is the second largest producer (994 million short tons in 1991—behind China) and our 1991 coal exports (109 million short tons) dwarfed coal imports (3 million short tons). Electric utilities consumed 87 percent of coal production in 1991. Because electric vehicles are more efficient, studies have shown you're going to use a lot less coal or other primary energy source. To quote Wang and DeLuchi:

"(in 1995). . . Electric vehicles will consume 30 percent less coal than internal combustion engine vehicles if coal is the primary energy source. In 2010, however, electric vehicles will consume less primary energy than internal combustion engine vehicles, regardless of the primary energy sources (50 percent less coal)."[7]

If you generate electrical energy directly—from renewable solar, geothermal, hydroelectric, wind, and fusion sources—you take nothing out of the ground at all and improve your efficiency still more. In the long term, renewables are our only source of energy.

Using electric vehicles moves our transportation away from total dependence on oil and supports our national energy security policy. It reduces our dependence on foreign oil—and substitutes relatively stable coal-pricing for unpredictable OPEC oil-pricing. It lowers the risk to the environment. Finally, it moves us over to consuming a nonrenewable energy resource (coal) in more plentiful supply than the oil energy resource.

Before electric vehicles roam our streets in significant numbers, internal combustion engine technology and fuel should be priced to reflect its true social cost—not just economic cost—because of the oil dependency and environmental problems it creates. Such incremental oil fuel price increases or taxes, directly used to subsidize electric vehicle technology and infrastructure development, could move us away from oil dependency, jumpstart the phaseover into electric vehicles, and put millions of Americans to work.

## GREENHOUSE EFFECT

The thin mantle of atmosphere around earth has for millennia protected its surface and inhabitants against excessive heating or cooling. Due to the widespread adoption of the internal combustion engine and its carbon dioxide ($CO_2$) emissions, mankind has managed to heat up the atmosphere—the greenhouse effect—and disturb its function in only a few generations. Let's look at the results of putting these enormous additional amounts of $CO_2$ and $CO_2$-equivalent greenhouse gases into the atmosphere and how electric vehicles can help.

In a glass greenhouse, the glass allows shorter light waves to pass through it much faster than longer infrared waves can. So heat inside a greenhouse cannot escape through the glass as fast as light can enter. Hence the term, "greenhouse effect" which applies equally well to the earth's atmosphere as it does to your closed car left parked in the sunlight with all the windows rolled up.

In the earth's case, shorter-wave light and ultraviolet radiation from the sun is absorbed and longer-wave infrared energy is radiated back into the atmosphere. Water vapor, $CO_2$ and other trace gases—the infrared-absorbing gases—absorb most of this outgoing energy (some is reradiated back to the earth's surface). If the concentration of infrared-absorbing gases is increased, the total energy stored in the atmosphere increases, thus heating it up. Atmospheric warming brings with it all manner of possible global consequences such as: melting the polar icecap, raising the mean sea level, changing rainfall and temperature patterns, altering weather patterns—increasing incidents of severe/unusual weather, disrupting established crop-growing regions and interfering with earth's natural growth (sea plankton) and cleansing processes (jetstream, tradewinds, ocean currents).

$CO_2$ is expected to be responsible for about half of future global warming because of the sheer volume of its emissions—the United States leads the world in the $CO_2$ released category with 5.8 tons per person per year.[8] The other infrared-absorbing greenhouse gases—methane ($CH_4$), nitrous oxide ($N_2O$), nitrogen dioxide ($NO_2$), etc.—while individually emitted in less volume than $CO_2$, collectively contribute as much to future global warming as $CO_2$ because of their greater heat-trapping abilities.

Don't get the mistaken impression that carbon dioxide is "bad." Carbon dioxide is both natural (as TABLE 2-1 shows,[9] it occurs naturally in air to the extent of .03 percent), and essential (it's inextricably involved in the carbon cycle that links the plant kingdoms—consuming $CO_2$—and animal kingdoms—producing $CO_2$—through photosynthesis). It is found wherever compounds of carbon are burned with an adequate supply of air. The amount of $CO_2$ varies widely: combustion produces $CO_2$ and increases its partial pressure in air over cities; photosynthesis consumes $CO_2$ and decreases its partial pressure in air over forests. Carbon dioxide is not bad in itself—it's bad only when its concentration in our atmosphere and over our cities is raised to unacceptable levels.

### Table 2-1
### Composition of air at sea level

| Compound | Partial pressures in dry air (in atm) |
|---|---|
| Nitrogen | 0.780 |
| Oxygen | 0.210 |
| Argon | 0.0094 |
| Carbon Dioxide | 0.0003 |
| Hydrogen | 0.0001 |
| Neon | 0.0000123 |
| Helium | 0.000004 |
| Krypton | 0.0000005 |
| Xenon | 0.00000006 |
| Radon | trace |

*(assuming total air pressure is exactly 1 atmosphere)

In chapter 1 you learned the internal combustion engine is just another variant on the combustion process. To continue the analogy with the match, if the match were pure carbon and burned completely, the chemical reaction would be:

$$C \text{ (solid)} + O_2 \text{ (gas)} \rightarrow CO_2 \text{ (gas)}$$

The result of combining the pure carbon (C) and the oxygen ($O_2$) is pure carbon dioxide ($CO_2$). The problem is that because of the ratios of the chemical weight of carbon dioxide (44.01 atomic mass units—amu) to the chemical weight of carbon (12.01 amu), burning *one* gram, ounce, pound, or ton of carbon gives you *three* (actually 44.01/12.01 = 3.66) grams, ounces, pounds, or tons of carbon dioxide. This also holds true for complete combustion of a complex hydrocarbon such as octane ($C_8H_{18}$) whose balanced reaction would be:

$$2\ C_8H_{18} \text{ (gas)} + 25\ O_2 \text{ (gas)} \rightarrow 16\ CO_2 \text{ (gas)} + 18\ H_2O \text{ (gas)}$$

The result here is almost the same: burning *one* gram, ounce, pound, or ton of carbon gives you *three* (actually $(8 \times 44.01)/114.22 = 3.08$) grams, ounces, pounds, or tons of carbon dioxide. The internal combustion engine carbon dioxide and water vapor ($H_2O$) emissions are not classified as pollutants (although you wouldn't live long in a room filled only with $CO_2$). The problem is that all this extra carbon dioxide goes up into the atmosphere and creates the "greenhouse effect" by trapping the sun's radiation and warming the earth.

How much carbon dioxide is generated? A recent study puts the total carbon dioxide-equivalent emissions at 455 grams per mile for light duty gasoline-powered vehicles and 2249 grams per mile for heavy duty diesel-powered vehicles.[10] That's *one pound per mile* and *five pounds per mile*, respectively. Stated another way, if you drive your 4000-pound internal combustion engine automobile 12,000 miles a year, you are putting three times its weight in carbon dioxide into the atmosphere every year! Multiply this by the nearly 200 million (actually 190,228,000 in 1990[11]) vehicles in use in the United States and you can see the magnitude of the carbon dioxide "greenhouse effect" problem.

## How electric vehicles can help

The bad news is that these carbon dioxide emissions will automatically increase with our increased transportation sector fuel consumption—predicted to *increase one third by the year 2010* as you read in the previous section. This is a dangerous situation.

How can electric vehicles reduce carbon dioxide emissions? Easy. No one disputes that an average mileage internal combustion engine vehicle emits thousands of pounds of $CO_2$ each year—by far its largest emission component—while all electric vehicles are zero-emission vehicles (ZEVs) that emit nothing. No meaningful data yet exists to substantiate the exact benefit of electric vehicles over internal combustion engine vehicles in terms of $CO_2$ generation. Until electric vehicles exist in greater quantities to allow such a meaningful study to be conducted, four obvious factors emerge:

1. Assuming that coal-powered electrical generating plants will be the primary source of electric vehicle energy in the near term, it's going to be many, many years—if not decades—before the energy drawn to recharge electric vehicles makes even a one percent dent in production. So studies that compare pollutant emissions at the tailpipe of internal combustion engine vehicles with emissions of coal-powered electrical generating plants providing

electricity to recharge electric vehicles are inaccurate. *There is no emission from an electric vehicle* and, until there exists an appreciable number of them, they do not impact in any way the emissions from the power plant used to generate the electricity.

2. It's far easier to control emissions from *two hundred* stationary electrical generating plants than it is to control emissions from *two hundred million* internal combustion engine vehicle tailpipes. Future studies should reflect the Clean Air Act of 1990 amendments-mandated emission levels for stationary electrical generating plants as well as emission levels mandated for internal combustion engine vehicles.

3. When you put a new electric vehicle into service versus a new internal combustion engine vehicle, you save on emission-generation right away—thousands of pounds annually. If you take an older (and by definition a more highly polluting) internal combustion engine vehicle out of service and reincarnate it as an electric vehicle, you more than double your emission savings proportionately.

4. Unlike internal combustion engine vehicles, *electric vehicle energy may be supplied directly from renewable energy sources* such as solar, wind, hydroelectric, geothermal, etc. All of these sources make no greenhouse contributions at all.

You cannot control a problem until you recognize it. While other nations at the 1991 United Nations Environmental Summit in Nairobi, Africa recognized the problem, the United States—the world's largest total emitter of $CO_2$—has refused to mandate any cuts. In other words, while $CO_2$ is by far the largest component emission of internal combustion vehicles, it isn't even regulated because regulators are still arm wrestling over whether putting a lot of extra $CO_2$ into the atmosphere is a bad thing. If analyzing data from our planetary probes of Venus are not enough (apparently its $CO_2$-laden atmosphere covers a too-hot-for-human-comfort surface), perhaps all our regulators could be ushered into a $CO_2$-filled room and we can observe how quickly they make a decision.

Even before we prove exactly how much electric vehicles will help us to solve the greenhouse effect problem, it should be obvious that they *will help*, and we should move ahead with electric vehicle technology and infrastructure development. After we help save the planet we'll have time to reflect on the intelligence of our decision, and to study exactly how much margin for error we had.

## TOXIC AIR POLLUTION

Of much more significant concern to the human beings and other life on the planet is the fact that we're going to run out of breathable air. In Los Angeles, California you can almost see the air. But Los Angeles air is clean compared to that of Mexico City. Let's look at toxic air pollution and how electric vehicles can help.

More than any other factor, the toxic, obnoxious emissions of internal combustion vehicles have brought electric vehicles into the foreground. The internal combustion engine is a just another variant on the combustion process, but gasoline, diesel, and the other fuels are highly-complex hydrocarbon compounds that burn incompletely, and you get a lot more out of them than just $CO_2$. While complete combustion produces carbon dioxide and water, the internal combustion engine fuel and oil that is not burned completely in the combustion chamber comes out the tailpipe as all kinds of derivative, unburned, and partially-burned hydrocarbons and byproducts, some of which are acted upon still further in the atmosphere by sunlight. Almost all internal combustion engine emissions are toxic to humans. To summarize the emission picture, there is:

*Carbon dioxide ($CO_2$).* This "greenhouse-problem" gas of the previous section is by far the largest-volume emission component. It's nonpoisonous but, unless you're from the plant kingdom, you can't breathe in a room filled with it.

*Carbon monoxide (CO).* This colorless, odorless and tasteless gas is poisonous, and a whiff of 0.3 percent (by volume) of it for 30 minutes—the level you'd get from an idling internal combustion engine vehicle in a closed garage—can kill you.

*Hydrocarbons.* This class includes hundreds of carbon-hydrogen compounds, usually broken into "does-not-react-to-form-ozone ($O_3$)," methane ($CH_4$), and the "reactive" nonmethane organic compounds (NMOG, shorthand for a broad class of hydrocarbon pollutants; its widely used equivalent forms are nonmethane hydrocarbons (NMHC), reactive organic gases (ROG), volatile organic compounds (VOC) and just hydrocarbons (HC). Hydrocarbons—some are carcinogenic—form oxidants that irritate the mucous membranes in the presence of sunlight and nitrogen oxide.

*Oxides of nitrogen ($NO_x$).* This includes nitrogen oxide (NO), a colorless, odorless and tasteless gas, which is rapidly converted to nitrogen dioxide ($NO_2$) in the presence of oxygen. Nitrogen dioxide is a reddish-brown, poisonous gas with a penetrating odor that destroys lung tissue.

*Particulate matter.* This includes all dust, ash, soot (carbon), and smoke present as solids or liquids in exhaust gas under two labels: *TSP*, or total suspended particulates (all particulates), and *PM10*, or particulate matter under 10 micrometers in diameter that is particularly unhealthy because of its ability to reach and affect the lower lungs.

*Ozone ($O_3$).* This is typically formed when $NO_x$ and NMOG react in the *lower* atmosphere in the presence of sunlight. Unlike the *upper* atmosphere ozone that protects us against the sun's ultraviolet rays, lower atmosphere ozone is the white haze that hangs over many cities as smog that causes nose and eye irritation, heart and lung problems, increased asthma, infections and problems for the elderly, along with crop, forest, wildlife, raw material, and structural damage.

*Sulfur dioxide ($SO_2$).* This is formed because gasoline typically contains 0.03 percent sulfur. In addition to being an irritating pollutant, when it combines with water vapor in the atmosphere and forms sulfuric acid, it also produces harmful "acid rain."

*Lead (Pb).* Lead in gasoline has been found so hazardous to health that its use has been drastically curtailed, but earlier vehicles using leaded gasolines produce lead emissions that continue the problem.

The relatively high amount of pollutant emissions in its exhaust gas is an inherent basic disadvantage of the internal combustion engine, in addition to its low efficiency. And this problem is not easily solved. If you adjust an internal combustion engine's fuel-air mixture setting so it's at maximum fuel economy, you get the least amount of carbon monoxide and hydrocarbons but the maximum output of nitrogen oxides. This setting is far past its peak torque setting, but if you richen it toward peak torque to get less nitrogen oxides, the carbon monoxide and hydrocarbon outputs increase. If you make it leaner, you get more hydrocarbons and increase misfires. In either case, you decrease fuel mileage.

So what does all this mean? Internal combustion engine vehicles in the Los Angeles area in 1985 produced 87 percent of the carbon monoxide, 60 percent of the nitrogen oxides, 46 percent of the hydrocarbons and 29 percent of the sulfur dioxide emissions.[12]

The Environmental Protection Agency (EPA) says that internal combustion engine vehicle emissions accounted for 24 percent of the hydrocarbons, 34 percent of the nitrogen oxides and 54 percent of the carbon monoxide released in the United States in 1987.[13] Not a good thing—especially when what harms you the most is in the air you breathe, and you can't immediately do anything about it.

## Southern California to the rescue

Southern California's love affair with internal combustion engine vehicles gave them an air pollution problem in severe and visible form as early as the 1950s (its days and weeks of haze, smoke, and fog coined the word *smog*). Future prospects look even worse (L. A. Basin vehicle miles travelled are expected to rise from 275 million to 400 million by the year 2010). This is why Southern California is also at the forefront of the air pollution solution.

In 1968, the California legislature formed the California Air Resources Board (CARB), the first body in the nation with legislative powers over its air quality standards. On May 6, 1988 (a day widely-celebrated by Electric Auto Association groups), a Los Angeles city council motion—pushed by Councilman Marvin Braude and Councilman John Ferraro—stated: "Widespread use of electric vehicles can contribute significantly to improvement of air quality in Southern California. Because these vehicles use a clean and easily transportable source of energy—electricity—they can provide mobility without resulting byproducts of combustion . . . [we] move that Department of Water and Power, on behalf of the city, serve as lead agency in sponsoring an international competition for development and sale in Southern California of at least 5000 Electric Vans and 5000 Electric Passenger Vehicles and seek participation of Southern California Edison as a co-equal sponsor of this competition."[14] In early 1989, what had become known as the *L. A. Initiative* was launched. Request For Proposal (RFP) packages were sent out to more than 200 companies and agencies worldwide for placing 10,000 electric vehicles in operation in the Los Angeles area by 1995. The rest, as they say, was history.

On September 28, 1990, new, more stringent regulations governing stationary and mobile pollution sources proposed by CARB were adopted. California is the only state in the nation that can set its own air quality standards because it was doing so before the EPA began setting them. Recognizing that ". . . automobiles contribute between 50 and 70 percent of the major components in the polluted air above Los Angeles," CARB's regulations mandated several new categories of alternative-fueled vehicles, allowable levels by pollutant, and milestone dates and quantities. Here are the new categories:

- Transitional low emission vehicle (TLEV).
- Low emissions vehicle (LEV).
- Ultra low emissions vehicle (ULEV).
- Zero emission vehicle (ZEV).

TABLE 2-2 shows the CARB emission standards. It says that emission standards become increasingly stricter in moving through the four categories from TLEV to ZEV. For example, TLEVs are allowed 90 pounds of carbon monoxide pollutants per 12,000 miles driven per year; ZEVs emit nothing at all.

TABLE 2-3 shows the CARB implementation dates, percentages and ZEV quantities. CARB mandated that two percent of all cars sold in California be ZEVs starting in 1998, rising to ten percent by 2003. Notice higher emission TLEVs and LEVs are phased

## Table 2-2  CARB emission standards

| Vehicle/Compound | grams/mile | lbs/mile | lbs per 12,000 miles |
|---|---|---|---|
| TLEV | | | |
| Carbon Monoxide (CO) | 3.4 | 0.0074956 | 89.9 |
| Hydrocarbons | 0.4 | 0.0008818 | 10.6 |
| Oxides of Nitrogen ($NO_X$) | 0.125 | 0.0002756 | 3.3 |
| LEV | | | |
| Carbon Monoxide (CO) | 3.4 | 0.0074956 | 89.9 |
| Hydrocarbons | 0.2 | 0.0004409 | 5.3 |
| Oxides of Nitrogen ($NO_X$) | 0.075 | 0.0001653 | 1.9 |
| ULEV | | | |
| Carbon Monoxide (CO) | 1.7 | 0.0037478 | 45.0 |
| Hydrocarbons | 0.2 | 0.0004409 | 5.3 |
| Oxides of Nitrogen ($NO_X$) | 0.040 | 0.0000881 | 1.1 |
| ZEV | | | |
| Carbon Monoxide (CO) | 0.0 | 0.0 | 0.0 |
| Hydrocarbons | 0.0 | 0.0 | 0.0 |
| Oxides of Nitrogen ($NO_X$) | 0.0 | 0.0 | 0.0 |

## Table 2-3
## CARB implementation percentages, dates and zero emission vehicle numbers

| Model Year | % TLEV | % LEV | % ULEV | % ZEV | # ZEV in CA |
|---|---|---|---|---|---|
| 1994 | 10 | - | - | - | - |
| 1995 | 15 | - | - | - | - |
| 1996 | 20 | - | - | - | - |
| 1997 | - | 25 | 2 | - | - |
| 1998 | - | 48 | 2 | 2 | 40,000 |
| 1999 | - | 73 | 2 | 2 | 40,000 |
| 2000 | 0 | 96 | 2 | 2 | 40,000 |
| 2001 | 0 | 90 | 5 | 5 | 100,000 |
| 2002 | 0 | 85 | 10 | 5 | 100,000 |
| 2003 | 0 | 75 | 15 | 10 | 200,000 |
| Totals | | | | | 520,000 |

down as ULEVs and ZEVs increase. The ZEV quantities were derived from the baseline new 1990 California automobile registrations (1,059,926),[15] and this figure was allowed to grow at a 5 percent per year rate. Using this assumption, California's ZEV—electric vehicle—population is at least 484,416 by the end of 2003. If 10 percent growth rate is assumed, ZEVs on hand at the end of 2003 rises to 833,854.

California is not alone. In 1992, Massachusetts became the first NESCAUM (Northeast States for Coordinated Air Use Management, which includes Massachusetts, New

York, New Jersey, Maine, New Hampshire, Vermont, and Rhode Island) state to officially adopt the California program. The governors of Delaware, Maryland, and Pennsylvania, and the Mayor of Washington, D.C. also announced their intention to adopt the California program, and Florida, Texas, Colorado and a number of other states are working on variants of it. This geographic lineup accounts for more than half the new vehicles sold in the United States and gives automakers a real incentive to get busy.

How far have we come? It's instructive to look at the earlier air pollution data. Prior to 1968 enactment of CARB and EPA Federal pollution control standards (in "precontrol" times), the Motor Vehicle Manufacturer's Association baseline 49-state standard data for passenger car exhaust emissions was enormous,[16] as shown on the top of TABLE 2-4. 2222 pounds of carbon monoxide pollutants were emitted per 12,000 miles driven every year! No wonder we were cautioned not to start the car with the garage door closed—instant asphyxiation by automotive exhaust! By 1980, the middle of TABLE 2-4 shows standards beginning to have a noticeable effect—only 397 pounds of carbon monoxide pollutants emitted per 12,000 miles driven per year. By the mid-1980s, another study of L. A. Basin air pollution,[17] summarized at the bottom of TABLE 2-4, showed still better results. The 2574 pounds of carbon monoxide emitted over a vehicle's 100,000 mile lifetime is a lot—but far better than the nearly equivalent amount emitted per year only two decades earlier!

## Table 2-4  Earlier emission levels and standards

| Date/Compound | grams/mile | lbs/mile | lbs per 12,000 miles |
|---|---|---|---|
| Prior to 1968 Levels | | | |
| Carbon Monoxide (CO) | 84.0 | 0.185 | 2222 |
| Hydrocarbons | 10.6 | 0.023 | 280 |
| Oxides of Nitrogen ($NO_x$) | 4.1 | 0.009 | 108 |
| 1980 Federal Levels | | | |
| Carbon Monoxide (CO) | 15.0 | 0.033 | 397 |
| Hydrocarbons | 1.5 | 0.003 | 40 |
| Oxides of Nitrogen ($NO_x$) | 2.0 | 0.004 | 53 |
| 1985 L.A. Basin Actuals[1] | | | |
| Carbon Monoxide (CO) | - | - | 2574 |
| Hydrocarbons | - | - | 262 |
| Oxides of Nitrogen ($NO_x$) | - | - | 172 |

[1]Pollutants generated in driving a gasoline-powered car over its typical 100,000 mile lifetime.

While it's great to have CARB and EPA Federal air quality regulations, the problem is that the newest standards apply only to the latest model year. Between 7 and 13 million vehicles have been sold in the United States during each of the last 30 years. In 1990, 10 million vehicles were sold in the United States (26 percent were imports), 11 million vehicles were scrapped and the average car age rose to 7.8 years.[18] The lion's share of the 200 million vehicles in the United States do not meet current emission regulations. People are keeping their vehicles longer, and new vehicles are replacing scrapped ones at a slow rate. Even the latest internal combustion engine vehicles exceed the emission limits with use, and must be periodically re-certified—a procedure which adds from $50 to $500 to annual vehicle maintenance cost.

### How electric vehicles can help

Maintaining stringent toxic air pollution emission levels along with conforming to increasingly higher mandated corporate average fuel economy levels puts an enormous burden on internal combustion engine vehicle technology and on your pocketbook. Automotive manufacturers have to work their technical staffs overtime to accomplish these feats, and the costs will be passed on to the new buyer. Pollution control equipment is a problem each internal combustion engine vehicle owner has to revisit every year: smog checks, emission certificates, replaced valves, pumps, filters, and parts—all extra cost-of-ownership expenses and inconveniences.

How can electric vehicles reduce toxic air pollution emissions? Easy. All electric vehicles are by definition zero-emission vehicles (ZEVs): they emit nothing. That's why California and other states have mandated electric vehicles to solve their air pollution problems. To quote Wang, DeLuchi, and Sperling, who studied the subject extensively:

"The unequivocal conclusion of this paper is that in California and the United States the substitution of electric vehicles for gasoline-powered vehicles will dramatically reduce carbon dioxide, hydrocarbons and to a lesser extent, nitrogen oxide emissions."[19]

The comments of the greenhouse section on power plant emissions associated with electricity production for electric vehicle use are equally applicable to air pollution. In addition, shifting the burden to coal-powered electrical generating plants for electric vehicle electricity production (often located out-of-state) has these effects:

- Focuses smokestack "scrubber" and other mandated controls[20] on stationary sites that are far more controllable than internal combustion engine vehicle tailpipes.
- Shifts automotive emissions from congested, populated areas to remote, less-populated areas where many coal-fired power plants are located.
- Shifts automotive emissions to night time (when most electric vehicles will be recharged) when fewer people are likely to be exposed and emissions are less likely to react in the atmosphere with sunlight to produce smog and other byproduct pollutants.

Electric vehicles generate no emissions whatsoever and, until you get an appreciable number of them, do not impact emissions from electrical generating plants.

## TOXIC SOLID WASTE POLLUTION

Almost everything going into and coming out of the the internal combustion engine is toxic. In addition to the internal combustion engine vehicle's greenhouse gas and toxic air pollution outputs, consider its liquid waste (fuel spills, oil, antifreeze, grease, etc.) and solid waste (oil-air-fuel filters, mufflers, catalytic converters, emission control system parts, radiators, pumps, spark plugs, etc.) byproducts. This does not bode well for our environment, our landfills, or anything else—especially not when multiplied by hundreds of millions of vehicles.

How can the electric vehicle help? The only waste elements of an electric vehicle are its batteries. Lead-acid batteries—the kind commonly available today—are 97 percent recyclable. This means 97 percent of all such batteries and the products that go into them (the sulfuric acid, the lead, and even the plastic of their cases) are recoverable.

## TOXIC INPUT FLUIDS POLLUTION

Remember, almost everything going into and coming out of the the internal combustion engine is toxic. The fuel and oil you put into an internal combustion engine vehicle, the fuel vapors at the pump (and those associated with extracting, refining, transporting and storing fuel), and the antifreeze you use in its cooling system are all toxic and/or carcinogenic, as a quick study of the pump and container labels will point out. On the output side, when burning coal, oil, gas, or any fossil fuel, you create more problems either by the amount of carbon dioxide ($CO_2$) or by the type of other toxic emissions produced.

Everything you pour into an internal combustion engine is toxic, but some chemicals are especially nasty. In addition to more than 200 compounds on its initial hazardous list, the Clean Air Act of 1990 amendments said:

". . . study shall focus on those categories of emissions that pose the greatest risk to human health or about which significant uncertainties remain, including emissions of benzene, formaldehyde and 1, 3 butadiene."

Fouling the environment as in the Exxon Valdez oil spill disaster is one thing. Poisoning your own drinking water is another. Those enormous holes in the ground near neighborhood gas stations everywhere (as they rush to be compliant with Federal regulations regarding acceptable levels of gasoline storage tank leakage) make the point. So does the recall of millions of bottled Perrier drinking water where only tiny levels of benzene contamination were involved.

How can the electric vehicle help? The only substance you pour into your electric vehicle occasionally is water (preferably distilled).

## WASTE HEAT DUE TO INEFFICIENCY

Although its present form represents its highest evolution to date, the gasoline-powered internal combustion engine is classified among the least efficient mechanical devices on the planet. In gasoline-powered vehicles, only 15 percent of the energy of combustion becomes mechanical energy. Eighty five percent becomes heat lost in the engine system, and of this 85 percent:

- One third overcomes aerodynamic drag (energy ends up as heat in air).
- One third overcomes rolling friction (energy ends up as heated tires).
- One third powers acceleration (energy ends up as heating the brakes).

In contrast to the hundreds of internal combustion engine moving parts, the electric motor has just one. That's why they're so efficient. Today's EV motor efficiencies are typically 90 percent or more. The same applies to today's solid-state controllers (with no moving parts), and today's lead-acid batteries come in at 75 percent or more. Combine all these and you have an electric vehicle efficiency far greater than anything possible with an internal combustion engine vehicle.

## ELECTRIC UTILITIES LOVE ELECTRIC VEHICLES

Even the most wildly optimistic electric vehicle projections show only a few million electric vehicles in use by early in the 21st century. Somewhere around that level, EVs will begin making a dent in the strategic oil, greenhouse, and air quality problems. But until you reach the 10 to 20 million or more EV population level, you're not going to require additional electrical generating capacity. This is due to the magic of load leveling. If electric vehicles are used during the day and recharged at night, they perform a great service for their local electrical utility, whose demand curves almost universally look like that shown in FIG. 2-9.

How electricity is generated varies widely from one geographic region to another, and even from city to city in a United States region. In 1991, the net electricity mix generated by electric utilities was 54.87 percent coal, 21.73 percent nuclear fission, 9.75

**2-9** Weekly peak power-demand curve for a large utility operating with a weekly load factor of about 80%.

percent hydroelectric, 9.36 percent natural gas, 3.93 percent oil, and 0.35 percent geothermal and other.[21] Electric utility plants producing electricity at the lowest cost (i.e., coal and hydro) are used to supply base-load demands, while peak demands are met by less economical generation facilities (i.e., gas and oil).

By owners recharging their electric vehicles in the evening hours (valley periods) they receive the benefit of an off-peak (typically lower) electric rate. By raising the valleys and bringing up its base-load demand, the electric utility is able to more efficiently utilize·its existing plant capacity. This is a tremendous near-term economic benefit to our electric utilities because it represents a new market for electricity sales with no additional associated capital asset expense.

## CHAPTER EPILOGUE

Hey, you only wanted to build an electric vehicle and this chapter has burdened you with environmental facts. Well, earth is your planet too. If you don't want to take action, that's certainly your prerogative. On the other hand, electric vehicle ownership is the best first step you can take to help save the planet. But there is still more you can do. Do your homework. Write your Senator or Congressperson. Voice your opinion. Get involved with the issues. But don't settle for an answer that says we'll study it and get back to you. Settle only for action—who is going to do what by when and why. I leave you with a restatement of the problem, a possible framework for a solution and some additional food for thought.

### Legacy of internal combustion engine is environmental problems

Internal combustion engine technology and fuel should be priced to reflect its true social cost, not just its economic cost, because of the environmental problems it creates:

- Our dependence on foreign oil security risk problem.
- The greenhouse problem.
- The air quality problems of our cities.
- Toxic waste problem.
- Toxic input fluids problem.
- Inefficiency problem.

No other nation prices its gasoline so cheaply. At the very least, our gasoline should have an associated tax with it that goes to rebuild our transportation infrastructure—just like other advanced nations do. Better still, our gasoline's cost should reflect our cost to defend foreign oil fields, reverse the greenhouse effect, and solve the air quality issues. Best of all, our gasoline's cost should include substantial funding to research solar energy generation (and other renewable sources) and electric vehicle technologies—the two most environmentally beneficial and technologically promising gifts we can give to our future generations.

## A proactive solution

People living in the United States have been extremely fortunate. We have been blessed by clean air, abundant natural resources, stable government, inexpensive energy costs, and a convenience and true standard of living second to no other country on the planet. But nothing guarantees our future generations will enjoy the same birthright. In fact, if we fold our hands behind our backs and walk away from today's environmental problems, we guarantee our children and children's children will not enjoy the same standard of living as we do. For the sake of our children, we cannot walk away, we must do something. We must attack the problem straight on, pull it up by its roots and replace it by a solution.

Figure 2-10 suggests a possible approach. We need to look at the results wanted in the mid-21st century and work backward—on both the supply and demand sides—to see what we must start doing today. Clearly, it's time for a sweeping change—but we all have to want it and work toward it for it to happen. No one has to be hurt by the change if they become part of the change. Auto makers can make more efficient vehicles. Suppliers can provide new parts in place of the old. The petrochemical industry can alter its mix to supply less crude as oil and gas and more as feedstock material used in making vehicles, homes, roads, and millions of other useful items. Long before any of these happen, you can do your part by building your own electric vehicle.

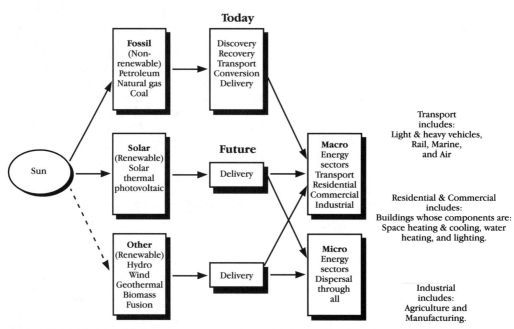

**2-10** Model of balanced future energy usage made possible by working from future desired goal back to today.

# 3
# Electric vehicle history

*Precedent said, it cannot be done. Experience said, it is done.*

Darwin Gross, *Universal Key*

While modern technology has made electric vehicles better, there is very little new electric vehicle technology. Today's EV components would be instantly recognizable in those that roamed our streets a century ago. As a potential EV builder or converter, you should be happy to know they have a long and distinguished heritage—you might even get some useful building ideas by looking at the earliest-vintage EVs in an automobile museum.

Ironically, EVs were around before internal combustion engine vehicles and will also be around after them. The two vehicle types will coexist for some time to come. In this chapter you'll learn about the history of EVs, the forces that shaped their demise, the trends that forced their resurgence, and the tremendous positive role awaiting them in the future.

## THREE STAGES OF VEHICLE HISTORY

Studying vehicle history is similar to looking at any economic phenomena. Hula hoops are a good example. The first hoola hoop is a novelty; the one hundredth creates a strong desire to own one. By the ten thousandth you own three; by the one millionth the novelty has worn off; and after the hundred millionth they're an eyesore on the landscape. The same with vehicles—past events shift the background climate, and affect current consumer wants and needs. The innovative Model T of the 1910s was an outdated clunker in the 1920s. The great finned wonders of the 1950s and muscle cars of the 1960s were an anachronism by the 1970s. A vehicle that was once in great demand is now only junkyard material because wants and needs change.

Figure 3-1 is rather busy, but studying it gives you the clues to the rise and fall of the three types of vehicles in one picture—steam, electric, internal combustion—plus the interrelationship between them during the three stages of vehicle history. Figure 3-1 shows that steam has been passed by as a vehicle power source but electric vehi-

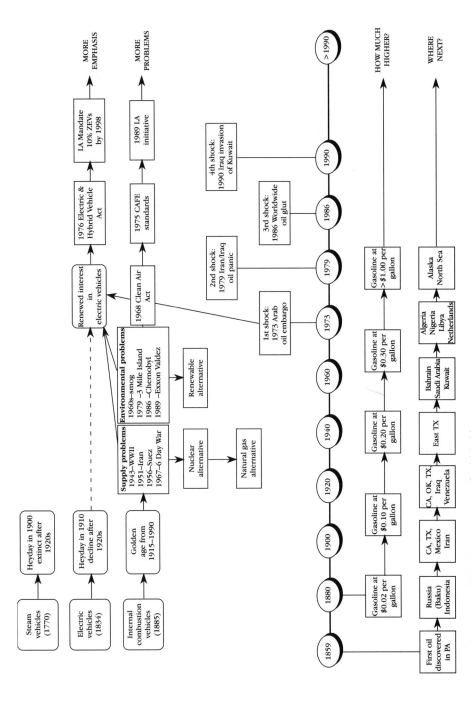

**3-1** Overview of the three streams of vehicle history.

cles, dominant in urban areas at the turn of the century, are again returning to favor as the majority of the world's industrial nations become "urbanized," with widespread availability of electrical power and roads. The 75 "golden years" of the internal combustion engine vehicle, which swamped the early steam and electric offerings in a wave of cheap gasoline prices and offered the ability to travel where there were no tracks, are over. The internal combustion engine loses attraction as mankind begins to choke on its toxic wastes, and as the primary supplies of oil are dominated by foreign interests. Vehicle history can be divided into three chronological periods or stages:

*Up until 1915.* This period marked the transition from the "novelty" era (most bystanders were amazed that these devices actually worked) to the "practical" era, where early buyers just wanted a vehicle to go from one point to another with minimum hassle, and finally to the "production" era. In the production era, after nearly 3000 vehicle manufacturers had come and gone, the survivors fell into two camps—those able to make a profit catering to custom needs at a high market price, and those offering a standard solution at a low market price. Steam and electric offerings were overwhelmed by the dominance of cheap oil and gasoline and virtually disappeared as competitors after 1915.

*1915 to 1990.* This 75-year period was the "golden age" of the internal combustion engine vehicle. With cheap, available gasoline prevailing as fuel, and basic internal combustion engine vehicle design fixed, manufacturing economies of scale brought the price within reach of every consumer. Expansion away from urban areas made vehicle ownership a necessity. The creation of an enormous highway infrastructure culminated in completion of the interstate highway system. This was accompanied by the destruction of urban non-internal-combustion-powered transit infrastructure by political maneuvering in the United States, and by damage during World War II in Japan and Europe, creating a highly complex and huge financial burden in the energy-efficient 1990s.

*Mid 1960s thru 1990s (and beyond).* This period marked a successively heightened awareness of problems with internal combustion vehicles. Smog problems of the mid-1960s made us aware we were polluting our environment and killing ourselves. Arab oil embargos, shortages, and gluts of the 1970s, 1980s, 1990s made us aware of our dependence on foreign oil. Nuclear and oil spill accidents of the 1970s and 1980s made us aware of the long-range consequences of our short-range energy decisions. The internal combustion engine and oil problems that started with a whimper in the mid 1960s turned into a groundswell of public opinion by the 1990s. The net result of new awareness in this still unfinished period has been the re-emergence of electric vehicles. When legislative action mandating zero emission vehicles in the 1990s forced rethinking of basic vehicle design, current technology applied to the EV concept emerged as the ideal solution.

A brief look at these three stages of vehicle history is helpful in understanding why electric vehicles came, went away, and are back again.


## THREE STREAMS OF THOUGHT

Steam engines came first, followed by electric motors, and finally by internal combustion engines. The close proximity of coal and iron deposits in the northern latitudes of what came to be known as the industrialized nations—United States, Europe/England and Asia—made the steam engine practical. The thriving postindustrial revolution economy provided by the steam engine created the climate for electrical invention. Electrical devices made the internal combustion engine possible. Vehicles powered by them followed the same development sequence.

In 1900, half of the 80 million people in the United States lived in a few large (mostly Eastern) cities with paved roads, and the other half in towns linked by dirt roads or in countryside with no roads at all. Less than 10 percent of the 2 million miles of roads were paved. More than 25 million horses and mules provided mobility for the masses. Electric lighting in the larger cities was dwarfed by the use of kerosene lamps—popularized by the discovery of plentiful amounts of oil—of the countryside. Coal-or wood-burning steam engine locomotives were high tech. Only 200,000 miles of railroad track existed, but some of it provided fast, efficient transportation between major locations—New York to Chicago on the "Twentieth Century Limited" took 20 hours. While a New York-to-California railroad ticket cost $50, vehicles of any kind cost around $5000 to $50,000 in today's dollars—putting them out of the hands of all but the well-to-do.

The three types of vehicles came into this turn-of-the-century United States environment. Almost 3000 manufacturers experimented with various combinations of propulsion (steam, electric, or internal combustion engine); fuel (water/kerosene, battery, gasoline/oil); cooling (air or liquid); mounting (front, rear or middle); drive (front or rear wheels via shaft, gear, chain or belt); chassis (three or four wheel; independent suspension or leaf/coil springs); and tires (pneumatic or solid). By the time World War I was over the internal combustion vehicle had emerged as the clear victor.

Figure 3-2 shows the steam, electric and internal combustion vehicle population in the United States from 1900 to 2000. Steam-powered vehicles, popular in the last part of the 1800s, declined in favor of the other two vehicle types after the early 1900s. Electric vehicles enjoyed rapid growth and popularity until about 1910, then a slow rise until their resurgence in the 1990s. Internal combustion engine-powered vehicles passed steam and electric early in the 1900s. More than any other factor, cheap and nearly unlimited amounts of domestic (and later foreign) oil, which kept gasoline prices between ten and twenty cents a gallon from 1900 through 1920, suppressed interest in alternatives to internal combustion engine vehicles until more than 50 years later (the 1970s).

## Huff and puff

The same phenomenon that makes the heated tea kettle on your stove whistle, when suitably harnessed, makes a steam engine go. The "huffing and puffing" associated with steam railroad locomotives is absent from the whisper silent steam automobiles, yet they are equally powerful pound for pound. However, you still need to heat the water, which means burning something, and if you don't continually monitor your boiler steam pressure, everything can blow up.

James Watt's steam engine of 1765—widely-acclaimed as responsible for the industrial revolution—was only an improvement of Thomas Newcomen's 1712 machine that, in turn, built on the more primitive 1690-vintage designs of Denis Papin, Christian Huygens, and Robert Boyle, and the initial patent of Thomas Savery in 1698. Steam technology was applied to the first land vehicle—Nicolas Cugnot's tractor—in 1770, to a steamboat by John Fitch in 1787, and to a rail locomotive by Richard Tevithick in 1804.

While the Cugnot steam tractor is a far cry from the Stanley Steamer automobiles of the early 1900s (a streamlined version of the latter set the land speed record at 122 mph in 1906), and still further removed from the high-performance Lear steam cars of a few decades ago, the problem with steam vehicles remains the steam. Water needs a lot of heat to become steam, and it freezes at cold temperatures. To get around these and the basic "time to startup" problems, technical complexity was introduced in the

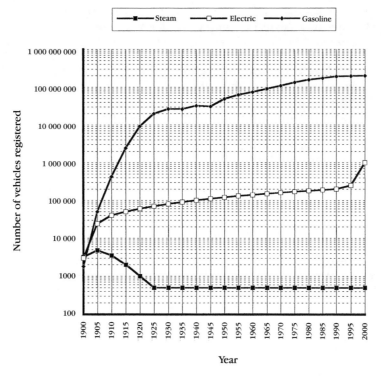

**3-2** Growth of the three vehicle types in the United States from 1900 to 2000.

form of exotic liquids to withstand repeated evaporation and condensation, and exotic metals for more sophisticated boilers, valves, piping, and reheaters.

In the early 1900s, steam vehicles unquestionably offered smoothness, silence, and acceleration. But stops for water were typically more frequent than stops for kerosene, and steamer designs required additional complexity and a lengthy startup sequence. While 40 percent of the vehicles sold in 1900 were steam (38 percent were electric),[1] electrics offered simplicity, reliability, and ease of operation, while gasoline vehicles offered greater range and fuel efficiency. Thus steamers declined, and only a handful operate today.

## Wheel goes round and round

Electricity is everywhere. In one place it lights a factory, in another it conveys a message, and in a third it drives an electric vehicle. Electricity is transportable—it can be generated at a low-cost location and conveniently shipped hundred of miles to where it is needed. A storage battery—charged from electricity provided by a convenient wall outlet—can reliably carry electricity to start a car anywhere, or power an electric vehicle. Who knows (or cares) that the wheels of your vehicle turn because the electricity was generated from water turning a turbine wheel linked to a generator? We take electricity for granted today and are continually developing new uses for it, primarily because of its advantages—it's clean, simple, available, and reliable. But our modern electrical heritage owes a great debt to many pioneers.

Alessandro Volta, building on the experiments of Luigi Galvani in 1782, invented the electric battery—his "Voltaic pile"—in 1800. Joseph Henry, building on the experiments of Han Christian Oersted in 1819 and Andre Ampere in 1820, created the first primitive direct current (dc) electric motor in 1830. Michael Faraday demonstrated the induction principle and the first electric dc generator in 1831. Battery-powered electric technology was applied to the first land vehicle by Thomas Davenport in 1834, to a small boat by M. H. Jacobi in 1834, and to the first battery-powered locomotive—the five-ton "Galvani"—by Robert Davidson in 1838.

Moses Farmer unveiled a two-passenger electric car in 1847, and Charles Page showed off a 20-mph electric car in 1851, but Gaston Plante's lead-acid "rechargeable" battery breakthrough of 1859—improved upon by Camille Favre in 1881 and H. Tudor in 1890—paved the way for extended electric vehicle use. Nikola Tesla's alternating current (ac) induction motor of 1882 and subsequent polyphase patents paved the way for the ac electrical power distribution infrastructure we use today. By the 1890s, dc power distribution via dynamos had been in use for a decade. ac power distribution began with the 1896 Niagara Falls power plant contract award to George Westinghouse (Nikola Tesla's patents) after a brief but intense battle with Thomas Edison's dc forces.

Figure 3-3 shows a cross-section of the more prominent United States electric vehicle manufacturers in operation from 1895 through the 1930s. By 1912, the peak pro-

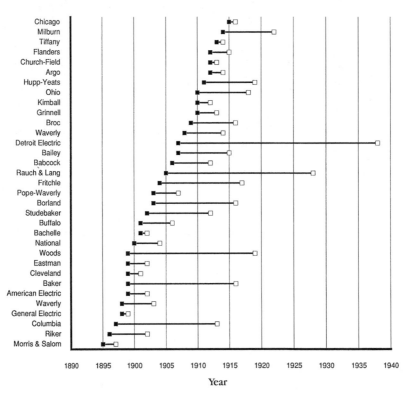

**3-3** Prominent electric vehicle manufacturers in the United States from 1895 through the 1930s.

duction year for early electrics, 34,000 cars were registered. The *Reader's Guide to Periodical Literature* listings tell the story. The half page of magazine articles listed in the 1890 through 1914 volumes dwindled to a quarter page in 1915–18 and disappeared altogether in the 1925–28 volume.

Early electric vehicle success in urban areas was easy to understand. Most paved roads were in urban areas; power was conveniently available; urban distances were short; speed limits were low; and safety, comfort, and convenience were primary purchase considerations. The quietness, ease of driving, and high reliability made EVs a natural with the wealthy urban set in general and well-to-do women in particular. Clara Bryant Ford (Mrs. Henry Ford) could have any automobile she wanted, but she chose the Detroit Electric now on display at the Henry Ford Museum shown in FIG. 3-4 for getting around the Ford Park Lane estate and running errands.[2] Thomas Edison's 1889-vintage electric vehicle was a test platform for his rechargeable nickel-iron battery experiments. Later, Edison production nickel-iron batteries went into the Bailey Electric and numerous other electrics. Edison had his own personal Studebaker electric vehicle, and both he and Henry Ford were strongly supportive of EVs. At one time these two planned to bring out a lightweight $750 electric auto that was to be called the Edison-Ford.[3]

**3-4** Mrs. Henry Ford's Detroit Electric now on display in the Henry Ford museum.

Electric vehicles also dominated the commercial delivery fleets in urban areas around the world. Department stores, express delivery companies, post offices, utility and taxicab companies in New York, Chicago, London, Paris, and Berlin used thousands of EVs. High reliability (99 percent of the 300-day work year availability) and low maintenance characterized commercial EVs and made them fleet favorites.

In the electric's performance department, typical 2500-pound cars went 20 to 30 mph, and got 50 to 60 miles on a battery charge. Half-ton trucks went 10 to 15 mph and had a 40- to 50-mile range. Ten-ton trucks went 5 to 10 mph and had a 30- to 40-mile range. An electric vehicle set the first land speed record. Camille Jenatzy's "Jamais Contente" (a streamlined vehicle powered by two 12-horsepower electric motors riding on narrow 25-inch diameter tires) went 66 mph in April, 1899—a record that stood for three years until broken by the Baker Electric "Torpedo" in 1902 at 78 mph, and later by the "Torpedo Kid" in 1904 at 104 mph. In 1900, the French B.G.S. Company's electric car set the world's electric distance record of 180 miles per charge.

A 1915 version of Clara Ford's car—a Detroit Electric powered by a 5.5-horsepower dc motor driven at 72 volts—when retested 60 years later (with new batteries) by Machine Design Magazine—still delivered 25 mph and an 80-mile range. It was still recommended by the magazine as a "best buy"—proving the point that we can routinely expect long lifetimes out of electrical machinery. While electric automobiles were a common sight until the mid-1910s, and commercial and industrial EVs have enjoyed continued growth and success on up through today, cheap oil combined with the nonelectrification of rural areas assured victory for internal combustion engine vehicles. Ironically, it was an electric vehicle's motor and battery—adapted as an electric starter for internal combustion engine vehicles by Charles Kettering in 1912—that delivered the crushing coup de grace to early electric autos.

## Ashes to ashes, dust to dust

The story of the internal combustion vehicle is inextricably linked to the story of oil itself. But the internal combustion engine's rise in popularity was due more to the great economic advantage of oil rather than any technical advantage of the internal combustion engine. Today, with the United States and other industrialized nations substantially dependent on foreign oil, the strategic economic disadvantage of oil coupled with the environmental disadvantage of the internal combustion engine has created strong arguments for alternative solutions. Let's examine how this situation was created.

Animal oils had been used for centuries to provide illumination. Rock oils (so called to indicate that they derived directly from the ground, and the original name for crude oil or petroleum) were envisioned in the 1850s only as superior alternatives for illumination and lubrication in the upcoming mechanical age. Earlier researchers had discovered that a quality illuminating oil—kerosene—could be extracted from coal or rock oil. Coal existed in plentiful quantities—all that remained was to discover a substantial source of crude oil/petroleum.

The discovery of oil in Western Pennsylvania by Edwin Drake in 1859, was the spark that ignited the oil revolution. Almost overnight, the boom in Pennsylvania oil, with its byproducts exported globally, became vitally important to the United States economy. The promise of fabulous wealth provided the impetus that attracted the best business minds of the age to the quest. Soon the oil business—dominated by kerosene—was controlled by the worldwide monopolies of John Rockefeller's Standard Oil (production/distribution from Pennsylvania in the United States), Ludwig and Robert Nobel (production from Baku on the Russian Caspian Sea), Alphonse and Edmond Rothchild (production from Baku, distribution from Batum on the Russian Black Sea), Shell (production and tanker distribution from Batum/Borneo to England and the Far East) and Royal Dutch (production from northeast Sumatra in Indonesia). These monopolies, securely in place before the 1900s, were all based on the markets for oil as kerosene and lubricating products. In the 1890s, gasoline—once thrown away after kerosene was obtained—was lucky to bring two cents a gallon, but that was about to change.[4]

Coal was the foundation for the industrial revolution, and the first internal combustion engine built in 1860 by Etienne Lenoir was fired by coal gas. Nikolaus Otto improved on the design with a four-cycle approach in 1876. But the discovery that gasoline was an even more "combustible" fuel that was also inexpensive, plentiful, and powerful was the spark that ignited the internal combustion engine revolution. All that remained was controlling the explosive gasoline-air mixture—solved by Gottlieb Daimler's carburetor design of 1885—and controlling the timing—solved by Karl Benz's enhanced battery-spark coil-spark plug ignition design of 1885—for the internal combustion engine as we know it today to emerge.

Early internal combustion vehicles were noisy, difficult to learn to drive, difficult to start, and prone to explosions (backfiring) that categorized them as dangerous in competing steam and electric advertisements. Internal combustion vehicle offerings from Daimler (Germany, 1886), Benz (Germany, 1888), Duryea (United States, 1893), Peugot (France, 1894), and Bremer (England, 1894)[5] were primitive engineering accomplishments in search of a marketing niche, while contemporary enclosed-body electrics—targeted at the elite urban carriage trade—sold briskly at $5000 a copy.

This changed quickly, perhaps due to inspiration of the Chicago Times-Herald Thanksgiving Day race of 1895,[6] an event held "with the desire to promote, encourage, stimulate invention, development and perfection and general adoption of the motor vehicle in the United States." Won by Frank Duryea, driving a Duryea brothers motor wagon, it brought instant fame to the brothers but, more importantly, brought most of the United States automotive pioneers together for the first time. Only three years later, more than 200 companies had been organized to manufacture motorcars.

Simultaneously, when internal combustion powered vehicles were still decades away from dominance, discovery of the Los Angeles field in the 1890s, the "Spindletop" field near Beaumont, Texas in January 1901, and the Oklahoma fields of the early 1900s saw boom and bust times that priced a 42-gallon barrel of crude oil (typically from 15 to 20 percent recoverable as gasoline) as low as three cents a barrel. While dc and ac electrical distribution systems guaranteed that electric lighting would replace the kerosene lamp, cheap domestic oil—which kept gasoline prices between two and ten cents a gallon between 1890 and 1910—guaranteed the success of internal combustion vehicles.

Like Rockefeller with oil, Henry Ford was the individual who was in the right place (Detroit) at the right time (October 1908) with the right idea (Model T Ford) at the right price ($850 FOB Detroit). Ford had attended neither the Chicago Times-Herald race nor the earlier World's Columbian Exposition of Chicago that opened May 1, 1893, but written information derived from these events doubtless inspired his first creation—the 1896 Quadricycle.[7] By 1908, Henry Ford had produced numerous designs. Hand-built 1899 and 1901 models followed the 1896 Quadricycle; Ford Motor Company models A, B, C, F, N, R, S, and K preceded the T; and Ford had won races with his Grosse Point racer of 1901 and famous "999" Barney Oldfield racer of 1902. But it was the innovation of the mass-produced, one-color-fits-all 1909 Model T at the $850 price that put the internal combustion vehicle on the map. The four-cylinder, 20-horsepower, 1200-pound 1909 Model T's instant success created an enormous demand that lasted nearly 19 years—more than 15,000,000 were manufactured until production ceased in May 1927. By producing nearly the same model, manufacturing economies of scale enabled the price to be dropped year after year until its all-time low of $290 in December 1924. In addition to low purchase price, the Model T's success was also due to its operating economy. Its two forward/one reverse speed planetary transmission and 30-inch wheels (with recommended tire pressures of 60 psi) drove the Model T's engine at 1000 rpm at 25 mph and 1800 rpm at 45 mph—producing a typical gas mileage of 20 miles per gallon and up.

Simultaneously, integration (along with the 1905 political problems in Russia) had consolidated the world oil market in the hands of two companies by 1907: Standard Oil and Royal Dutch/Shell. But by 1911, the investigation of Standard Oil launched by president Teddy Roosevelt in 1904 resulted in the United States Federal court finding Standard Oil guilty of antitrust violations and ordering its breakup into the companies we recognize today: Standard Oil of New Jersey (Exxon), Standard Oil of New York (Mobil), Standard Oil of California (Chevron), Standard Oil of Ohio (Sohio—BP/America), Standard Oil of Indiana (Amoco), Continental Oil (Conoco) and Atlantic (ARCO/Sun). While this breakup initially led to a decade of peaceful coexistence among the former allied parts, it also paved the way for the oil industry as we know it today, dominated by multiple, large, fiercely competitive, multinational corporations.

Other events of the period also contributed to oil's rise and dominance: the introduction of "thermal cracking" by Standard Oil of Indiana in 1913 (a process that more than doubled the amount of gasoline recoverable from a barrel of crude oil, up to 45 percent); discovery of oil near Tampico, Mexico in 1910; discovery of oil in Persia (Iran) that led to construction of an Abadan refinery in 1912 by Anglo-Persian (a pre-World War I "strategic" decision by Winston Churchill gave the British government—British Petroleum—51 percent ownership of Anglo-Persian after it ran into financial difficulties); and World War I itself.

Meanwhile, other internal combustion engine vehicle innovators were busy too: Walter Chrysler, John and Horace Dodge (the brothers who began as captive suppliers to Ford), and numerous others provided innovations that survive to the present day. William Durant incorporated General Motors in September 1908,[8] and by the 1920s its major divisions (Buick, Oldsmobile, Cadillac, Oakland, Chevrolet, GM Truck) and its supporting divisions (Fisher Body, Harrison Radiator, Champion Spark Plug, DELCO, Hyatt Roller Bearing, et al) were household names in the United States. While Durant acquired valuable assets in assembling GM's acquisitions under one holding-company umbrella, it was the talent he obtained (such as Alfred Sloan from Hyatt and Charles Kettering from Cadillac) that paved the way for GM's later rise to dominance. GM innovations such as color, streamlining, smoother, more-powerful six-cylinder engines, and annual model styling changes made obsolete Ford's Model T—despite its $290 price in 1924. The internal combustion vehicles were now on their way.

## THE GOLDEN AGE OF INTERNAL COMBUSTION

Internal combustion engine vehicle growth in the United States exploded with World War I. After World War II, world internal combustion engine automotive growth was even more dramatic. What made all this possible was the unprecedented oil availability, and the relative price stability shown in FIG. 3-5, which allowed United States gasoline prices to move from roughly $0.10 to $0.30 during this 50-year period (FIG. 3-1). Yet in terms of inflation-adjusted "real dollars" the cost of gasoline actually went down. Is it any wonder that no one cared how large the cars were in the 1950s or how much gas they guzzled in the 1960s? Gasoline was cheaper than water.

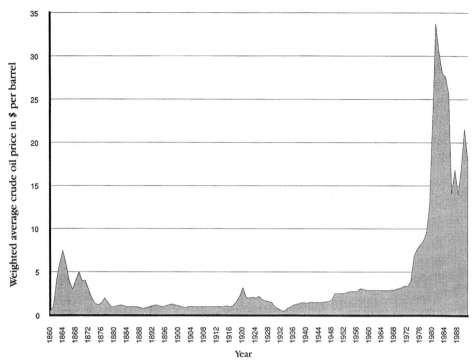

**3-5** Oil prices from 1860 to 1990.

But FIG. 3-1 also shows the problems: no major discoveries since the Alaskan and North Sea fields of two decades ago; the increasing concerns with oil supplies beginning during World War II; the introduction of nuclear and natural gas energy alternatives in the 1950s; and the hardening of public opinion with the increasing frequency of smog and air-quality problems, oil and nuclear environmental accidents, and the foreign oil shocks. Already forced to comply with more stringent emission standards by the Clean Air Act of 1968, the first oil shock of 1973 caught the "big three" United States automobile manufacturers with their pants down. Japanese and European auto manufacturers had smaller, more fuel-efficient internal combustion vehicle solutions as a result of years of higher gasoline prices (due to higher taxes earmarked for infrastructure rebuilding). The market share lost by the big three to foreign automakers has never been regained.

By the early 1990s, the wild oil party of the preceding 75 years was over. Environmental problems, the need for energy conservation, and the instability of foreign oil supply all signal that the sun is setting on the internal combustion vehicle. It will not happen overnight. In the near term the industrialized nations of the world and emerging Third World nations will consume ever greater amounts of foreign oil. But it's inevitable that a replacement of the fossil-fuel-burning internal combustion vehicle will be found.

## Internal combustion vehicle growth fueled by cheap and plentiful oil

The lessons of World War I were simple. Flexible, oil-powered internal combustion engine cars, trucks, tanks, and airplanes were superior to fixed, coal-powered railway transportation; and those who controlled the supply of oil won the war. The allies had Standard Oil, Royal Dutch/Shell, and Anglo-Persian Oil. The Germans did not have access to vast amounts of oil; the destruction of the Ploesti refinery in Rumania, and their belated, failed attempts at capturing Baku cost them the war.

Meanwhile, as FIG. 3-2 showed, internal combustion engine vehicle registrations in the United States exploded from one-half million in 1910, to 9 million in 1920, to 27 million in 1930 and—slowed by the depression—to 33 million in 1940. Gasoline that was sold by the local blacksmith in containers in the early 1900s gave way to 10,000 wooden "filling stations" with gravity-feed tanks at the beginning of the 1920s, to more than 150,000 buried tank/electric-pump-driven "service stations" in the 1930s. More and more paved roads were built; the landscape was changed forever. More oil continued to be found in the 1920s: California (Signal Hill), Oklahoma (greater Seminole and Oklahoma City), West Texas (Yates Field), Venezuela (Maracaibo Basin), and Iraq (Kirkuk). Then the biggest of them all was discovered in October 1930—the giant East Texas oil reservoir that later proved to measure 45 miles long and up to 10 miles wide. Crude oil that sold for around $2 a barrel in the mid 1920s dropped to less than 10 cents a barrel in the early 1930s (the low was 4 cents a barrel in May 1933), and gasoline prices that had been chugging along between 10 and 20 cents per gallon from 1910 through the 1920s dropped accordingly. Now the problem was too much oil, and the United States government had to enter the picture to control prices.

## World War II oil lessons are learned by all

Oil was the one resource Japan did not have at all. In retrospect, Japan's war was easy to understand. It needed the oil resources of Indonesia, Malaysia, and Indochina. After an oil embargo against Japan was set up in mid-1941 by blocking the use of Japanese funds held in the United States, Japan was desperate for oil, and did what it had to do to get it. The Pearl Harbor attack was an effort to protect its Eastern flank, but poor timing made it an infamous event (Japan's "declaration of war" didn't get delivered until after the attack). Japan's early loss of planes and ships at Midway meant it was never able to provide adequate protection for its oil tanker convoys from Indonesia. Dwindling oil reserves and a nonfunctioning synthetic fuels program meant new pilots couldn't be trained and ship fleets couldn't maneuver. While Japan "lost" World War II long before 1945, it learned its oil lesson well and converted to the oil standard soon after the war.

On the other hand I.G. Farben, the huge German chemical combine, had mastered synthetic fuel recovery from coal by the early 1920s—*hydrogenation* was the most popular method—and Germany had plenty of coal. But most of Germany's oil imports came from the West, and increasing demand was causing a foreign exchange hemorrhage. Hitler believed if Russian Baku oil reserves could be added to those of its ally Rumania, along with Germany's own 1940 synthetic fuel reserves, the "Thousand Year Reich" was a cinch via a blitzkrieg-tactics war that didn't consume much fuel. Unfortunately for Hitler, his blitzkrieg advances frequently outran their fuel supply trucks, he never got to Baku, his Rumanian oil supply at Ploesti was destroyed early in the war, and the German advance in North Africa came to a halt when its oil tankers couldn't make it across the Mediterranean. In addition, German synthetic fuel aviation gasoline was never quite as "hot" as that produced from real crude oil, and the entire German war machine came to a halt when systematic bombings of its synthetic fuel plants later in the war reduced them to rubble. Germany also lost World War II long before 1945, but learned its oil lesson well and converted to the oil standard soon after the war despite massive reserves of coal.

What did the allies learn from World War II? They relearned the lesson from World War I: Whoever controls the supply of oil wins the war. They also learned the value of a strategic petroleum reserve. Up until 1943, they nearly lost the war to the Germans in the North Atlantic—the success of submarine wolf packs made it nearly impossible for allied oil tankers to resupply England, Europe, and Africa. Even in 1945, countless lives were wasted and the Russians moved toward Berlin while Patton's tanks sat without fuel in France, giving the Germans time to regroup and resupply.

While the United States provided six out of every seven barrels of allied oil during World War II, it was recognized by many in government that it would soon become a net importer of oil. More oil had been discovered in Bahrain in 1932, and in Kuwait (Burgan field) and Saudi Arabia (Damman field) in 1938. In 1943, as all eyes turned toward the mideast with its reserves variously estimated at around 600 billion barrels, the United States government proposed the "solidification" plan to assist the oil companies (i.e., share the financial risk) in Saudi Arabian oil development. While this plan and subsequent revisions to it were rejected with much indignation by the oil companies, decades later with 20/20 hindsight, they would all gladly change their votes.

## A world awash in oil after World War II

After World War II, pent up consumer demand released another type of blitzkrieg—the internal combustion engine automobile. While FIG. 3-6 shows United States automobile registrations were almost quadrupling, the rest of the world's auto population grew more than twentyfold: from 13 million in 1950 to nearly 300 million in 1990. Oil exploration in this period was in high gear. A nearly inexhaustible supply had apparently been found in the Middle East—gasoline prices bounced between 20 and 30 cents per gallon until the early 1970s. Aided by the convenience of the internal combustion automobile, America moved to the suburbs, where distances were measured in commuting minutes, not miles. Fuel-efficient automobiles were the last item on anyone's mind during this period—gasoline was plentiful and cheap (reflecting underlying oil prices) and regular local retail price wars made it even cheaper. *Environment*

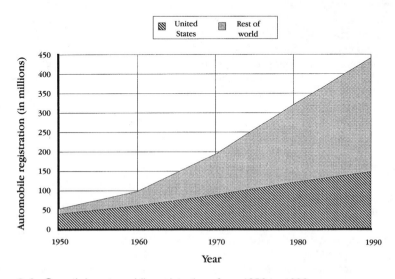

**3-6** Growth in automobile registrations from 1950 to 1990.

was an infrequently used word with an unclear meaning. Highway construction proceeded at an unprecedented pace, culminating with the Interstate Highway Bill signed by President Eisenhower in 1956, authorizing a 42,500-mile superhighway system. Public transportation and the railroads—the big losers in Japan and Europe due to World War II damage—also became the big losers in the United States as the U.S. government formally finished the job that major industrial corporations—acting in conspiratorial secrecy and convicted of violating the Sherman Antitrust Act—had started in the 1930s and 1940s: ripping up the tracks, dismantling the infrastructure, and scrapping intercity and intracity light rail and trolley systems that could have saved consumers, cities, and the environment the expenditure of billions of dollars today.[9]

The post World War II rebuilding of European and Japanese infrastructures made them more modern than the United States. Germany and Japan (and most of the rest of the industrialized world) rapidly converted from coal to oil economies after World War II, and underwent an unprecedented period of economic and industrial expansion as the surge in automobile registrations outside of the United States (FIG. 3-6) attests. All the industrialized economies of the world were now dependent on internal combustion engine vehicles and oil.

## Twilight of the oil gods

By the middle of the 1960s, many in government and industry around the globe became aware that something was very wrong with this picture. Although United States movement towards energy alternatives such as natural gas and nuclear fission (started decades earlier) was rapidly bringing new alternative capacity online, dependence on oil was becoming worse, and smog and environmental issues began coming into the foreground. Passage of the Clean Air Act of 1968 was one result. The 1975 passage of a corporate average fuel economy (CAFE) standards bill was another. The problem was obvious to some, but most of the public chugged merrily along in their internal combustion powered vehicles.

After the "first shock"—the Arab embargo that followed the October 1973 Yom Kippur War—everyone knew there was a problem. By cutting production 5 percent per month from September levels, and cutting an additional 5 percent each succeeding month until their price objectives were met, the Organization for Petroleum Exporting Countries (OPEC) effectively panicked, strangled, and subverted the industrialized nations of the world to their will. The panic was exacerbated by nations and oil companies scrambling for supplies on the world market—overbuying at any price to make sure they had enough—and consumers doing the same by waiting in lines to "top off their tanks" when weeks before they would have thought nothing of driving around with their gas gauges on empty. When the dust settled, the United States consumer, who had paid about $0.30 a gallon for all the gas he could get a few months earlier, now paid $1 a gallon or more at the height of the crisis, and waited in line to get a rationed amount. Figure 3-7 shows the drastic change.

How could it happen? Easy. The oil crises in 1951 (Iran's shutdown of Abadan), 1956 (Egypt's shutdown of Suez canal), and 1967 (Arab embargo following June Six-Day War) were effectively managed by joint government and oil industry redirection of surplus United States capacity. But by the early 1970s there was no longer any surplus capacity to redirect—the United States production peak of 11.3 million barrels of oil per day occurred in 1970. Up until the 1970s, the oil industry focused on restraining production to support prices. Collateral to this action, relatively low oil prices forced low investment and discovery rates, and import quotas kept a lid on supplies. But rising demand erased the need for production-restraint tactics and the surplus capacity along with it.

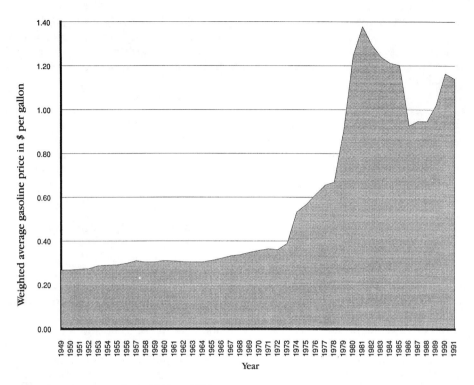

**3-7** Gasoline prices from 1949 to 1991.

The "second shock" occurred after the Iranian oil strikes began in November 1978. Iran was the second largest oil producer, exporting 4.5 million barrels per day. The strikes reduced this to 1 million in mid-November, and exports ceased entirely by the end of December. While loss of this supply was partially offset by other suppliers within OPEC, burgeoning TV-network coverage—simultaneously broadcast throughout the world—of the Shah's mid-January 1979 departure and the Ayatollah Khomeini's arrival (along with other internal Iranian events) convinced the world that Iran would never return to its pro-Western ways, and initiated a 1973-style hoarding and panic buying spree. What started as a 2 million barrel per day shortfall became 5 million barrels per day as governments, oil companies, and consumers scrambled for supplies. Hoarding at all levels exacerbated the problem, gas lines appeared again, and oil prices went from $13 to $34 a barrel.

Although Iranian exports returned to the market by March 1979, the ill-timed Three Mile Island nuclear accident of March 28, 1979 further intensified the energy-awareness panic mood, in addition to forever altering public opinion on nuclear power. Several other factors contributed to making the gasoline crisis that occurred during 1979 in most industrialized nations of the free world more severe than any previous crisis. Many refineries set up to process light Iranian crude could not deliver as much gasoline from the alternate heavier crude oil they were forced to accommodate. Uncooperative (and in some cases, conflicting) policies by federal, state, and local governments and oil companies disrupted the orderly distribution of the gasoline supplies that were available. The appearance of oil commodity traders—who could make huge profits on the play between the long-term contract and spot prices for oil—artificially bid up the price of spot oil in response to the prevailing buy-all-you-can-get-at-any-price mentality. Lengthy gasoline lines and rationing interfered with all levels of business and personal life. By the time the Iran hostage crisis began, a state of anarchy existed in the world oil market that president Carter's subsequent embargo of Iranian oil and freeze on Iranian assets did little to ease.

The "third shock" occurred in the opposite direction—prices went down. The self-correcting market forces that swing into action after any shortage or glut accomplished what the world leadership could not—this time with a vengeance. At the OPEC meeting of June 1980, the "official" price averaged $32 per barrel, but OPEC inventories were high, and approaching economic recession caused price and demand to fall in consuming countries, further swelling their inventory surplus. When Iraq attacked Iran, one of the first steps in its war plan was to bomb the Abadan refinery (September 22, 1980). The net result after reprisals was that Iran oil exports were reduced during the war, but Iraq exports almost ceased. After Saudi Arabia raised prices from $32 to $34 a barrel in October 1981, an unprecedented boom was created, and drillers came out of the woodwork.

Earlier fears of shortage at the beginning of the 1920s and mid-1940s had ended in surplus and glut because rising prices had stimulated new technology and development of new areas. This pattern was repeated with $34 a barrel oil. High OPEC oil prices created a major non-OPEC production buildup in Mexico, Alaska, the North Sea, as well as in Egypt, Malaysia, Angola and China. Technological innovations improved exploration, production, and transportation. Meanwhile demand was reversed by the economic recession, higher prices, government policies, and the growth of nuclear power and natural gas alternatives. Not only did the oil share of the world energy market decrease from 53 percent in 1978 to 43 percent by 1985, but conservation shrank the entire market as well—the 1975 United States CAFE legislation doubled auto fleet mileage to 27.5 miles per gallon by 1985, and this alone removed nearly 2 million barrels per day from the demand side of the world oil ledger. By 1985, the collapse in demand combined with the relentless buildup of non-OPEC supply (and everybody dumping inventory) reduced demand on OPEC by 13 million barrels per day. By the May, 1985 Bonn economic summit, excess oil capacity around the world exceeded supply by 10 million barrels per day—the exact opposite of the 1979 situation, but twice as bad!

West Texas Intermediate crude on the futures market plummeted from its all time high of $31.75 a barrel in November 1985, to $10 a barrel in only a few months. Some Persian Gulf cargoes sold for as little as $6 a barrel. The ensuing "buyers market" saw oil commodity traders sell oil to anyone at any price as spot prices plummeted and sellers scrambled to get out of their long-term contracts. By the time the dust settled and OPEC established a new "official" oil price of $18 in December 1986, consumers were again jubilant, and all fears of permanent oil shortage had been laid to rest. Meanwhile, the Chernobyl nuclear accident in April 1986 gave the now energy-aware public another boost of environmental awareness. The 1989 Exxon Valdez tanker accident in Alaska further heightened this awareness. As oil and internal combustion engine vehicles headed into the 1990s, they faced a tripartite alliance of issues: CAFE standards, environmental standards, and unstable supply dominated by foreign interests.

When the "fourth shock" occurred, the invasion of Kuwait by 100,000 Iraqi troops in August 1990, it removed both the Kuwait and Iraq oil supply from market and again sent oil prices climbing. Iraq's Saddam Hussein had problems—his eight year war with Iran and enormous weapons purchases had run him out of money, his oil production was in disarray, and his political popularity at home was potentially at risk. He needed money, oil, and a new external threat upon which to focus Iraqi citizens' attention. Kuwait had money, oil and was conveniently located at the border. Hussein need only get assurances from the U.S. State Department that it would look the other way, and mock up a few sovereignty claims (Iraq had claimed ownership of Kuwait before OPEC in 1961, right after Kuwait became independent of Britain, but in fact Kuwait's origins go back to 1756, predating Iraq's origin; Iraq was formed out of the three former provinces of the Turkish empire by the British in 1920). Having taken care of both these steps, Saddam Hussein's military might had no problem in quickly dispatching Kuwait's defense and occupying it. Unfortunately for Saddam Hussein, he had stepped on the industrialized world's oil jugular, the implied threat being first Kuwait, then Saudi Arabia and the rest of the peninsula, and total domination of the world's oil supply. Never before had the industrialized world acted so quickly (and in concert) before, and Iraq was dispatched in short order. Fortunately for Saddam Hussein, he remained to fight another day.

The Iraq-Kuwait war and its aftermath brought the real priorities of all the industrialized world's citizens—now linked together by real-time TV network news coverage—into clear focus. The supply of foreign (mainly Middle Eastern) oil that makes everything work is—and will continue to be—highly vulnerable. While the majority of Earth's citizenry continue to drive their internal combustion engine vehicles as if the supply of oil was secure and inexhaustible, fouling the air with pollutants and soiling the lands, rivers, and seas with toxic byproducts, more and more individuals are waking up. The stage is set for rising individual responsibility and the reemergence of the electric vehicle.

## ELECTRIC VEHICLES RISE AGAIN

While electric vehicles of all kinds continued to be built in the United States and overseas during the 1920s through the 1950s, the resurgence of interest in EVs directly coincides with the environmental problems of 1960s, as well as the first, second, and fourth oil shocks (FIG. 3-1). Unfortunately, the lull between these four waves—particularly the third oil shock glut—was equally responsible for retarding further EV development after each interest peak. What guarantees the lasting impact of the fourth wave are the following universal perceptions:

- The security of our oil sources is a serious problem affecting the whole world.
- Our environment is at risk.
- There is a real need to conserve our scarce planetary resources and nonrenewable fossil fuel supplies.

Thanks to the miracle of instant global TV and telecommunications, almost everyone in the industrialized world now knows this. The individual citizens of the advanced industrialized nations, nagged by the increasingly insistent urgings of their conscience, find it harder to conduct "business as usual" if it involves polluting the air, land, or water, or wasting a suddenly precious, precarious, and limited commodity—oil. The handwriting is on the wall. Gas-guzzling cars are gone forever, as are smoke-belching junkers. Even so, these beg the ultimate question—our proven oil reserves should last us 40 years as shown in chapter 2, but what then? Isn't there a better way than oil today?

Of course there is: Electric vehicles and renewable energy sources. But before EVs could reappear in quantity on our streets and highways, consumers had to believe in them, and companies had to believe they would be profitable to invest in. All this took time. Now, at the beginning of the 1990s, the foundation is firmly in place. This section will explore how the interaction of five diverse areas set the stage for EVs to rise again:

- Interest in electric vehicle speed records.
- Interest in electric vehicle distance records.
- Development of electric vehicle associations.
- Development of U.S. legislation.
- Increasing frequency of electric vehicle events.

The remainder of the chapter will explore how the four waves influenced the production history of EVs in the United States, Europe, and Japan. Manufacturers, in response to growing public interest and awareness, gradually increased their own awareness of electric vehicle feasibility, and established a more open development climate. This fostered technological innovation and, fueled by global competition between United States, European, and Japanese companies, led to the actual rise of today's modern, technically sophisticated electric vehicles.

## Setting the stage for electric vehicles

You don't have to be a marketeer to know about the power of "tell a friend" marketing. People talk. They tell others about what they like and don't like. In the automotive field, they talk about speed and distance records and automobiles. They go to gatherings to see what's going on and to read about the gatherings that they missed. When enough people get together and talk to the government about something, the government listens. Speed and distance records, associations, legislation, and events put EVs back on the map. Figure 3-8 is rather busy, but it gives you a snapshot overview of how it happened.

## The need for speed

In chapter 1 you read about the "speed myth" and learned that you can make electric vehicles go as fast as you want. The first line in FIG. 3-8 shows the evolution. Each step involved a bigger motor, more batteries, a better power-to-weight ratio, and a more streamlined design.

The very first land speed record of any kind was set by an electric vehicle in 1899 at 66 mph. In the same year the first speeding ticket awarded any sort of vehicle went to a Manhattan electric cab zipping along at 12 mph![10] Just after the turn of the century, a Baker Electric went 104 mph.

After interest in electrics resumed, Autolite reached 138 mph in 1968, Eagle-Pitcher bumped it to 152 mph in 1972, and Roger Hedlund's "Battery Box" pushed it to its present 175 mph in 1974. In 1992, Satoru Sugiyama, in the Kenwood-sponsored "Clean Liner" was going to try for 250 mph at the Bonneville salt flats, using a 650-hp motor Fuji Electric motor from a Japanese bullet train and 113 Panasonic lead-acid batteries. Unfortunately, a simple component failure prevented the first day's run, and wind or weather wiped out the next six days of his Speed Week window.[11]

When you consider that GM's Impact dusted off a Mazda Miata and Nissan 300ZX in 0-to-60 mph standing-start races—the film of this feat appeared on the CNN News—there should be no question in your mind that today's electric vehicles can go fast and get there quickly. If you have difficulty with the concept, remember that France's T.G.V. electric train routinely goes 186 mph en route (it can go 223 mph) and the

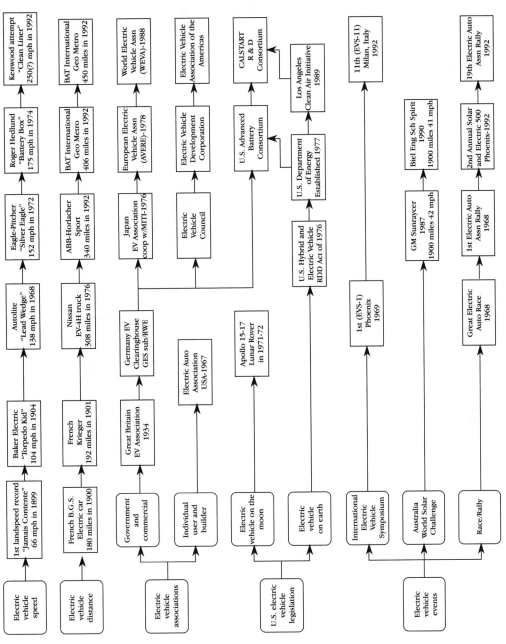

**3-8** How speed, distance, associations, legislation, and events put EVs back on the map.

Swedish-built X2000 electric train will ride the Washington, D.C. to New York corridor at 125 mph (it can go 155 mph) starting in early 1993. An electric train is nothing more than an electric vehicle on tracks (and with a long cord!). Meanwhile, no one is saying electric vehicles are wimpy in the speed department anymore.

## The need for distance

You also heard the "range myth" debunked in chapter 1; you can make electric vehicles go as far as you want. The second line in FIG. 3-8 shows the evolution. In 1900, a French B.G.S. Electric went 180 miles on a single battery charge. A Nissan EV-4H truck went 308 miles on lead-acid batteries in the 1970s; a Horlacher Sport pushed that to 340 miles using Asea Brown Boveri sodium-sulphur batteries in 1992;[12] and BAT International bumped it to 450 miles later in the year using proprietary batteries/electrolyte in a converted Geo Metro. The Horlacher, by the way, averaged nearly 75 mph during its 10-hour test period! The BAT Geo Metro was driven at a more sedate 35 mph.

There should also be no question in your mind that electric vehicles can go far. Given the fact that most Americans average less than nine miles a trip, and are probably less than 100 yards from an electric outlet in any American urban area, the likelihood of an electric vehicle running out of juice would appear to be less than an internal combustion vehicle running out of gasoline; certainly less than running out of diesel or any of the new alternate fuels that aren't carried in every filling station!

## The need for association

The third line in FIG. 3-8 shows two types of associations: those for government and commercial interests, and those for individuals. The government/commercial types can be further divided into associations for sharing information (the British formed the first in 1934); associations for lobbying (like the Electric Vehicle Council, Electric Vehicle Development Corporation, and Electric Vehicle Association of the Americas); and consortium associations for advancing research like the U.S. Advanced Battery Consortium and Calstart.

Not all government/commercial associations for sharing information are created equal. The British association was very "loose," but had the world's largest EV population by the 1970s—approximately 150,000 vehicles of all kinds. The German Gesellschaft fur Electrischen Strassenzerkehr (GBR) clearinghouse—a subsidiary of their largest electric utility—was instrumental in focusing German EV efforts; and Japan's Electric Vehicle Association, under the guidance of its Ministry of International Trade and Industry (MITI), gave them a definitive EV mission statement—focus on urban acceleration and range. In contrast to the earlier efforts of the British, and the focused EV efforts of the Germans and the Japanese in the 1970s, the United States didn't have any focused industry effort until the USABC and Calstart consortia of the 1990s.

Although individual members of the Electric Auto Association (formed in 1967) were a "voice in the wilderness" during this period, they made it increasingly difficult for people to say, "it couldn't be done" after they were already doing it.

## The need for U.S. legislation

The fourth line in FIG. 3-8 tells an interesting story. It is an enigma that we had the legislation (the U.S. government-funded National Aeronautics and Space Agency-guided Apollo program) to put an electric vehicle on the Moon before we had the legislation to develop EVs on Earth.

When the very first bills were being introduced in Congress in 1966 to sponsor electric vehicles as a means of reducing air pollution, government-funded contractors had Lunar Rover EV prototypes already working at their facilities. While the pros and cons of EVs were argued during various Congressional hearings in the early 1970s (leading to the passage of the landmark Public Law 94-413, the "Electric and Hybrid Vehicle Research, Development and Demonstration Act of 1976")—Lunar Rover electric vehicles had already performed flawlessly on the Moon during three separate 1971 and 1972 Apollo missions.

The Lunar Rover, shown in FIG. 3-9, was optimized for light weight. It had four 0.25-hp series dc motors—one in each wheel—and woven wire wheels and aluminum frame that gave it a mere 462-pound weight (which could carry a 1606-pound payload). Its silver-zinc nonrechargeable batteries gave it an adequate one-time 57-mile range and 8-mph speed. Future moon travellers only have to bring a new set of batteries with them—three Lunar Rovers are already there!

**3-9** Apollo 17 Lunar Rover vehicle on Moon with Astronaut Eugene A. Cernan at the controls.

The point is, legislation and subsequent funding, along with the focus and emphasis provided by consolidating the major Federal energy functions into one Cabinet-level Department of Energy in 1977, were vital components. Until these pieces were in place, EV development and renewal could not proceed.

But while the government-funded early ETV-I and ETV-II programs in 1977 provided the impetus to jumpstart electric vehicle activity again, it was clear that government legislation and funding provided the merest beginning, not the path for introducing electric vehicles in widespread numbers. The testimony of one General Motors executive at the June 1975 Congressional hearings drives home the point: "General

Motors does not believe that much can be gained by subsidizing the sale of electric vehicles," and ". . . we feel that building more electric vehicles is a waste of resources."[13]

Just how far thinking has progressed since then is summarized by an August 1991 Motor Trend article, "Among all the auto makers, General Motors is by far the most vocal in support of the electric car as a real alternative in the U.S. market."[14] Just how effective legislation can be is shown by the existence of the U.S. Advanced Battery Consortium (whose principals are General Motors, Ford, and Chrysler, with Department of Energy participation), the CALSTART consortium (involving utilities, large and small aerospace/high-tech companies including Hughes, a subsidiary of General Motors) and research institutions driven by the Los Angeles Clean Air Initiative.

### The need for events

Follow the fifth line and its branches in FIG. 3-8. Some people are thinkers, others are doers, some just like to tinker around—and EVs provide a fertile field for all three types. Thinkers can attend symposia; these typically alternate between Eastern and Western hemisphere locations at two or three year intervals—Paris in 1984, Washington, DC in 1986, Toronto in 1988, Hong Kong in 1990, Milan in 1992, and Anaheim (planned for 1994). Doers can go to races; what started out as the simple MIT versus Cal Tech Great electric vehicle Race of 1968 evolved to the huge crowds and multiple classes of the Phoenix 500 race by the 1990s. Doers can also go to rallies—the modest first Electric Auto Association rally of 1968 had grown to substantial proportions by the 1990s, with speed and distance records and TV coverage along the way. Tinkerers have their hands full with the greatest of all challenges (and the greatest of all publicity stunts): going 1900 miles across the Australian outback using a motor the size of a coffee can powered only by sunlight captured people's imagination via the covers and pages of numerous newspapers and magazines.

## FIRST WAVE OF THE 1960s—THE SLEEPER AWAKENS

While electric vehicle automobile development languished since the 1920s (except for Detroit Electric's efforts), commercial and industrial EV activities continued to flourish—perhaps best exemplified by Great Britain's electric milk trucks (called "floats") and its total electric vehicle population of more than 100,000.

The heightened environmental concerns of the 1960s—specifically air pollution—were the first wave upon which electric vehicles rose again. While numerous 1960s visionaries were correctly touting EVs as a solution, the manufacturing technology was, unfortunately, not up to the vision.

Figure 3-10 shows a chronological summary of what was being done by the primary electric vehicle developers in the United States, Europe, and Japan during the four waves. EV interest during the first wave fell into two distinctly different perceptual camps:

- Individuals who successfully converted existing internal combustion vehicles.
- Manufacturers who could not figure out how to make the existing technology justify the financial figures, let alone figure out how to market EVs to an American public that wanted quick and large internal combustion engine automobiles.

### Individuals lead the way

Converting was easy enough, and also inexpensive. Then, as today, you picked a vehicle shell—hopefully light in weight and/or easy to modify—added motor, controller,

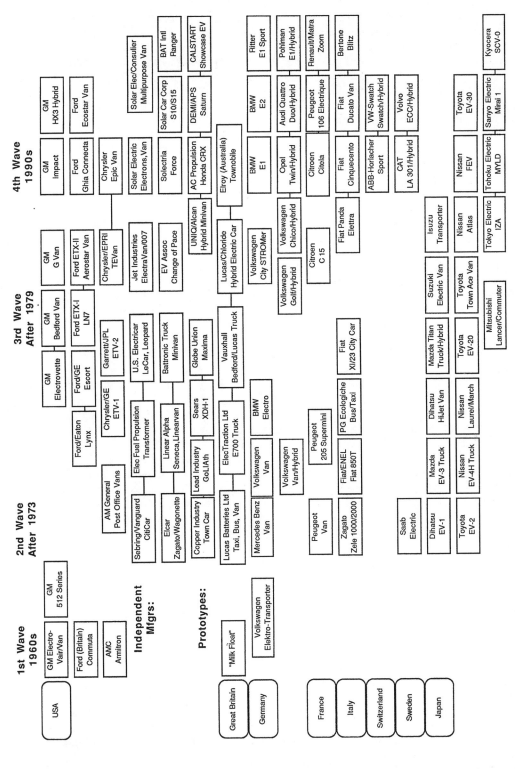

**3-10** Chronological summary of electric vehicles' four development waves.

and batteries, and went. Unfortunately, the most available motors in the appropriate size were decades-old war surplus aircraft starter dc motors; do-it-yourself controllers were barely more sophisticated than their turn-of-the-century counterparts; and battery technology—although cosmetically improved by modern manufacturing and packaging techniques—was virtually unchanged from 1900. As the most readily available controllers came from golf carts that typically used six 6-volt batteries (36 volts), and aircraft starter motors were typically rated at 24 to 48 volts, many first time do-it-yourself EV attempts suffered from poor performance, and contemporary internal combustion muscle car owners of the 1960s just laughed at them.

Then someone discovered motors were actually underrated to ensure long life, and began driving them at 72 to 96 volts. Some early owners found they could make simple, non-current-limiting controllers, and create vehicles that could easily embarrass any internal combustion muscle car at a stoplight. For a conceptual picture of this, imagine a subway traction motor in a dune buggy. In fact, these owners simply left the starting resistance out of a series dc motor, or equivalently diddled a shunt or compound motor. A series dc motor delivers peak torque at stall, and while starting currents were enormous, these early innovators just made sure they had a load attached when they switched on the juice. The immediate result was a rush. Predictably, the longer-term results were burned-out motors and, occasionally, broken drive shafts or axles. But the sanely driven and controlled 72-to 96-volt EV conversions were not bad at all. This was the 1960s, and the Electric Auto Association was founded in 1967.

## Manufacturers don't

Established internal combustion engine vehicle manufacturers in the late 1960s did not produce much in the way of electric vehicles. The General Motor's XP512E Series (GM's ElectroVairs—converted 1964 and 1966 Corvairs—and ElectroVan—a converted 1966 GMC HandiVan)—could have been easily replicated by any individual except for the ElectroVair's high-cost silver-zinc batteries, and ElectroVan's high-cost hydrogen-oxygen fuel cells. Ford of Britain's Comuta (even more easily replicated by anyone) and American Motor's Amitron, like GM's offerings, all resembled souped-up golf carts (though the Amitron was in a class by itself—it featured Gulton's lithium batteries, a solid state controller, 50-mph speed and a 150-mile range). It was sad that numerous individuals could develop EV solutions far superior to anything put forth by the giant industrial corporations that had helped to put a man on the Moon in the same decade.

The problem was not that these corporations lacked talent, money, or technology. The problem was that corporate thinking of this era was locked into a mode that presumed their current successes would continue forever, and they were committed to maintaining the status quo to assure it.

## SECOND WAVE AFTER 1973—PHOENIX RISING, QUICKLY

The late 1960s policies of the major American automobile manufacturers put them in a poor position to respond to the crisis of the early 1970s—the oil shock of 1973. A huge inventory of stylish but large, gas-guzzling cars along with four- to five-year new car development cycles put them in an impossible situation. All they could do was wait out the crisis and import smaller, more fuel-efficient cars from their foreign subsidiaries. The higher European and Japanese gasoline prices had, over the years, forced them to develop lighter, more compact automobiles with economical drivetrains. While this helped the Europeans and the Japanese only a little at home after the crisis—their already-high gasoline prices just rose proportionately higher, negating any advantage—the European/Japanese automotive solution was ideal for the United States market of that time. Thus imports, unimportant in the United States until the early 1970s, gained a foothold that was to become a significant factor over time. It was against this backdrop that the electric vehicle rose again, like the Phoenix from the ashes. Five trends (FIG. 3-10) highlight electric vehicle development during this second wave:

- The inactivity of GM in distinct contrast to Ford, Chrysler, and American Motors.
- A period of frantic activity by the independent manufacturers.
- A period of strong prototype promotion by industry associations and suppliers.
- Resumption of serious overseas development.
- Continuation of individuals converting existing internal combustion vehicles.

## GM leads, the world does not follow

Under the "2nd wave" heading in FIG. 3-10, GM's inactivity is in marked contrast to what others are doing in electric vehicles during this period. The GM executive's remarks at the June 1975 Congressional hearings—quoted earlier in the chapter—clearly convey the reason for GM's electric vehicle nonactivity.

While GM's actions are inconsequential today—the times have changed and GM has changed along with them—GM's strong stance against electric vehicles caused much grief among the pro-EV industry forces of the early 1970s.

Late in the decade GM re-entered the electric vehicle world with its ElectroVette (a converted Chevette) and Bedford Van (a converted GM-United Kingdom van). But GM did nothing technically innovative, and both conversion efforts became self-fulfilling prophecies: neither vehicle's performance specs were spectacular, and the economics just didn't make sense for a manufacturer/marketer.

As mentioned in chapter 1, an 8000-pound van would never be my first conversion choice. The Chevette had twenty 12-volt maintenance-free batteries, a 53-mph top speed, a 50-mile range at 30 mph, and weighed in at 2950 pounds—maybe a marginal conversion choice. A contemporary individual electric vehicle converter's outlay might have been $5000 for the whole package—perhaps only $2000 with used parts and heavy scrounging—while GM would have been hard pressed to wring a profit out of a $20,000 retail price. The van performance and pricing were even worse. GM could honestly tout their conclusion without mentioning what they hadn't done (i.e., a total systems design such as their 1990 Impact EV).

## Ford, Chrysler, and American Motors move ahead

Ford's direction was totally different than GM's during this wave. They took a hard look at the problem and decided the two critical areas were battery technology and drivetrain efficiency. This period marked their planting of seeds in sodium-sulfur battery technology (they invented it in 1965) and integrated ac induction motor drivetrains during the Ford/Eaton Lynx and Ford/GE Escort projects that would later bear sweet fruit.

The Chrysler/GE team took advantage of Federal government funding of the Electric Test Vehicle One (ETV-1) program in 1977 to couple a lightweight, low-rolling resistance, streamlined body with front-wheel drive to a unique dc motor, transistorized controller and 108 volts worth of advanced lead-acid batteries. At the same time, a Garrett/JPL team was working on ETV-2 and learning the benefits of a similarly constructed and identically powered vehicle using rear-wheel drive and aided by flywheel energy storage.

In 1971, the United States Post Office studied electric vehicles in a pilot program implemented at the Cupertino, California Post Office, using British Harbilt electric vans. The program was a resounding success. Because of it, the United States Post Office ordered 350 converted "Jeep" type electric vans from A. M. General Corp, a division of American Motors, for the next phase of the program. The A.M. General vans were also a resounding success. Both the Harbilt and A.M. General vans had enormously high uptimes and low cost per mile while being driven almost continuously during their evaluation periods. The program had strong support inside the Post Office and was cancelled only after the "third oil shock" made it economically unattractive.

## Independent EV manufacturers rise and fall

Numerous independent electric vehicle manufacturers came out of the woodwork after the 1973 oil shock; it was a repetition of turn-of-the-century vehicle development with the good, the bad, and the ugly. There were many EVs to choose from, but most were not technically innovative, manufacturing quality was inconsistent, and component quality was occasionally poor.

Prevailing designs used either conversions of existing internal combustion vehicles or unsophisticated new chassis, and many were poorly engineered. In addition, most firms were severely undercapitalized. While the automotive industry marketed its internal combustion powered vehicles via public relations, lobbying to blunt legislation, measuring public taste through survey and prototype programs, product advertising, and distribution through a dealer network, EV manufacturers of the 1970s used few of these, and haphazardly.

The Sebring/Vanguard CitiCar was the most famous and infamous electric vehicle of this period. Sebring/Vanguard was the first manufacturer out of the blocks in 1974, and for a brief time sold all it could make. Eventually, more than 2000 CitiCars were produced. It was very popular in its day; owners were fiercely loyal, and it received much publicity. Unfortunately, although it was well-built (many are still on the road today), efficient, and practical, its design and styling gave it the appearance of a glorified golf cart—similar to the late 1960s engineering prototypes previously mentioned. When they were unjustly crucified by a Consumer's Union report (along with Elcar's golf-cart-sized Zagato imported from Italy), and an unfavorable article went out over UPI news, even a letter from the chief attorney of the Department of Transportation saying the criticisms were false failed to undo the damage. The public would always associate EVs with golf carts and some nebulous stigma.

But this painful lesson was well learned by other electric vehicle manufacturers, and later models avoided golf-cart-looking designs like the plague. Another well-known manufacturer of this period was Electric Fuel Propulsion. Their early Renault R 10 and Hornet conversions led to their original and innovative Transformer (featuring 180-volt tripolar lead-cobalt batteries, a 70-mph top speed, and 100-mile range) with its range-extending Mobile Power Plant trailer. Linear Alpha produced the Seneca (Ford) and Linearvan (Dodge) conversions. U.S. Electricar produced the Electricar (Renault LeCar) and Lectric Leopard conversions. Finally, there was Battronic Truck's Minivan, co-produced with the Electric Vehicle Council; more than 60 utilities received production versions of this 6800-pound van whose 18 6-volt lead-acid batteries pushed it to 60 mph.

## Electric vehicle industry closes ranks to show support

One of the more innovative promotions of this period involved the development of prototypes by industry associations and individual manufacturers who stood to gain from the sale of their product(s) in electric vehicles. The prototypes were used for all sorts of public relations event-style marketing. The result was a raising of the level of public awareness about EVs—so much so that individuals, thinking these were production products, frequently called these organizations to place orders.

Copper Development Association's Electric Town Car, one of the most well-known prototypes of this period, squeezed 59 mph and 103 miles out of its 18 Exide EV-106 6-volt lead-acid batteries. Another well-known prototype was Lead Industry Association's GoLIAth; co-sponsored by the electric utility industry and its Electric Vehicle Council, this lightweight, lead-acid battery powered truck was a familiar sight at events around the country in the mid-1970s. Two others were Sears and Roebuck's XDH-1, a converted Fiat 128 with DieHard lead-acid batteries, and Globe-Union's Maxima (a 4350-pound station wagon that got 75 mph out of its 20 deep-cycle, 12-volt lead-acid batteries).

## Europe gets serious about EVs

Great Britain led the world in electric vehicle production for many years, and had more than 150,000 EVs of all kinds in service by the mid-1970s. They optimized their EVs to overcome air pollution, noise pollution, and cramped urban road conditions. The British were technically strong in vehicles for fleet use—the best known was their "milk float" electric vehicle—and they had countless other vans, trucks, and taxis in daily operation. British manufacturers such as Lucas and Chloride were known around the world, but British EVs were not optimized for speed or range, featured right-hand drive, and used 220-volt/50-Hz recharging services—criteria that created barriers to their effective export to other world markets.

The British example was not lost on the Germans. German history in small car optimization was legendary, with more than 20 million Volkswagen Beetle cars produced in more than 20 countries around the world in its unprecedented 54-year (and still going) production history.[15] Volkswagen had also done conversions based on its familiar Transporter chassis in the late 1960s (the Elektro-Transporter), and joint electric utility and government sponsorship had placed 30 Mercedes-Benz and 20 Volkswagen electric van test vehicles on the roads by the early 1970s. Under the guidance of Gesellschaft fur Electrischen Strassenzerkehr (GES, an arm of Germany's largest electric utility), coordinated German industry and government cooperation resulted in several bus programs: Electrobus (Mercedes-Benz/Bosch/Varta) and MBB-Elektrotransporter (Messerschmitt Bolkow Blohm/Bosch/Varta/Bayer); the Volkswagen Commercial (Volkswagen/Bosch/Varta); an automobile program using the Audi 100 (Volkswagen/Bosch/Varta), and various hybrid programs.

Looking over Germany's shoulder, electric vehicle efforts by Peugeot, Citroen, and Renault in France, Zagato and Fiat in Italy, Saab in Sweden, and activities elsewhere on the continent were pushing developments along.

## Japan gets more serious about EVs

Meanwhile, the European example was not lost on the Japanese. The Japanese rise to world-class automotive power is a remarkable and fascinating story,[16] but even before they achieved world dominance in the internal combustion engine vehicle in the

1980s, leading visionaries at Japan's state planning agencies had seen the future, and it was electric vehicles. Japan needed little incentive—it was the world's largest importer of oil, had dangerous levels of pollution, and high speeds on its narrow, urban streets were a fact of life.

While the Japanese Electric Vehicle Association and its tight coordination with MITI directives did not arrive on the scene until 1976, Japanese government funding of EV programs began in 1971 with Phase I basic research into batteries, motors, control systems, and components across the spectrum of car, truck, and bus platforms.

The fruits of its labors—augmented by MITI directives to focus on urban acceleration and range—appeared in Phase II. As TABLE 3-1 attests, Japan's 1970s Phase II offerings from Daihatsu, Toyota, Mazda and Nissan put it into a world-leadership class. The Nissan EV-4P truck's 188 miles before recharging was a record for lead-acid battery-powered vehicles, and its EV-4H truck's 308 miles was the world record for that period.

### Table 3-1 Comparison of Japan's 2nd wave phase II electric vehicles

|  | Daihatsu EV-1 car | Toyota EV-2 car | Mazda EV-3 truck | Nissan EV-4P truck | Nissan EV-4H truck |
|---|---|---|---|---|---|
| Range (miles) | 109 | 283 | 127 | 188 | 308 |
| Top Speed (mph) | 55 | 53 | 45 | 54 | 56 |
| Battery Pack | lead-acid | zinc-air/lead-acid | lead-acid | lead-acid | zinc-air/lead-acid |
| Curb Weight (lbs) | 2500 | 2770 | 1720 | 5000 | 5490 |

Throughout the rest of the 1970s all Japan's big nine automakers—Daihatsu, Honda, Isuzu, Mazda (Toyo Kogyo), Mitsubishi, Nissan, Subaru, Suzuki and Toyota—were involved in EV activities, although some to a greater extent than others (FIG. 3-10).

## Individuals assisted by more and better everything

The best news of the 1970s was for individuals wanting to do EV conversions. More of what was needed was available for conversions, and how-to books even started to appear. Other than the fact that components—particularly the controllers—were still unsophisticated, individual converters enjoyed relating their conversion experiences at regular Electric Auto Association meetings and "pushed the outside of the speed and distance envelope" at rallies and events. The greatest irony of this period is that at the same time General Motors was providing extremely negative information to the Congressional hearings, the individuals who had actually done a conversion to an electric vehicle were reporting high degrees of satisfaction, with operating costs in the 2 cents per mile range, and most had yet to replace their first set of batteries.

## THIRD WAVE AFTER 1979—EVs ENTER A BLACK HOLE

While the second "oil shock" of 1979 and ensuing shortage further spurred electric vehicle development onward, the "oil shock glut" of 1986 and events leading up to it nearly shut development down. While the larger internal combustion automobile manufacturers were "whipsawed"—their crash programs of the late 1970s were now bringing lighter, smaller cars to market that (temporarily at least) no one wanted—the independent electric vehicle manufacturers were simply wiped out. With oil and gasoline prices again approaching their 1970s levels, everyone lost interest in EVs, and the capital coffers of the smaller EV manufacturers were simply not large enough to weather the storm. Even research programs were affected. From mid-1983 until the early 1990s, it was as if everything having to do with EVs suddenly fell into a black hole. No manufacturers, no books, not even many magazine articles. The EV survivors were the prototype builders and converters, the parts suppliers (who typically had other lines of business such as batteries, motors, electrical components) and EV associations—although their membership ranks thinned somewhat. Four trends (FIG. 3-10) highlight EV development during this third wave:

- Low levels of activity at GM, Ford, and Chrysler.
- The best independent manufacturers arrive and then depart.
- Low levels of activity overseas.
- Continuation of individuals converting existing internal combustion vehicles.

## Lack of EV activity at GM, Ford, and Chrysler

In retrospect, given all the other problems the big three had to deal with during this period, it's amazing that electric vehicle programs survived at all. But survive they did, to emerge triumphant in the 1990s.

The GM Bedford van project became the GM Griffon van—the G-Van. With a broad base of participation from the Electric Power Research Institute (EPRI), Chloride EV Systems, and Southern California Edison, the General Motors G-Van, actually an OEM aftermarket conversion by Vehma International of Canada, was widely tested for fleet use. While it was humorous to read numerous complaints about the G-Van's 53-mph top speed, 60-mile range and 0 to 30 mph in 12 seconds acceleration, one has to wonder how many report readers correctly associated this data with G-Van's 8120-pound weight, 36 batteries, and huge frontal area.

Ford's direction was to continue to build on its sodium-sulfur battery and integrated propulsion system technology using government funding. Teamed with General Electric, Ford's ETX-I program adapted sodium-sulfur batteries and an integrated ac induction motor propulsion system to a front-wheel drive LN7 automobile test bed. The follow-on Ford/GE ETX-II program utilized sodium-sulfur batteries and a permanent magnet synchronous motor propulsion system in a rear-wheel drive Aerostar van. Meanwhile Chrysler, under the sponsorship of EPRI, used their standard Caravan/Voyager minivan platform, a GE dc solid-state motor, and 30 Eagle Picher NIF 200 6-volt nickel-iron batteries to achieve 65 mph and a 120-mile range in their 6200-pound TEVan.

## Arrival and departure of independent manufacturers

Numerous independent electric vehicle manufacturers had already come and gone during the previous wave. As an independent EV manufacturer in the third wave, you either were doing something good, or you had come out with something better. Electric Vehicle Associates of Cleveland, Ohio is the best example of the first type. While their Renault 12 conversion and ElectroVan project with Chloride were interesting diversions, they are best known for their Change of Pace wagons and sedans built on AMC Pacer platforms. The Change of Pace 4 passenger sedan weighed in at around

3990 pounds, and used 20 Globe-Union 6-volt lead-acid batteries driving a dc motor via an SCR chopper to achieve 55 mph and a 53-mile range.

Jet Industries of Austin, Texas, unquestionably the largest and the best of the independent EV manufacturers, was also the last to arrive on the scene. Jet's most popular products were its ElectraVan 600 (based on a Subaru chassis) and 007 Coupe (based on a Dodge Omni chassis). It also offered larger 8-passenger vans and pickups. The 2690-pound ElectraVan 600 had a GE 20-hp or Prestolite 22-hp dc series motor, SCR controller, and 17 6-volt lead-acid batteries that could push it to 55 mph with a 100-mile range. Hundreds of ElectraVan 600s and 007 Coupes are prized possessions among Electric Auto Association members today—attesting to their outstanding quality and durability. Jet Industries, alas, is no more.

Needless to say, industry association support of independent EV manufacturers—at its zenith during the previous wave—moved to its nadir during this one. There were no longer any independent electric vehicle manufacturers to support.

### Electric vehicle activity overseas barely moves forward

Europe and Japan, plagued by the same problems as the United States during this period—an oil shortage followed by a glut—were not focused on EV development either. Germany's Volkswagen occupied itself with the City STROMer and Hybrid Golf projects, and little else notable came out of the continent. Meanwhile, Japan made slow but steady progress in its overall EV plan. Electric vehicle developments increasingly centered on activities of Nissan, with its Laurel, March and Atlas projects, and Toyota, with its EV-20 and Town Ace Van projects.

### Individual conversions continue

Individuals assisted by more and better everything during the last wave now had to make do with more modest resource levels. But EV conversions by individuals continued throughout this wave—albeit at a slower pace. The best news of the 1980s was that the resources of the 1970s could still be found and used! During this wave, individual converters still enjoyed relating their conversion experiences at regular Electric Auto Association meetings; they still "pushed the outside of the speed and distance envelope" at rallies and events; and they still reported high degrees of satisfaction with what they had done.

## FOURTH WAVE OF THE 1990s—EVs ARE HERE TO STAY

The fourth wave was a new ballgame entirely. Oil security, environment, and conservation concerns put real teeth back into EV efforts, and even General Motors got the message. Indeed, GM did a complete about-face and led the parade to electric vehicles. Resumption of interest in EVs during this wave was led by unprecedented legislative, cooperative, and technological developments.

Legislation provided both the carrot and the stick to jumpstart EV development. California started it all by mandating that 2 percent of each automaker's new-car fleet be comprised of zero-emission vehicles (and only electric vehicle technology can meet this rule) beginning in 1998, rising to 10 percent by the year 2003. As TABLE 2-3 of chapter 2 showed, this means 40,000 electric vehicles in California by 1998, and more than 500,000 by 2003. California was quickly joined in its action by nearly all the Northeast states (and at this time, states representing more than half the market for vehicles in the United States have California-style mandates in place)—quite a stick! In addition, for CAFE purposes, every electric vehicle sold will count as a 200- to 400-mpg car under the 1988 Alternative Fuels Act. But legislation also provided the carrot. California EV builders and buyers can claim up to a $1000 credit on their state income taxes. The recently passed National Energy Policy Act of 1992 allows a 10 percent Federal tax credit up to $4000 on the purchase price of an EV—quite a carrot! And it's happening everywhere, not just the United States: Japan's MITI set a target of 200,000 domestic EVs in use by 2000; and both France and Holland enacted similar tax incentives to encourage electric vehicle purchase.

The stakes are high, and volume EV deliveries will start relatively soon—so companies have literally been forced to work together to meet their common goals. The best example of this is the formation of the U.S. Advanced Battery Consortium in 1991 by former protagonists General Motors, Ford, and Chrysler—followed by announcement in late 1992 that they intended to cooperate more closely in development of EVs. This was such an unprecedented step that Japan was forced to raise and call—Nissan and Toyota, the two premier Japanese EV developers (but also protagonists), also pledged to work together on technology and components. A lot is happening—and quickly—with EVs in the 1990s, because cooperation builds synergy.

Electric vehicles of the 1990s also benefit from improvements in electronics technology, because the 1980s mileage and emission requirements increasingly forced automotive manufacturers to seek solutions via electronics. Although EV interest was in a lull during the 1980s, that same decade saw a hundredfold improvement in the capabilities of solid-state electronics devices. Tiny integrated circuits replaced a computer that took up a whole room with a computer on your desktop at the beginning of the 1980s, and by one that could be held in the palm of your hand by the beginning of the 1990s. Development at the other end of the spectrum—high power devices—was just as dramatic. Anything mechanical that could be replaced by electronics was, in order to save weight and power (energy). Solid-state devices grew ever more muscular in response to this onrushing need. Batteries became the focused targets of well-funded government-industry partnerships around the globe; lead-acid, sodium-sulfur, nickel-iron, and nickel-cadmium advanced to new levels of performance. Everyone began quietly looking at the mouth-watering possibilities of lithium-polymer batteries.

Meanwhile, EVs were the beneficiary of all this technological advance, and the new technology guarantees EVs are here to stay. TABLE 3-2 compares electric vehicle specs from the major automotive manufacturers at the beginning of the fourth wave of the 1990s. At first glance, TABLE 3-2 appears to be a step backward from TABLE 3-1—Japan's accomplishments of the 1970s. But first impressions are misleading. The 1990s electric vehicles offer substantial technology improvements under the hood, are more energy-efficient, and are closer to being manufacturable products than engineering test platforms. Six trends (FIG. 3-10)—highlight EV development during this fourth wave:

- High levels of activity at GM, Ford and Chrysler.
- New independent manufacturers bring vigor.
- New prototypes are even better.
- High levels of activity overseas.
- High levels of hybrid activity.
- A boom in individual internal combustion vehicle conversions.

## Table 3-2  Leading 4th wave electric vehicles compared

|  | GM Impact car | Ford Ecostar van | Chrysler Epic van | BMW E2 car | Nissan FEV car |
|---|---|---|---|---|---|
| Range (miles) | 120 | 100 | 120 | 150 | 150 |
| Top speed (mph) | 75 | 65 | 65 | 75 | 80 |
| Power train | 2 ea 54 hp ac | 75 hp ac | 65 hp dc | 45 hp dc | 2 ea 27 hp ac |
| Battery pack | Lead-acid | Sodium-sulfur | Nickel-iron | Sodium-sulfur | Nickel-cadmium[1] |
| Curb weight (lbs) | 2200 | 3100 | 3200 | 2020 | 1980 |

[1]Augmented by crystal-silicon solar battery.

## General Motors, the follower, now leads

Two developments changed General Motors forever, and they had a common thread. Like the person holding on to the edge of the cliff who stuggles to hold on, and finally gives up and lets go, only to fall six inches to the ledge that was always there, GM's struggles finally forced it to go outside itself for help. Once it did, the results were simply amazing.

GM's problem was the 1987 Australian World Solar Challenge. They needed a winner but hadn't a clue. They turned to a little company called AeroVironment for help. The founder of the company, Ph.D. Paul MacCready, who had already won a prize for the longest human-powered flight with his ultra-lightweight and efficient "Gossamer Condor" in 1977, and another for the first human-powered plane to cross the English channel with his "Gossamer Albatross" in 1979, and whose solar-powered "Solar Challenger" flew 163 miles from Paris to the English coast in 1981, was now given the ultimate challenge from GM—can you go 1900 miles across the Australian outback on solar power only? MacCready could and did. GM's winning "Sunraycer" set the course record that still stands today.

Now it was MacCready's turn. Would GM be interested in an electric car? MacCready solicited the assistance of Hughes Vice President Howard Wilson, with whom he had worked on the SunRaycer team. Howard Wilson was able to present the question personally to Bob Stemple, the formal plan was approved later in 1988 by then-chairman Roger Smith and the rest, as they say, is history.

GM's Impact electric vehicle, shown in FIG. 3-11, appropriately debuted at the January 1990 Los Angeles Auto Show, and promptly set the automotive world on its ear. Offering 100-mph capability (it's governor limited to 75 mph), 120-mile range, 8 second 0-to-60 elapsed time, in a slippery package that had a 0.19 coefficient of drag, weighed only 2200 pounds and used off-the-shelf lead-acid batteries, the Impact was not your grandparent's electric vehicle. GM now showed the rest of the world the way.

How did it happen? MacCready himself provides us the key: "No one had ever tried to build a superefficient car from scratch. That's because no one had ever needed to. Energy had always been cheap and pollution controls were relatively recent, so automakers never needed to pay fanatical attention to efficiency."[17]

The latest word is that GM plans to produce 50 Impacts in 1993—based on the prototype Impact—for use by utilities, local governments and others to provide technical feedback to GM about performance capabilities and user requirements. Meanwhile, GM has hedged its bet by working on the hybrid HX3 concept minivan that uses a gasoline-powered 40-kW generator to extend the range of its two 60-hp ac front-wheel drive electric motors.

**3-11** GM Impact electric vehicle.

## Ford has a better idea

Ford's electric vehicle perseverance is finally close to paying off. By using its European Escort van as the platform, building on the sodium-sulphur battery technology it invented in 1965 and applying its 1980s ETX-I and ETX-II drivetrain experience, Ford is hoping to leapfrog its competition and be the first to production with 80 Ford Ecostar vans planned for distribution to fleet customers in 1994. Ford is serious—management, resources, tooling, and facilities have already been put in place.[18] A Ford Ecostar van is shown in FIG. 3-12 and featured on the cover. Weighing in at 3100 pounds and driven by a solid-state-controlled 75-hp ac induction motor coupled to an integrated front-wheel drive and powered from sodium-sulfur batteries, the Ford Ecostar's specs are impressive—75 mph and 100-mile range. Plus the ability to carry an 850-pound payload. All of which led automotive writer Dennis Simanaitis to comment in February 1993 *Automobile*, "The first electric vehicle you're likely to see is the most transparent we've driven so far."[19]

**3-12** Ford Ecostar van electric vehicle.

## Chrysler arrives late but great

Chrysler's TEVan (*T* for T-115 minivan platform, *E* for Electric) efforts have paid a real dividend. Their TEVan will earn the distinction of being one of the first Federally certified production electric vehicles to ship when shipments of 50 units to fleet customers begin in early 1993. The already-proven electronics, drivetrain, and batteries lifted from their TEVan when used in their next generation lighter-weight Epic van, shown in FIG. 3-13, give them an immediate parity in the performance specs (TABLE 3-2).

**3-13** Chrysler Epic van electric vehicle.

## Independents return with gusto

In contrast to the independent EV manufacturers of the early 1970s, who often ran financially shaky operations, the best-known 1990s independents are financially snug, well-run operations that have carefully evolved from custom shops into manufacturing.

Solar Electric (Santa Rosa, CA) is unquestionably the largest and best-financed of the new breed of independents. Although originally known for its alternative energy products, it has manufactured EVs since 1984, and has sold several hundred of its various models in the 1990s alone. It offers an economical Ford Escort conversion; its well-known "Electron" series features Ford Tempo, GM Fiero, and Ford Aerostar van conversions; and its Destiny 2000—a highly modified Fiero conversion (FIG. 3-14) was featured in the movie *Naked Gun 2 ½*. Its most current offering, the "Aesop," goes 85 mph and has a 70-mile range—all on lead-acid batteries. Recently, Solar Electric was joined by Consulier, a well-known automotive racing circuit participant and composite monocoque sports car manufacturer, in a joint venture to produce a lightweight, multi-purpose, all-electric van.

Solectria Corporation (Arlington, MA), founded as Solectron in 1989, is another high-profile manufacturer. A frequent participant at electric vehicle events in the 1990s, its solar-assisted Force, a Geo Metro conversion (FIG. 3-15), goes 60 mph and has a range of more than 60 miles. Its Lightspeed, a ground-up design now in prototype form, pushes the envelope to a solar-assisted 85 mph and 150-mile range.

Solar Car Corporation (Melbourne, FL) began offering a converted Ford Fiesta with hybrid option to the public in 1991. Today it is best-known for its Chevy S-10 (FIG. 3-16) and GMC S-15 pickup truck conversions. These 3700-pound vehicles, powered by 20 lead-acid 6-volt batteries, attain 75 mph with a daily range of 50 to 80 miles, as numerous fleet owners attest.

Battery Automated Transportation (Salt Lake City, UT) offers Geo Metro car and Ford Ranger pickup truck (FIG. 3-17) conversions. A BAT Geo Metro, powered by their proprietary "Ultra Force" lead-acid batteries, is the current world's record holder at 450 miles on a single recharge.

**3-14** Solar Electric Engineering Destiny 2000 electric vehicle.

**3-15** Solectria Force electric vehicle.

## New prototypes are grabbers

Understandably, the resurgence in EVs has again brought out associations, utilities, and manufacturers, with an all-new crop of prototype electric vehicles. All of them are attention grabbers that have again raised public interest.

**3-16** Solar Car Corp Chevrolet S-10 electric vehicle.

**3-17** BAT Ford Ranger electric vehicle.

Unique Mobility (Englewood, CO) goes back to the late 1960s. They produced dune buggies early on and the Elek-Trek electric car in the late 1970s. Today they are best known for their efficient motor and drivetrain designs, and their Alcan Aluminum Limited-sponsored hybrid minivan—a Chrysler minivan conversion that goes 70 mph on batteries and gets 100-mile range via its natural gas, liquid propane, or methanol-fueled onboard internal combustion generator.

Calstart's (Burbank, CA) Showcase Electric Vehicle made its debut at the Los Angeles Auto show in January 1993. The public/private consortium that combined the best of high-tech aerospace firms with advanced university and electric utility thinking, made good on its promise to deliver mouth-watering electric vehicle technology today, and points the way to even greater possibilities tomorrow.

Dreisbach Electromotive's (Santa Barbara, CA) Arizona Public Service Company-sponsored Saturn conversion (FIG. 3-18) has been a particular attention-getter on the show and rally circuit. Powered by zinc-air/nickel-cadmium batteries, with dual dc motors and a custom Motorola controller, it holds the lap record of 42.06 seconds at Phoenix International Raceway's one-mile oval.

**3-18** DEMI Saturn electric vehicle.

AC Propulsion's (San Dimas, CA) founders, who developed the ac motors and controller that power GM's Impact, have brought further positive contributions to the EV movement on their own via consistent event attendance and test rides for the automotive media. Virtually all automotive writers swoon when they stomp on the throttle of AC Propulsion's converted Honda CRX and feel the blood rush from their temples. The most famous article about it is probably *Road and Track*'s October 1992 issue that features a "smoking the tires" photograph.[20] You'll meet this vehicle during chapter 8's ac controller discussions.

Motorola's (Phoenix, AX) converted 1993 Dodge Dakota pickup (FIG. 3-19) is a test platform for Motorola's latest solid-state control and power technology devices. This vehicle wowed audiences at the 1993 Phoenix Solar and Electric 500 with its Motorola-processor-powered control module capable of handling 900 amps from a 150-volt battery source.

**3-19** Motorola Dodge Dakota electric vehicle.

Westinghouse's Electronic Systems (Baltimore, MD) has also chosen a Dodge Dakota pickup platform for its propulsion system test platform. Its advanced WPT-100 ac controller system delivers 182 foot-pounds of torque (more than 100 bhp) and drives the 4500-pound pickup from 0 to 60 mph in under 12 seconds, with a top speed of 85 mph.

## BMW appropriately leads the Europeans

Germany's electric vehicle efforts likewise exploded in the 1990s. But while Volkswagen (Golf, Chico, and the secretive co-venture on the Swiss Swatch), Audi (Quattro Duo), Opel (Twin) and even independent Pohlman (E1) all concentrated their energies on hybrids (except for Volkswagen's converted-Jetta "City STROMer"), BMW became Germany's designated EV driver with its E1 (European version) and E2 (USA version) programs. *Outstanding* was the superlative used when BMW's E2 was unveiled at the 1992 Los Angeles Auto show. Featuring a body by Design Works, a California design studio, the E2 represents the best of the E1 drivetrain accomplishments mated to an automobile tailored to the United States—specifically Los Angeles—tastes and needs. TABLE 3-2 shows it can hold its own against the competition.

Speaking of competition, France provides it via Citroen's Citela, Peugeot's 106 Electrique, and Renault's Zoom. In Italy, Fiat's Panda Eletta, Cinquecento, and Ducato are all outstanding EV solutions, and Bertone's Blitz is hands down today's fastest EV sports coupe—0 to 62 mph in 6 seconds! It weighs only 1496 pounds—572 for the lead-acid batteries—yet is driven by two 36-hp motors that apply 145 pound-feet of torque to each rear wheel when you punch it! And a premonition of possible wonderful things to come occurred when the Italian policeman wouldn't give one writer/driver a ticket for speeding but admonished him, "Don't forget to drive slowly, despite being in an electric car."[21]

## Nissan leads the Japanese

Japan just possibly leads the world's effort in electric vehicles. No one has studied vehicles, batteries, and infrastructure longer or more systematically than the Japanese—thanks to the strong and wise guiding hand of MITI—nor does anyone else enjoy the broad technical, manufacturing, and financial base provided by Japan's big nine automakers. The results are predictable.

Nissan's nickel-cadmium-powered, crystal silicon solar battery-augmented FEV (Future Electric Vehicle) four-seater and Toyota's zinc-bromine-powered EV-30 two-seater are the slickest, quickest, and longest hitters in the EV universe. Innovative prototypes such as Tokyo Electric's IZA, Tohoku's Electric's MYLD, Sanyo Electric's Mirai 1, and Kyocera's SCV-0 confirm the enormous breadth and depth of Japanese EV understanding.

Just a brief look back at TABLE 3-1 (which featured the specs on Japan's 1970s-vintage EV-1 through EV-4 models), and the fact that Toyota is now touting an EV-30 model, should convince you that Japanese excellence in EVs is poised to come on strong.

## Hybrids don't measure up to power boost trailers

Many automotive manufacturers have sought to cover their EV range bets via hybrids: GM's HX3; Volkswagen's Golf, Chico, and Swatch; Audi's Quattro Duo; Clean Air Transport's LA 301; Volvo's ECC; Mazda's Titan Truck; Sanyo's Mirai; UNIQ's Minivan; and the Lucas Hybrid Electric Car, Chloride XREV, and California Energy Commission XA-100 studies are the most visible. But hybrids, whatever the combination, are basically a marketing concession to a perceived automotive industry problem with electric vehicles—that they don't have enough range.

General Motor's own surveys proved that most people don't drive very far. More than 40 percent of the trips were under 5 miles, and only 8 percent of the people took a weekday trip of greater than 25 miles. Electricity to recharge an electric vehicle is readily available everywhere—provided you carry an on-board recharger. Is the hybrid a solution to a nonexistent problem?

Considering that you pay an added weight/efficiency penalty by carrying another engine, generator, fuel source, drivetrain, etc., around with you that you don't use all the time when using a hybrid—versus a relatively lightweight emergency recharger—it would appear that hybrids contribute more to the problem than to the solution:

- With more parts they have to cost more.
- Higher parts count just means something more to go wrong later.
- You get worse performance than by using either pure energy source by itself.

Few automotive industry people writing about the limited range problem of electric vehicles have ever driven one, much less owned one and used it regularly. On the other hand, more than 99 percent of the Electric Auto Association members who own and regularly do drive their electric vehicles report that range is simply not an issue. When EAA members need to take longer trips they either use another vehicle, another mode of transportation, or they use a power boost trailer.

A far better solution is a trailer connected to your electric vehicle that converts it into a hybrid—but only when you need it. This "power-boost trailer" can utilize any type of power source (gasoline with alternator, diesel with generator, turbine, extra battery pack, etc.). The net result is that it extends your range, enabling you to recharge your batteries on the fly, etc. You disconnect it and leave it in the garage most of the time while you are taking short trips and tooling around town. In the future, maybe there will even be places where you can rent them for your occasional need.

### Individual electric vehicle conversion for fun, enjoyment and profit

As you have seen, numerous individuals have been successfully converting internal combustion vehicles to EVs for at least the past 25 years. This entire period has been marked by an almost total absence of comments about this activity from the naysayers. To the contrary, builders and converters are highly pleased with the results, hang on to their electric vehicles for years (if not decades), and are more than proud to say, "Why, yes I built it myself" hundreds of times (if not thousands) to interested onlookers during their trips. You can join this elite group. The rest of the book shows you how.

## ELECTRIC VEHICLES FOR THE 21ST CENTURY

What goes around comes around, as the saying goes. Although it's ironic that EVs were around before internal combustion engine vehicles and will also be around after them, the logic of 1990s' needs makes it just as sensible to switch to electrics as the logic of 1910s' needs made it to switch from them. Internal combustion engine vehicles will unquestionably be with us for a long time to come. You don't implement sweeping change overnight. But change we must. Electric vehicles are the most environmentally beneficial and technologically promising gift we can give to future generations. In the near term, electric vehicles powered by renewable energy sources solve greenhouse and air pollution problems, and make us independent of foreign oil. For the long term, rethinking our future transportation infrastructure around electric vehicles, Roadway-Powered electric vehicles (RPEV), electric railroad engines, and maglev systems is an urgent imperative to ensure the survivability of species Homo Sapiens on spaceship earth.

# 4

# The best
# electric vehicle
# for you

*Put an electric motor in your chassis and save a bundle.*

You learned from earlier chapters that electric vehicles are fun to drive, and save you money as you use them, and help save the planet. The benefit of building/converting your own electric vehicle is that you get this capability at the best possible price with the greatest flexibility.

In this chapter you'll learn about electric vehicle purchase tradeoffs, conversion trade-offs, and conversion costs. You'll be able to pick the best electric vehicle for yourself today—whether buying, converting, or building.

## ELECTRIC VEHICLE PURCHASE DECISIONS

When you go out to buy an electric vehicle today, you have three choices: buy ready-to-run, convert an existing internal combustion engine vehicle, or build from scratch. In addition, each choice has two broad subcategories: new or used. Your two most basic considerations are how much money you can spend and how much time you have. Another dimension of your purchase trade-off is where (and from whom) you can obtain an electric vehicle. This section will overview your options and highlight why conversion is your best choice at this time.

### Conversions save you money and time

Saving money is an important objective for most people. It's also logical to assume you want to arive at a working electric vehicle in a reasonably short time. When you combine these two considerations, conversion emerges as the best alternative—FIG. 4-1 shows you why at a glance. A conversion:

- Costs less money than either buying ready-to-run or building-from-scratch.
- Takes less time than building-from-scratch and only a little more time than buying ready-to-run.

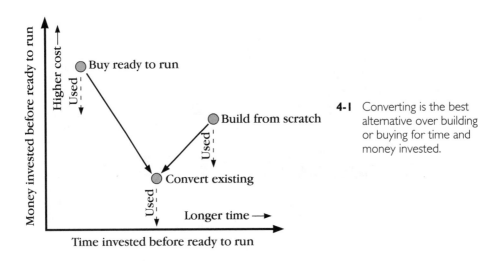

**4-1** Converting is the best alternative over building or buying for time and money invested.

Buying "used" in any category—the downward pointing arrows in FIG. 4-1—clearly lowers the cost for that category, but doesn't alter the overall relationship. When you compare equivalent used pricing across all three choices, conversion is still the best alternative.

A few actual figures make this easy to see. While you might spend upwards of $100,000 to obtain a new electric vehicle from the big three automakers, and could typically spend from $20,000 to $30,000 for a new EV at an independent dealership, in only a month or two, you can be driving around in your own EV conversion for under $10,000. When you build an EV from scratch, on the other hand, your electrical components cost the same as a conversion's, but your frame and body costs start where buying a conversion body leaves off, and you still have to do the work—pushing your time way up. Now you know the reason for this book.

Let's look at the buy-build-convert alternatives more closely, then look at the who-where-when trade-off.

## Buy ready-to-run from big three

Ready-to-run electric vehicles are available today from the major manufacturers. But if you want a brand new EV today from one of the big three automakers or any major automaker on the planet, you're going to spend a lot of money. In addition, because there are only a limited number of units available from any manufacturer, you must still meet "additional qualifications" or you're going to be way at the end of the receiving line. If you are able to offer the major manufacturers something of value to them (publicity, an in-depth driver opinion study, a scientifically conducted vehicle performance study, etc.), this might just earn you a better place in line but, realistically speaking, unless you represent one of the Fortune 500 firms, a major utility, or a major university, you are not likely to get one of 1993's crop of newly manufactured electric vehicles. Personal ownership of one of these vehicles is not their highest and best use at this time anyway. Let's quickly check out your chances with the big three.

*Ford.* Seeing one of the Ford Ecostar vans, shown in FIG. 4-2, pass you on the highway is one thing. Seeing it stopped with its hood raised, as also shown in FIG. 4-2, is something entirely different. Rolling around town draws a "Yeah, that's neat" re-

**4-2** Ford Ecostar van electric vehicle.

sponse. When it's parked and you pop the hood and passers-by are able to get a better look at what's present (or not present!) in the engine compartment, the barrage of questions starts with "Hey, where's the engine?" and inevitably leads to "Where can I get one?"

Ford plans to deliver about 100 of its Ecostar vans, priced around $100,000, to selected customers on 3-year leases starting in late 1993. Ford gets the data during the lease and gets the vehicle back at the end so it can see what's happened. Ford has also set up dealerships around the country in the key metropolitan areas where it intends to "seed" its initial van fleet to handle any maintenance. As maintenance is expected to be low, these dealerships will mostly visit their assigned customer units and swap out the computer diskettes that log results from the Ecostar's computers.[1] Ford is serious about its electric vehicle program.[2] Get your proposal in the mail. But don't grab for your checkbook until you hear from them first.

*Chrysler.* The Chrysler TEVan, shown in FIG. 4-3, is also not likely to turn heads on the highway. But when it's stopped with its hood raised, this isn't your average minivan and the questions will come pouring out. You can almost hear the first two: "Hey, what's in the box?" followed by "Well then, where is the engine?"

**4-3** Chrysler TEVan electric vehicle.

Chrysler plans to deliver about 50 electric TEVans, priced around $120,000, starting in late 1993, and they are already booked up with orders—again from the larger utilities and commercial customers. But get your proposal in the mail anyway—you might have a better idea. Though Chrysler had only 50 people working on electric vehicles in late 1992 (compared to hundreds at Ford and GM), many of those had come over from the highly successful Viper program, so Chrysler is also serious about its electric vehicles.

*General Motors.* Since 1990, GM has maintained the highest profile of any of the big three in EVs. GM also owns part of its Impact electric vehicle designer (AeroVironment, Inc.), and all of its battery/motor (Delco Remy) and power/charging electronics (Hughes) suppliers, so it's the most vertically integrated of the big three. GM's Impact EV—unlike the Ford and Chrysler vans—is more likely to turn heads just driving down the highway, and just as likely to draw a crowd when displayed with its hood up (FIG. 4-4).

**4-4** GM Impact electric vehicle.

Unfortunately, General Motors has fallen on hard economic times, and reduced budgets have shelved its electric vehicle mass production plans. Instead, GM announced plans to work more closely with Ford and Chrysler—they are already partnered in the U.S. Advanced Battery Consortium—to develop EV technology and possibly even build a car together. GM will deliver only 50 Impact two-seater electric vehicles, priced around $100,000, to its fleet customers for evaluation starting in late 1993.[3] So get your proposal in the mail to GM also—yours might offer just the solution they need.

## Buy ready-to-run from independent manufacturer

Ready-to-run electric vehicles are available from independent manufacturers and their dealers today. But these EVs—typically conversions that use internal combustion engine chassis—cost $20,000 to $30,000. EVs will come down in price as the numbers manufactured increase—their nonrecurring costs can then be amortized over a greater number of units. EVs have the potential to be far lower in cost than internal combustion powered vehicles in the future—when economies of scale from increased production kick in—because there are far fewer components that go into an electric vehicle than go into an internal combustion powered vehicle.

Buying a ready-to-run electric vehicle from an independent manufacturer or dealer means haggling and doing all the other things you are accustomed to doing when you buy a vehicle, with one important caveat: It's a long walk to the next nearest dealership! So keep things on a cordial business level while angling for the best price, and think about buying a used or demonstrator model, or working a variation of the publicity/advertising theme, or buying an older or discontinued model they still happen to have in stock. On balance, you'll find independent electric vehicle manufacturers and their dealers to be knowledgeable and enthusiastic on the subject of EVs, and willing to go halfway to meet your unique needs.

Solar Electric is easily the best and the largest of the independent EV manufacturers. Their offerings, such as the EV2 converted Fiero (FIG. 4-5) are among the nicest available. Coupled with the wide range of conversion products they offer (including cars, trucks, and minivans) is their commitment to quality and customer satisfaction. Typically, you can be driving your EV within a week after visiting their Santa Rosa, California showroom. You can also relax, knowing exactly who to call if ever there's a service problem. Of course, for all this you pay a price.

Green Motor Works is a Solar Electric dealer and conversion specialist that has gotten the entire industry's attention for its classy North Hollywood, California showroom and shop operation.[4] You would be the envy of all your friends driving its mouth-watering Porsche look-alike (FIG. 4-6), and no one would know why it was so quiet until they looked inside its engine compartment (FIG. 4-7). You could drive it away the same day you visit the showroom, but you'd have to come up with the $33,200 (approximate) sticker price.

## Building from scratch

If your desire is to custom-build a "Starship Electrocruiser" prototype from the ground up, there is certainly nothing to stop you. But you are certainly going to spend a lot

**4-5A**   Solar Electric Fiero electric vehicle.

**4-5B** Neatness counts—Solar Electric Fiero rear compartment.

**4-6** Green Motorworks Porsche look-alike electric vehicle.

more time and money than in buying ready-to-run (unless you put a zero value on your labor time).

There is also a hidden problem here—the safety aspect. If you buy a vehicle from any manufacturer or convert an existing internal combustion powered vehicle chassis, the safety aspects have already been handled for you. When you roll your own design, dealing with the safety issues becomes your responsibility. While licensing

**4-7** Green Motorworks Porsche "engine" compartment.

your one-of-a-kind "Starship Electrocruiser" will consume less time with the local Motor Vehicle authorities than would getting the certifications for a production run from multiple agencies, you are still going to have to convince someone that the wheels aren't going to come off before they permit you to cruise anywhere except in your driveway.

Building to someone's plans is a logical step towards saving time and hassle. You still have to provide the muscle and the bucks, but there's someone who has already blazed a trail before you in terms of licensing and construction shortcuts. In addition, you have someone to write or call if you get hung-up on a construction detail.

Putting a custom kit on an existing internal combustion powered vehicle chassis saves you yet more building time and hassle, but still requires an extensive labor investment on your part and additional out-of-pocket cost. The advantage is that you not only have someone to call or write to for help, but better instructions typically make that help less necessary. Pre-fab parts also go a long way toward ensuring that your finished product looks like a professional job.

There's also a hidden problem with custom-, plan-, or kit-built electric vehicles; it occurs if you decide to sell your creation. With new and do-it-yourself conversions using internal combustion engine vehicles—even if quite radical in the electrical department—the prospective purchaser comes to look at it and the first impression is, "Yeah, it's a Ford" or "Ok, it's a Honda" and so forth. The prospective buyer mentally catalogs the make and model of the body, then moves on. With custom-built electrics, you might deal with a lot of questions on the body, such as "How safe is it?" or "Does it leak much when it rains?" or "Will this crack here get much wider?" It can go on and on.

Building an EV from scratch today involves thoroughly planning what you are going to do in advance (will you make or buy parts, and from what vendors; when will parts be needed; where will assemby happen; which subcontractors will help, etc.) then pursuing your plan (leaving room for contingencies, problems, and any bargains that come your way). If you can scrounge, barter, or scavenge used parts, so much the better.

## Custom-built electric vehicles

The advantage to custom-building an electric vehicle is that you can "go where no one has gone before." In practical terms, it means you can build something like Don Moriarity's custom sport racer (FIG. 4-8), which is an excellent example of the meticulous attention to design and construction detail that results in winning entries. With the custom-built approach, ideal for high-speed or long-distance race vehicles, you are free to make the design and component trade-offs that optimize your vehicle in the direction of your choice. But this approach also assumes you have the skills, talent, resources, and money to accomplish it.

**4-8** Don Moriarity's custom sports racer electric vehicle.

## Electric vehicles built from plans

The best example of building from plans is the Doran three-wheeler electric (shown front and rear in FIG. 4-9), whose plans are typically offered mail-order for $39.95. With the Doran construction manual and plans you get excellent body, mechanical, and electrical instruction. They result in a less than 1500-pound vehicle (motorcycle classification) that gets 80 mph and an around-town 60-mile range from its recommended Prestolite 28-hp dc motor and 108-volt (nine 12-volt lead-acid) battery string. But to build even this small, simple vehicle requires mastery of fiberglass-over-foam-core body construction techniques—or at least making haste slowly while you learn them—and is, again, not for everyone.

**4-9** Doran three-wheeler electric vehicle.

## Electric vehicles built from kits

The best example of building from kits is the Bradley GT II kit car (FIG. 4-10), of which there are several versions around. This combination of 96-volt electrics on a Volkswagen chassis with a lightweight body delivers impressive performance along with a snazzy, classic body-style, but you're going to have to part with a substantial amount of cash and labor before you make it happen. A kit-built EV can be a highly reward-

**4-10** Bradley GT II kit car electric vehicle.

ing and satisfying showpiece project for those who have the experience, enthusiasm, and persistence to pull it off—but (for the third and final time) it's not for everyone.

## Converting existing vehicles

Conversion is the best alternative because it costs less than either buying ready-made or building from scratch, takes only a little more time than buying ready-made, and is technically within everyone's reach (certainly with the help of a local mechanic, and absolutely with the help of an EV conversion shop).

Conversion is also easiest from the labor standpoint. You buy the existing internal combustion vehicle chassis you like (certain chassis types are easier and better to convert than others), put an electric motor in your chassis, and save a bundle. It's really quite simple; chapter 11 covers the steps in detail.

To do a smart EV conversion, the first step is to buy a clean, straight, used, internal combustion vehicle chassis. A used model is also to your advantage (as you'll read in chapter 6) because its already-broken-in parts are smooth and the friction losses are minimized. Then you add well-priced electrical parts—or a whole kit from vendors you trust—and do as much of the simple labor as possible, while farming out the tough jobs (machining, bracket-making, etc.). Whether you do the work yourself and just subcontract a few jobs, or elect to have someone handle the entire conversion for you, you can convert to an electric vehicle for a very attractive price compared to buying a new EV.

## Converting existing vans

While large vans, such as SMUD's converted GMC van (FIG. 4-11), make great test bed vehicles for the utility companies, they are heavy, more expensive to buy, take longer

**4-11** SMUD converted GMC van electric vehicle.

to convert, give less-than-adequate after-conversion performance, and cost you more to own and operate. For these reasons and others, you will never find an 8000-pound van recommended in this book as a potential conversion candidate. On the other hand, minivans—particularly the newer, lighter models—offer intriguing prospects for conversion, and you can look further into them as your needs require. But vans in general—even minivans—are usually more expensive, heavier, or take longer to convert than other chassis styles, so investigate before you invest.

## Converting existing cars

This is a rich, fertile topic and can only be lightly skimmed here. Even a partial listing of car conversions done would fill far more pages than are in this book. We'll just hit a few high points and move on.

Scott Cornell's 1971 Karman Ghia conversion (shown front and rear in FIG. 4-12), is an excellent example of conversion technique and component placement in a Volkswagen chassis. The 16 6-volt batteries (96 volts total) give you excellent performance and range from this body style and weight combination. But 96 volts worth of batteries is about the limit for this type of vehicle, and most of the batteries are placed in the back seat area inside the passenger compartment—not a particularly safe or convenient arrangement.

Vic Jager's 1971 Mazda RX-2 conversion (shown front and rear in FIG. 4-13), is also a 96-volt system, but takes advantage of a little more chassis room to conveniently stow four batteries in the front under-hood area and 12 batteries in the rear trunk area—keeping them out of the passenger compartment entirely. Notice the neatly laid out components and tied-off wiring runs in the engine compartment.

Conversion manual author Bill Williams' ultra-clean 1971 Datsun 1200 (FIG. 4-14) takes neatness in the engine compartment to a new standard of excellence—but at the expense of being only an 84-volt system.

Lyle Burresci's 1977 Plymouth Arrow (shown front and rear in FIG. 4-15), was the basis for Mike Brown's conversion manual. It also uses a 96-volt system (six batteries

**4-12** Scott Cornell's 1971 Karman Ghia conversion electric vehicle.

in front, 10 in rear) and shows how good chassis selection, construction techniques, and "battery boxes" can transform a potential safety hazard.

John Wasylina's 1983 Renault Alliance (shown front and rear in FIG. 4-16), offers another solution—use 12-volt instead of 6-volt batteries. Four batteries up front and six in the rear take up less room, yet deliver 120 volts.

**4-13** Vic Jager's 1971 Mazda RX-2 conversion electric vehicle.

Conversion manual author Michael Hackleman and Otto Ebenhoech's 1991 Honda CRX (shown front and rear in FIG. 4-17) is today's state-of-the-art conversion. It offers breathtaking performance and range from its 120 volts of 12-volt batteries, rear battery box, and under-hood area, which is full of performance enhancements.

**4-14** Bill Williams' 1971 Datsun 1200 conversion electric vehicle.

**4-15A** Lyle Burresci's 1977 Plymouth Arrow conversion electric vehicle.

## Converting existing pickup trucks

Until recently, not many pickup truck conversions were done. But when EV converters took a closer look at the newer, lighter pickups, this trend reversed itself in a resounding way.

**4-15B** Lyle Burresci's 1977 Plymouth Arrow rear compartment.

**4-16A** John Wasylina's 1983 Renault Alliance front compartment.

Solar Car Corporation's SMUD-owned Chevrolet S-10 pickup truck conversion (shown front and rear in FIG. 4-18), provides a lot more room under the hood and even more space in the rear.

**4-16B** John Wasylina's Renault Alliance rear compartment.

**4-17A** Michael Hackleman's and Otto Ebenhoech's 1991 Honda CRX front compartment.

The same is true for BAT's Ford Ranger pickup truck conversion (shown front and rear in FIG. 4-19), and note BAT's fantastic treatment of the rear battery box.

Pickup trucks are a great EV conversion choice—one you'll get a chance to examine in detail in chapter 11.

**4-17B** Michael Hackleman's and Otto Ebenhoech's 1991 Honda CRX rear compartment.

**4-18A** Solar Car Corporation's Chevrolet S-10 pickup truck front compartment.

## What, who, where, when

TABLE 4-1 adds the other dimensions to the buy, convert, or build trade-off: What vehicles or parts are available, who do you get them from, where they are located, and when will you get/assemble them? There are not many places to buy new EVs today.

**4-18B** Solar Car Corporation's Chevrolet S-10 pickup truck rear compartment.

### Table 4-1 Electric vehicle purchase decisions compared

| Who | Where | When | |
|---|---|---|---|
| Buy New | Independent Manufacturers & Dealers | Regional Locations | Buy - Now Shipping - Weeks |
| Buy Used | EAA Classifieds Manufacturers & Dealers | Anywhere | Buy - Now Shipping - Weeks |
| Convert New | New Auto Dealers Electric Vehicle Parts Dealers | Local Auto Dealer Mail Order EV Parts | Buy - Weeks Assy - 80-100 Hours |
| Convert Used | Used Auto Dealers Electric Vehicle Parts Dealers | Local Auto Dealer Mail Order EV Parts | Buy - Weeks Assy - 100-200 Hours |
| Build From Scratch | Manufacturers & Dealers Electric Vehicle Parts Dealers | Local Raw Material Mail Order EV Parts | Buy - Months Assy - 200-??? Hours |

If you don't reside in one of those places, adding the weeks it might take to get your new EV shipped to you puts the conversion option—based on an internal combustion vehicle chassis obtained locally—into an even more attractive light. The weeks in the conversion option "When" boxes are primarily consumed by finding (locally) the vehicle chassis you like. Conversion assembly time is then measured in hours or days. Starting with a used chassis will take you longer for cleaning, preparation and so on. The months it takes to build from scratch are consumed by your finding the vehicle kit, chassis, and parts you like and having them shipped to you. Build-from-scratch assembly times start where conversion times leave off, and can range up into the thousands of hours, depending on how exotic you get.

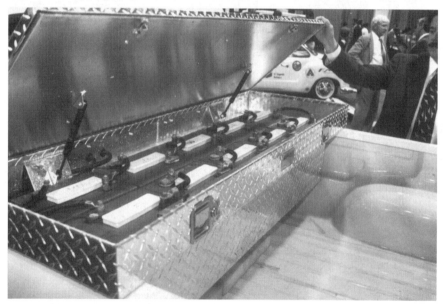

**4-19** BAT's Ford Ranger pickup truck conversion electric vehicle.

## ELECTRIC VEHICLE CONVERSION DECISIONS

When you do an electric vehicle conversion today, you have three chassis choices: van, pickup truck, or car. You have additional choices of motor, controller, batteries and charger. This section will look at these choices and prepare you for the guidance given by the rest of the book.

## Your chassis makes a difference

If you're going ahead with the conversion alternative, your most important choice is the chassis you select. TABLE 4-2's comparison of the leading vehicles sold in 1992 and 1991 gives you a hint of what's coming. Pickup trucks were the the number 1, 2, and 7 best selling vehicles. Why? Pickup trucks are the most popular personal transport vehicle for commuter use, and have a high utility for carrying loads. Study after study has shown that most commuter vehicles are occupied by only one person, and pickup trucks are ideal for carrying one person loads in style. According to the trend of the last 10 years (FIG. 4-20), pickup trucks are gaining on passenger cars in terms of percent of all light vehicles sold in the United States—up from 20 percent in 1982 to 35 percent in 1992.

### Table 4-2 Top 10 leading
### 1992 vs. 1991 vehicle sales compared

| Vehicle | 1992 rank | 1992 sales | 1991 rank | 1991 sales |
|---|---|---|---|---|
| Ford F-series pickups | 1 | 488,539 | 1 | 432,122 |
| Chevrolet C-K pickups | 2 | 455,250 | 2 | 424,181 |
| Ford Taurus | 3 | 409,751 | 4 | 299,659 |
| Honda Accord | 4 | 393,477 | 3 | 399,297 |
| Ford Explorer | 5 | 306,681 | 7 | 250,059 |
| Toyota Camry | 6 | 286,602 | 5 | 263,818 |
| Ford Ranger | 7 | 247,777 | 9 | 233,503 |
| Dodge Caravan | 8 | 244,149 | - | - |
| Ford Escort | 9 | 236,622 | 8 | 247,864 |
| Honda Civic | 10 | 219,228 | - | - |
| Chevy Cavalier | - | - | 6 | 259,385 |
| Chevy Corsica/Beretta | - | - | 10 | 231,227 |

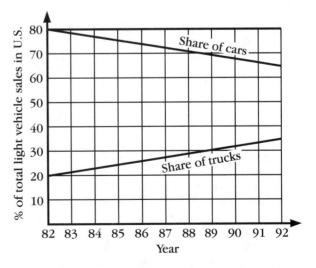

**4-20** Pickup truck sales gaining on passenger car sales in the United States during the last 10 years.

Minimizing weight is always the number one objective of any EV conversion. When added to the criteria of minimizing conversion time and maximizing the odds for get-it-right-the-first-time success, the trade-off points squarely in the direction of the pickup truck (FIG. 4-21). The van weighs more, and the car is typically the most time-consuming conversion—less room to mount EV options increases the problem of getting parts to fit.

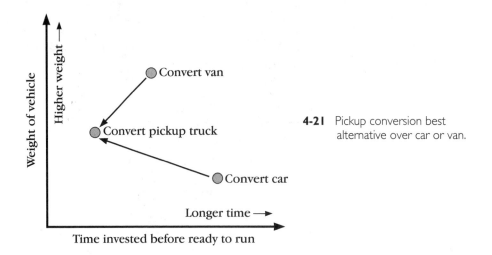

**4-21** Pickup conversion best alternative over car or van.

The internal combustion vehicle pickup truck chassis is actually an outstanding electric vehicle conversion choice because:

- A single cab model pickup truck's curb weight, minus its internal combustion engine components, is typically not much more than a car, yet it has considerable room to add the extra batteries that translate to better performance.
- A manual transmission, no-frills, 4- or 6-cylinder internal combustion engine pickup truck is one of the least expensive new or used conversion platforms you can buy.
- The additional battery weight presents no problems for the pickup's structure—its sturdy frame is specifically designed to carry extra weight and extra or heavier springs are readily available if you need them.
- The pickup isolates the batteries from the passenger compartment very easily—an important safety criteria not found in car or van conversion platforms.
- The pickup is much roomier—the engine compartment and pickup box or bed space offer flexibility for your component design and layout, and a front-wheel-drive model gives you additional flexibility for battery mounting.
- A late-model, compact, or intermediate pickup offers frontal-area comparable to equivalent-sized cars, yet front grill and engine areas can be more easily blocked or covered to reduce wind resistance and engine compartment turbulence.
- The pickup truck can be made into an instant hybrid—just load a portable emergency electrical generator and one or more five-gallon gas cans into the back.

## Your batteries make a difference

Of all the pickup truck advantages, its extra room makes the biggest difference because it makes more space available to mount batteries. With car conversions, you choose a larger chassis or go to 12-volt batteries. With pickup truck chassis, you can use more powerful 6-volt batteries to produce 120 volts and more. Both the 6-volt (more energy storage) and 12-volt (less weight) options are available with the pickup conversion.

More batteries—higher voltage—dramatically improves your performance (FIG. 4-22). 1970s-vintage EV designs found 72 volts acceptable. Today, you would be unhappy with anything less than 96 volts, and 120 volts—the setup used for chapter 11's conversion—is even better. Beyond that, a 144-volt battery setup is still better, but you start to encounter diminishing returns with this particular dc motor and controller combination.

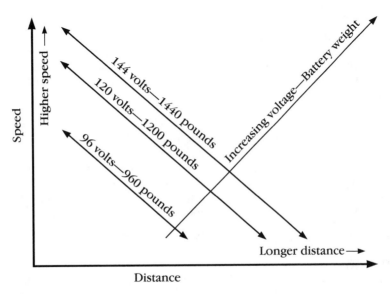

**4-22** More batteries always better than less—up to a point.

A 120-volt battery string comprising 20 6-volt batteries (about 1200 pounds) typically delivers a top speed of 60 mph or more, and a 60-mile range (at reduced speed) in a 3000-pound curb weight pickup truck with 4-speed transmission. You might get more or less depending on your design and components. Jim Harris' chapter 11 Ford Ranger pickup conversion goes 75 mph, and his ranges were still increasing at press time. But whatever you're getting at 120 volts, you'll get more of at 144 volts.

## All your component trade-offs make a difference

TABLE 4-3 presents all the dimensions of the conversion trade-off: chassis, motor, controller, batteries, and charger versus OK, ideal, best, and unlimited money alternatives. If money is no object, you customize your chassis, add the latest powerful ac induction motor with custom controller and tool around the countryside powered from nickel-cadmium, sodium-sulfur, or (soon to come) lithium polymer batteries towing

**Table 4-3  Electric vehicle conversion decisions compared**

| What | OK | Ideal | Best | Unlimited money |
|------|-----|-------|------|-----------------|
| Chassis | Van | Pickup truck | Car | Custom-built Starship Electrocruiser |
| Motor | dc series 19 hp | dc series 22 hp | dc compound 30 hp | ac induction 50 hp |
| Controller | Curtis MOSFET PWM | Custom IGBT PWM | Custom IGBT PWM + Regen | Custom ac IGBT PWM + Regen |
| Battery Pack | 96 volts - 16ea 12 volt Lead-Acid | 120 volts - 20ea 6 volt Lead-Acid | 144 volts - 6 or 12 volt Lead-Acid | 120+ volts - NiCd, NaS, Lithium Polymer |
| Charger | 120 volts, 20 amps | 120 volts, 20 amps + Onboard | 240 volts, 50 amps + Onboard | 240 volts, 50 amps + Power boost trailer |

your power boost trailer. The rest of us have to take it a bit slower; our movement toward the best category can only proceed as fast as our pocketbooks allow: a pickup, readily-available series dc motor, Curtis PWM controller, lead-acid batteries, and an onboard 120-volt 20-amp charger. Actually, anything from the OK or ideal is acceptable. And while I am obviously prejudiced towards pickups, that doesn't mean you have to be.

The point is, use what's available. dc motors, Curtis controllers, and lead-acid batteries are with us in abundance today. While not state of the art, they are proven to work, available from many sources, and you can get lots of help if something goes wrong. By starting with known quantities, your chances of initial success and satisfaction are very high. You can get up and running quickly. When you're ready for your next conversion, you can experiment a bit, "push the outside of the envelope," etc. You know what it's all about, what's working, and where you'd like to make the changes.

## The procedure

Chapter 5 provides sources to get you started; subsequent chapters introduce you to chassis, motors, controllers, batteries and chargers. In chapter 11, you'll look over my shoulder while I convert a Ford Ranger step-by-step—instructions you can adapt to nearly any conversion you want. Chapter 12 shows you how to maximize the enjoyment of your EV once it's up and running.

Use the chapter 5 sources—don't just take my word for it. Join the EAA and subscribe to their newsletter. Read all the books, magazines, and research material you can. Go to meetings, shows, and rallies. Most of all, talk to people who have already done it. If you listen to what they say, you will soon discover there are more correct ways to do an electric vehicle conversion than there are snowflakes in the known universe. Then integrate all this information and make your own decision. After you've done your first conversion, you'll notice a new phenomenon—people will start listening to you.

## HOW MUCH IS THIS GOING TO COST?

Now we come to the part where, as they say, the rubber meets the road—your wallet. Let's look at an actual quote, then add vehicle and battery costs and analyze the results. While you should not consider these costs the last word, you can consider them typical for today's EV conversion efforts. In any event, they will give you a good idea of what to expect.

Figure 4-23 shows you an actual quote provided from one of the established, long-time suppliers—KTA—in mid-1992. (KTA's address and phone number have since changed; check their new listing in chapter 5.) Notice the KTA quote is dominated by four items: motor, controller, charger, and the costs of making a custom motor-to-ve-

## KTA SERVICES
### 12531 BREEZY WAY
### ORANGE, CA 92669
### 714/639-9799

July 27, 1992

Thanks for the opportunity to furnish an EV components quotation on your 1983 Mazda RX-7 conversion. We'll assume that the vehicle will be using a 108-volt battery system comprised of 18 ea. 6-volt batteries with L-posts. Top speed should be about 70+ MPH, and maximum range would be 50-65 miles. The components recommended are listed below, and are the same as our California certified Kit #4:

| DESCRIPTION | PRICE | DELIVERY | REMARKS |
|---|---|---|---|
| PROPULSION MOTOR & ADAPTER: | | | |
| ADVANCED DC #FB1-4001 SERIES-WOUND MOTOR & DRAWING | $1500.00 | 1 WEEK | |
| PROTOTYPE EVCC ADAPTER PLATE, COUPLING, AND HARDWARE KIT | 650.00 | 2-3 WEEKS | FURNISHED BY EVCC |
| EVCC 9" CRADLE MOTOR MOUNT WITH 4" X 8" BASE PLATE | 150.00 | IN STOCK | PROTO. CROSS-MEMBER REQ'D |
| TOTAL MOTOR & ADAPTER | 2300.00 | | |
| MAJOR PROPULSION COMPONENTS & INSTRUMENTATION: | | | |
| CURTIS-PMC #1221B-7401 MOTOR CONTROLLER & MANUAL | $ 750.00 | IN STOCK | |
| CURTIS-PMC #PB-6 POTBOX | 60.00 | IN STOCK | |
| ALBRIGHT #SW-200B MAIN CONTACTOR | 130.00 | IN STOCK | |
| BUSSMAN #KAA-400 SAFETY FUSE | 22.00 | IN STOCK | |
| GENERAL ELECTRIC #TQD-200 MAIN CIRCUIT BREAKER | 110.00 | IN STOCK | |
| BLACK PLASTIC SWITCHPLATE & MOUNTING HDWE. FOR #TQD-200 | 20.00 | IN STOCK | |
| SEVCON #622-11014 13.5 VOLT/25 AMP DC-DC CONVERTER | 420.00 | IN STOCK | REPLACES AUX. 12-V BATTERY |
| KTA SVCS. #V50150-914 DUAL-SCALE VOLTMETER | 60.00 | IN STOCK | |
| KTA SVCS. #A500-50 DUAL-SCALE AMMETER | 40.00 | IN STOCK | |
| DELTEC #MKA-50-50 AUXILIARY CURRENT METER SHUNT | 17.50 | IN STOCK | |
| DELTEC #MKB-500-50 PROPULSION CURRENT METER SHUNT | 22.50 | IN STOCK | |
| ROTARY SELECTOR SWITCH & KNOB FOR METER SELECTION | 7.50 | IN STOCK | |
| CURTIS INST. #900R-108-BN BATTERY FUEL GAUGE | 225.00 | IN STOCK | OPTION TO KIT #4 |
| AUXILIARY WIRING 10-POS. 25 A. BARRIER STRIP | 10.00 | IN STOCK | |
| K & W ENG. #BC-20 ONBOARD BATTERY CHARGER | 550.00 | IN STOCK | |
| PHILLIPS #RLY8845F CHARGER INTERLOCK RELAY | 15.00 | IN STOCK | |
| 48 FT. #2/0 AWG CAROL ULTRA-FLEX WELDING CABLE | 172.80 | IN STOCK | |
| 54 #54111 THOMAS & BETTS #2/0 x 3/8 WELDING CABLE LUGS | 162.00 | IN STOCK | |
| 8 FT. HEAT SHRINK TUBING | 15.00 | IN STOCK | |
| WIRING DIAGRAMS, INSTALLATION INSTRUCTIONS, TAX CREDIT KIT | 50.00 | IN STOCK | INCLUDES BOOK "CONVERT IT" |
| COMPLETE SET OF VACUUM RESTORATION COMPONENTS | 303.00 | 1 WEEK | OPTION TO KIT #4 |
| TOTAL COMPONENTS | 3162.30 | | |
| TOTAL MOTOR & ADAPTER | 2300.00 | | |
| 8.25% CA SALES TAX - EXEMPT PER TAX CODE SUBDIV. (b), SECT. 6356.5 | 0.00 | | |
| FREIGHT & INSURANCE | 0.00 | | PICKUP XMSN/FLYWHEEL/ETC. APPROX. 8/9, DELIVER COMPLETE KIT APPROX. 9/19. |
| TOTAL KIT #4 ORDER | 5462.30 | | |

The above quotation covers all the major components except for the batteries and a few minor things. Additional components are needed to restore vacuum if the vehicle has power brakes. Also, battery 'fuel gauges' are available as an option, as is the LB-20 booster for 114 or 120 volt systems. Prices shown here will be in effect for 60 days.

With kind regards,

*KWKoch*

Ken Koch - KTA Services
KTAMEM175.WRI

**4-23** Sample KTA Services quotation.

hicle adapter. Notice also that the professionals at KTA tell you what performance you can expect from what you get, when you're going to get it, how much it's going to cost, and how long the quoted prices are valid. Be sure you get the same information, in writing, out of any supplier you choose.

Figure 4-24 shows the several kits KTA offers, each capable of powering different vehicle weights. Again, KTA gives you the performance you should expect per weight classification. It also points out that these kits are exempt from the California State sales tax and qualify for a California tax credit of $1000 after being installed and inspected; you might have the same or similar savings available in your state, so check it out. From the KTA kit list, it should be apparent that you will mostly be interested in the two largest kit options, for powering vehicles up to 4000 and 5000 pounds, respectively. The largest kit—with the 22-hp dc motor—was the basis for the quote of FIG. 4-23 and the analysis of TABLE 4-4.

**KTA SERVICES**
12531 BREEZY WAY
ORANGE, CA 92669
714/639-9799

**AVAILABLE NOW....5 CERTIFIED PURE ELECTRIC VEHICLE RETROFIT/CONVERSION KITS**
That qualify for the California State $1000 tax credit and sales tax exemption..a savings of up to $1433!

KIT NO. 1:    For 2, 3, or 4-wheeled vehicles up to 1500 lbs. with optional electrical reversing if no transmission is employed. May be used on any motorcycle or small custom kit car of 1992 manufacture or older. Suitable for propelling 500 to 1500 lb. vehicle at speeds up to 55 MPH................................................**Basic Kit $2450**

KIT NO. 2:    For 3 or 4-wheeled vehicles up to 3000 lbs. with optional electrical reversing if no transmission is employed. May be used on any custom kit car, compact passenger car, or compact pickup truck of 1992 manufacture or older. Suitable for propelling 1000 to 3000 lb. vehicle at speeds up to 50 MPH.......................................... ...........................................................................................................**Basic Kit $3450 to $4250**

KIT NO. 3:    For 3 or 4-wheeled vehicles up to 3300 lbs. May be used on any custom kit car, compact passenger car, or compact pickup truck of 1992 manufacture or older. Suitable for propelling 1000 to 3300 lb. vehicle at speeds up to 85 MPH.........................................................................................**Basic Kit $4450**

KIT NO. 4:    For 4-wheeled vehicles up to 4000 lbs. May be used on any custom kit car, passenger car, or pickup truck of 1992 manufacture or older. Suitable for propelling 1000 to 4000 lb. vehicle at speeds up to 85 MPH......... ...........................................................................................................................**Basic Kit $4950**

KIT NO. 5:    For 4-wheeled vehicles up to 5000 lbs. May be used on any full-sized passenger car, full-sized pickup truck, or mini-van of 1992 manufacture or older. Suitable for propelling 2000 to 5000 lb. vehicle at speeds up to 85 MPH.........................................................................................**Basic Kit $5250**

Kits do not include price of batteries, which can be purchased from a local battery supplier at a system cost of between $200 and $1300, depending on battery type and number. All *complete* kits are exempt from California State sales tax, and qualify for the State tax credit of $1000 after being installed and inspected in a *pure* electric conversion or retrofit. All kits delivered or shipped in Southern California *free of charge*....an additional savings of up to $100!

**4-24**  KTA Services kit offerings.

TABLE 4-4 adds the pickup truck chassis and battery costs to the FIG. 4-23 quote (the *Typical* column); shows what savings you might expect with a used and older chassis (the *Economy* column); and shows what extra costs to expect when using the latest new chassis and a few extra bells and whistles (the *High or custom* column). The

## Table 4-4 Electric vehicle conversion costs compared

| Item | Economy | Typical | High or Custom | Percent Allocated |
|---|---|---|---|---|
| Chassis | $1500 | $3000 | $10,000 | 28 - 31 - 57 |
| Pickup truck | Earlier model used | Late model used | Latest year new | |
| Motor adapter plate | $400 | $800 | $800 | 8-8-4 |
| Custom | Local or do-it-yourself | Professionally done | Professionally done | |
| Motor | $800 | $1500 | $1500 | 15 - 16 - 10 |
| Advanced dc 22 hp | Used | New | New | |
| Controller | $400 | $750 | $750 | 8 - 8 - 4 |
| Curtis PWM | Used | New | New | |
| Wiring & components | $600 | $1850 | $2500 | 12 - 19 - 14 |
| Switchers, meters, wire | Used | New | New | |
| Battery pack | $1200 | $1200 | $1200 | 23 - 12 - 7 |
| 20ea 6 volt lead-acid | New | New | New | |
| Charger | $300 | $550 | $800 | 6 - 6 - 4 |
| 120V, 20A, onboard | Used | New | New | |
| Total | $5200 | $9650 | $17,550 | 100 |

amounts you might obtain for selling off the internal combustion engine components were omitted from the comparisons; you can expect the vehicle costs to be lower if you do sell them. The *Typical* column summarizes the 1987 Ford Ranger pickup EV conversion of chapter 11.

## CONVERSION FOR FUN & PROFIT

Darwin Gross drew a picture of a two-seater sportscar EV on a napkin over lunch one day and said, "You could sell that for $4995." My own scribbling on the napkin (aluminum tubular frame, plastic body, thin/hard tires, no power steering, heater (fabric top optional), motor, controller, batteries, etc.), led me to a more sedate $9995. I'm talking about a TR3-sized sportscar that could whip a TR3 ala Bertone's Blitz—a mouth-watering idea. In thinking more about it later, if a person applied the Dr. Paul MacCready technique and optimized on cost, such a vehicle is not only energy efficient and high-performance, but very reachable. Think of it as a "poor man's Impact." And if a MacCready-style team was brought together to accomplish it, you'd have a working model on the streets within a year or so. Somebody's going to make a lot of money on this, or something just like it . . . you heard it here first.

# 5
# Sources

*Knowledge is but the beginning of wisdom.*

Darwin Gross, *Universal Key*

Most books put the sources in the back. In this book, sources follow the choices and trade-offs chapter because sources are the very next thing you should investigate *before* you buy, convert, or build anything. The reason is simple: A little bit of knowledge is a dangerous thing.

One of the most valuable benefits this book provides is information about where to go and who to see. While there are thousands of sources, you'll find your needs best served by confining your initial search to just a handful—until you get acquainted with the EV field. Like the ripples moving outward from a rock dropped into a pond, you'll find your sphere of contacts, written information, and events widening outward with the passage of time.

In this chapter you'll find out about the clubs and associations, manufacturers and suppliers, converters and consultants and books, articles and papers that can help you. But the electric vehicle field is changing rapidly, so consider what you read about here only as a starting frame of reference. You must do your own reading, searching, listening, talking before shopping to buy, convert, or build in order to get the best products and price.

No one ever became a doctor, astronaut or automobile mechanic by reading a book. You've got to get the experience yourself. No one will become an electric vehicle buyer, builder, or converter by reading a book either. Sure, it helps, but you owe it to yourself to do more. When it comes to talking about cars and trucks, everyone's got opinions and preferences. Four articles about the same vehicle in the automotive press will give you four different opinions for four different (and apparently logical) reasons. This is especially true of electric vehicles—a field in which opinions have been shaped and molded by powerful forces for years.

The best guides for individuals interested in EVs are usually found in, around, and through the Electric Auto Association. You are fortunate if you live near a local chapter—if not, their newsletter and a few phone calls to people listed inside will get you started instead. Even within the Electric Auto Association membership, you will encounter strong and differing individual opinions. Some members will tell you a certain vehicle chassis, motor, controller, and charger is best. The facts are that the dc motors, PWM controllers, lead-acid batteries, and unsophisticated chargers most readily available today are not the best, but are not too bad. EV'rs of the next century will unquestionably use ac, smarter controllers, and alternative batteries. My advice is to listen carefully to what everyone has to say, then go and do what makes the most sense for you and your wallet.

Other than purchasing your vehicle/chassis locally, most of the time you're going to be dealing over the phone when ordering, and your merchandise will come a few days later via UPS, or a few weeks later via freight (if it's a heavy item like an electric motor). You'll find many suppliers of parts and services in the pages that follow. But not all follow the same standards for customer satisfaction, so it's sensible to take certain steps to protect yourself:

- Remember it's healthy to question, challenge and doubt all claims until proven.
- Talk to several vendors—at least three—for all major parts; get literature, etc.
- Get everything in writing—price/delivery quotes, specs, and "promises."
- Pay by credit card if possible—this gives you additional protection.
- If it's a large purchase, see if you can negotiate progress payments.
- Be sure you understand vendor's warranty and return privileges before you buy.

People seem to tell many friends if they've been "wronged." Remember to tell at least one friend when you find an outstanding parts or service provider. EVs are a growing field, and this simple action ensures that "the cream rises to the top."

### Less is more

I have chosen to give you a few sources in each category to get you started rather than attempting to list everything and get you confused. The rest of this chapter is divided into four sections:

- Clubs, associations, and organizations.
- Manufacturers, converters, and consultants.
- Suppliers.
- Books, articles, and literature.

Start with the associations and articles, and then work your way through the consultants, converters, suppliers, etc.

## CLUBS, ASSOCIATIONS, & ORGANIZATIONS

The original Electric Auto Association, whose logo appears in FIG. 5-1, has numerous local chapters. There are offshoots from the original, and new local entities having no

**Electric Auto Association**

**5-1** The Electric Auto Association's logo says it all.

connection with the original. There are also associations and organizations designed to serve corporate and commercial interests rather than individuals. Each one of these has its own meetings, events, and newsletter. Here is the list:

ELECTRIC AUTO ASSOCIATION
2710 St. Giles Lane
Mountain View, CA 94040
800-537-2882

Founded in 1967, this is the oldest, largest organization, and has consistently been the best source of EV information for the individual. Its $35 membership dues/newsletter subscription is well worth the price. Recent newletters have averaged 16 to 20 pages and are a goldmine of current EV news and happenings. When you collect a year's worth of these, you literally have another electric vehicle book. I can't emphasize enough the invaluable knowledge on tap in the members of this organization.

Here are some EAA local chapters—maybe there's one near you.

ARIZONA—PHOENIX
Lee Clouse
PO Box 11371
Phoenix, AZ 85061—4th Saturday
(602) 943-7950

CALIFORNIA—EAST BAY
Jim Danaher
1986 Gouldin Road
Oakland, CA 94611—2nd Saturday
(510) 339-1984

CALIFORNIA—LOS ANGELES
Irv L. Weiss
2034 N. Brighton "C"
Burbank, CA 91504—1st Saturday
(818) 841-5994

CALIFORNIA—NORTH BAY
Preston McCoy
117 Birch Avenue
Corte Madera, CA 94925—2nd Saturday
(415) 492-7973

CALIFORNIA—PENINSULA
Jean Bardon
540 Moana Way
Pacifica Way
Pacifica, CA 94044—1st Saturday
(415) 355-3060

CALIFORNIA—SACRAMENTO
Richard Minner
5201 Dover Avenue
Sacramento, CA 95819-3824—4th Saturday
(916) 454-5524

CALIFORNIA—SAN DIEGO
Ron Larrea
9011 Los Coches Road
Lakeside, CA 92040—1st Tuesday
(619) 443-3017

CALIFORNIA—SAN JOSE
Don Gillis
5820 Herma
San Jose, CA 95123—2nd Saturday
(408) 225-5446

CALIFORNIA—SAN LUIS OBISPO
James Donnell
PO Box 1000
Morro Bay, CA 93443

CALIFORNIA—SANTA BARBARA
Dale Ross
PO Box 91327
Santa Barbara, CA 91327
(805) 687-3919

CALIFORNIA—SILICON VALLEY (FOUNDING CHAPTER)
Paul Brasch
1968 Elden Drive
San Jose, CA 95124-1313—3rd Saturday
(408) 371-5969

FLORIDA—NORTH
Bill Young
PO Box 156
Titusville, FL 32781-0156
(407) 269-4609

FLORIDA—SOUTH
Steve McCrea
101 SE 15th Avenue #5
Ft. Lauderdale, FL 33301
(305) 463-0158

MASSACHUSETTS
Bob Batson
1 Fletcher Street
Maynard, MA 01754
(508) 897-8288

NEVADA—LAS VEGAS
Gail Lucas
Desert Research Institute
PO Box 19040
Las Vegas, NV 89132-0040
(702) 736-1910

NEW JERSEY
Kasimir Wysocki
293 Hudson Street
Hackensack, NJ 07601
(201) 342-3684

NEW MEXICO—DEMING
Dr. Jack Hedger
PO Box 1077
Deming, NM 88031-1077
(505) 546-0288

NORTH CAROLINA
Lawson Huntley
PO Box 1025
Monroe, NC 28111-1025
(704) 283-1025

TEXAS—HOUSTON
Ken Bancroft
4301 Kingfisher Street
Houston, TX 77035
(713) 729-8668

WASHINGTON, D.C.
David Goldstein
Electric Vehicle Industry Association
9140 Centerway Road
Gaithersburg, MD 20879
(301) 231-3990, (301) 869-4954

WASHINGTON—SEATTLE
Ray Nadreau
19547 23rd NW
Seattle, WA 98177
(206) 542-5612

## Other EV associations & organizations

These are organizations focused on government and/or industry-related goals.

CALSTART
3601 Empire Avenue
Burbank, CA 91505
(818) 565-5600

California statewide public/private sector consortium promoting Showcase Electric Vehicle Program (SEVP) and prototype electric vehicles.

ELECTRIC VEHICLE ASSOCIATION OF THE AMERICAS (formerly electric vehicle Development Corp.)
20823 Stevens Creek Boulevard, Suite 440
Cupertino, CA 95014
(408) 253-5262

A trade association for electric utilities, manufacturers and government agencies.

## Electric vehicle events

The events listed here might be better known, but there are literally hundreds of these held all over the world annually. A little research ought to uncover an event held relatively close to your own backyard.

NORTHEAST SUSTAINABLE ENERGY ASSOCIATION (NESEA)
23 Ames Street
Greenfield, MA 01301
(413) 774-6051
Organizes the annual "American Tour de Sol" and electric vehicle symposium.

SOLAR ENERGY EXPO AND RALLY (SEER)
239 S. Main Street
Willits, CA 95490
(707) 459-1256
Host for annual "Tour de Mendo," when Willets temporarily becomes the solar capitol of the world.

SOLAR AND ELECTRIC RACING ASSOCIATION (500)
11811 N. Tatum Boulevard., Suite 301
Phoenix, AZ 85028
(602) 953-6672
Organizes annual Solar and Electric 500 in Phoenix and promotes electric vehicles.

## Electric utilities & power associations

Any of the following organizations can provide you with information.

AMERICAN PUBLIC POWER ASSOCIATION
2301 M Street, N.W.
Washington, DC 20202
(202) 775-8300

ARIZONA PUBLIC SERVICE COMPANY
PO Box 53999
Phoenix, AZ 85072-3999
(602) 250-2200

CALIFORNIA ENERGY COMMISSION
1516 9th Street
Sacramento, CA 95814
(916) 654-4001

ELECTRIC POWER RESEARCH INSTITUTE
412 Hillview Avenue
PO Box 10412
Palo Alto, CA 94303
(415) 855-2580

DIRECTOR OF ELECTRIC TRANSPORTATION
Department of Water and Power
City of Los Angeles
111 N. Hope Street, Room 1141
Los Angeles, CA 90012-2694
(213) 481-4725

PUBLIC SERVICE CO. OF COLORADO
2701 W. 7th Avenue
Denver, CO 80204
(303) 571-7511

SACRAMENTO MUNICIPAL UTILITY DISTRICT
PO Box 15830
Sacramento, CA 95852-1830
(916) 732-6557

SOUTHERN CALIFORNIA EDISON
2244 Walnut Grove Avenue
PO Box 800
Rosemead, CA 91770
(818) 302-2255

## Government

The following are agencies involved with EVs directly or indirectly at city, state, or federal government levels.

CALIFORNIA AIR RESOURCES BOARD
1012 Q Street
PO Box 2815
Sacramento, CA 95812
(916) 322-2990

ADMINISTRATOR
Environmental Protection Agency
401 M Street S.W.
Washington, DC 20460
(202) 260-2090

THE HONORABLE MARVIN BRAUDE
Los Angeles City Council
City Hall
Los Angeles, CA 90012
(213) 485-5700

ADMINISTRATOR
National Highway Traffic Safety Administration
400 7th Street S.W.
Washington, DC 20590
(202) 366-1836

J. MICHAEL DAVIS
Assistant Secretary, Conservation and Renewable Energy
U.S. Department of Energy
1000 Independence Avenue S.W.
Washington, DC 20585
(202) 586-9220

## MANUFACTURERS, CONVERTERS, & CONSULTANTS

There is a sudden abundance of people and firms doing EV work. This category is an attempt to present you with the firms and individuals from whom you can expect either a completed EV or assistance with completing one.

## Manufacturers

This category includes the household names plus the major independents you already met in chapters 3 and 4. When contacting the larger companies, it is best to go through the switchboard or a public affairs person who can direct your call after finding out your specific needs.

ENGINEERING & TECHNOLOGY
PUBLIC RELATIONS
Chrysler Corporation
12000 Chrysler Drive
Highland Park, MI 48288
(313) 956-5741

ENVIRONMENTAL COMMUNICATIONS
Ford Motor Company
Ford World Headquarters
The American Road, Room 904
Dearborn, MI 48121
(800) ALT-FUEL, (800) 258-3835

GENERAL MOTORS ELECTRIC VEHICLES
432 North Saginaw Street, Suite 801
Flint, MI 48502
(800) 25-ELECTRIC, (800) 253-5328

AUDI OF AMERICA
3800 Hamlin Road
Auburn Hills, MI 48326
(313) 340-5000

BMW OF NORTH AMERICA
300 Chestnut Ridge Road
Woodcliff Lake, NJ 07675
(201) 307-4000

MERCEDES-BENZ OF NORTH AMERICA
1 Mercedes Drive
Montvale, NJ 07645
(201) 573-0600

VOLKSWAGEN OF AMERICA
3800 Hamlin Road
Auburn Hills, MI 48326
(313) 340-5000

SOLAR ELECTRIC
116 Fourth Street
Santa Rosa, CA 95401
(800) 832-1986, (707) 542-1990

NISSAN MOTOR CO USA
18501 Figueroa Street
Carson, CA 90248
(213) 532-3111

TOYOTA MOTOR CO USA
19001 S Western Avenue
Torrance, CA 90509
(213) 618-4000

AUTOMOBILES PEUGEOT
75 Avenue de la Grande Armee
F-75116 Paris FRANCE

AUTOMOBILES CITROEN
62 Boulevard Victor Hugo
F-92200 Neuilly-sur-Seine
Hauts-de-Seine FRANCE

RENAULT (REGIE NATIONALE DES USINES)
8/10 Avenue Emile Zola
F-92100 Boulogne-Billancourt
Hauts-de-Seine FRANCE

FIAT AUTO S.p.A.
Corso Giovanni Agnelli 200
I-10125 Torino 016 ITALY

BERTONE S.p.A.
Corso Canonico Allamano 46
I-10095 Grugliasco Torino 016 ITALY

SMH—SWATCH
Seevorstadt 6
Bienne, CH-2501 SWITZERLAND

By far the largest of the independent vehicle manufacturers, they offer a wide range of vehicles and conversions both direct and through dealerships.

GREEN MOTOR WORKS (Also a Solar Electric dealer)
5228 Vineland
North Hollywood, CA 91601
(818) 766-3800
Selling, servicing, and converting EVs is their livelihood.

SOLAR CAR CORPORATION
1300 Lake Washington Road
Melbourne, FL 32935
(407) 254-2997

Best known for its Chevy S-10 and GMC S-15 pickup truck conversions.

SOLECTRIA CORP
68 Industrial Way
Wilmington, MA 01887
(508) 658-2231

Best known for its "Force" Geo Metro conversion, they are very active in racing and events as well.

BATTERY AUTOMATED TRANSPORTATION
2471 S. 2570 W.
West Valley City, UT 84119
(801) 977-0119

Best known for its proprietary "Ultra Force" lead-acid batteries and Ford Ranger pickup truck conversions.

CALIFORNIA ELECTRIC CARS
1669 Del Monte Boulevard
Seaside, CA 93955
(408) 899-2012

Best known for its "Monterey" electric vehicle.

CLEAN AIR TRANSPORT OF NORTH AMERICA
23030 Lake Forest Drive, Suite 206
Laguna Hills, CA 92653
(714) 951-3983

Best known for its "LA301" electric vehicle.

BOB BEAUMONT
Columbia Auto Sales
9720 Owen Brown Road
Columbia, MD 21045
(301) 799-3550

Best known as the former head of Seebring-Vanguard and its CitiCar offering, Bob Beaumont is beginning production again with another two-seater electric vehicle that he says will not repeat the mistakes of its predecessor. If true, it will be a well-priced, well-positioned offering for the 1990s.

CONCEPTOR INDUSTRIES
521 Newpark Boulevard
PO Box 149
New Market, ON L3Y 4X7 CANADA
(416) 836-4611

A subsidiary of Vehma International, best known for its "G-Van" EV conversions of the General Motors Vandura van.

CUSHMAN
900 North 21st Street
Lincoln, NE 68503
(402) 475-9581
Manufactures three-wheeler industrial and commercial electric carts.

ELECTRIC FUEL PROPULSION CORP.
4747 N. Ocean Drive, #223
Ft. Lauderdale, FL 33308
(305) 785-2228
Robert R. Aronson, a long-timer in the EV field, now offers the luxury "Silver Volt," with tri-power lead-cobalt batteries.

ELECTRIC MOBILITY
591 Mantua Boulevard
Sewell, NJ 08080
(800) 257-7955
Manufactures electric carts, bicycles, etc.

PALMER INDUSTRIES
PO Box 707
Endicott, NY 13760
(800) 847-1304
Manufactures an electric bicycle.

SEBRING AUTO CYCLE
PO Box 1479
Sebring, FL 33871
(813) 655-2131
The latest incarnation of the original Sebring-Vanguard operation, best known for its three-wheeled Zzipper electric vehicle.

## Conversion specialists

In this category, the line between those who provide parts and those who provide completed vehicles is blurred.

JIM HARRIS
Zero Emissions Motorcar Company
1031 Bay Boulevard, Suite T
Chula Vista, CA 91911
(619) 425-4221  Fax: (619) 425-2312
Jim's Ford Ranger pickup truck conversion is the centerpiece feature of chapter 11. His company offers prepackaged motor adapter and controller kits, and complete vehicles, components, literature, and expertise. His word is his bond.

JEFF SHUMWAY
Ecotech Autoworks
1524-V Springhill Road, Box 9262
McLean, VA 22102
(703) 893-3045
Vehicles and components.

BILL KUEHL
Electric Auto Conversions
4504 W. Alexander Road
Las Vegas, NV 89030
(702) 645-2132
Vehicles and components.

KEN BANCROFT
Electric Motor Cars Sales and Service
4301 Kingfisher
Houston, TX 77035
(713) 729-8668
Vehicles and components.

DON KARNER
Electric Transportation Applications
PO Box 10303
Glendale, AZ 85318
(602) 978-1373
Vehicles and components.

LARRY FOSTER
Electric Vehicle Custom Conversion
1712 Nausika Avenue
Rowland Heights, CA 91748
(818) 913-8579
Vehicles and components.

STAN SKOKAN
Electric Vehicles, Inc.
1020 Parkwood Way
Redwood City, CA 94060
(415) 366-0643
Experienced EV conversions and
consulting.

ED RANBERG
Eyeball Engineering
16738 Foothill Boulevard
Fontana, CA 92336
(714) 829-2011
Experienced EV conversion professional;
components and consulting.

LON GILLAS
E-Motion
515 W. 25th Street
McMinnville, OR 97128
(503) 434-4332
Vehicles and components.

GENE HITNEY
Hitney Solar Products
655 N. Highway 89
Chino Valley, AZ 86323
(602) 636-2201
Vehicles and components.

FRANK KELLY
Interesting Transportation
2362 Southridge Drive
Palm Springs, CA 92264-4960
(619) 327-2864
Experienced EV conversions and consulting.

W. D. MITCHELL
20 Victoria Drive
Rowlett, TX 75055
(214) 475-0361
Vehicles and components.

RON LARREA
San Diego Electric Auto
9011 Los Coches Road
Lakeside, CA 92040
(619) 443-3017
Vehicles and components.

## Consultants

Companies and individuals who are more likely to provide advice, literature or components—rather than completed vehicles—are listed.

PAUL McCREADY
Aerovironment
PO Box 5031
Monrovia, CA 91017-7131
(818) 359-9983
Developer of the GM Impact, Paul McCready and Aerovironment need no further introduction.

3E VEHICLES
Box 19409
San Diego, CA 92119

Another experienced participant in the EV field, 3E offers an outstanding line of conversion booklets that (although somewhat dated today) are still highly useful.

MICHAEL HACKLEMAN
Earthmind
PO Box 743
Mariposa, CA 95338
(310) 396-1527

Author, editor of Alternative Transportation News, experienced EV participant and a consultant.

MICHAEL BROWN AND SHARI PRANGE
Electro Automotive
PO Box 1113
Felton, CA 95018-1113
(408) 429-1989

This organization, experienced participant in the EV field, offers books, videos, seminars, consulting, and components.

CLARENCE ELLERS
Electronic Transportation Design
PO Box 111
Yachats, OR 97498
(503) 547-3506

Another experienced participant in the EV field, ETD offers books, a newsletter, consulting, and components.

MIKE KIMBALL
18820 Roscoe Boulevard
Northridge, CA 91324
(818) 998-1677

EV technician and maintenance mechanic extraordinaire; Mike has probably forgotten more about EVs than most people will ever know.

CARL TAYLOR
3871 S.W. 31st Street
Hollywood, FL 33023
(305) 981-9462

EV maintenance, repair and troubleshooting.

BILL WILLIAMS
Williams Enterprises
Box 1548
Cupertino, CA 95015

Experienced participant in the EV field, conversion specialist and consultant Williams offers an outstanding conversion guide that (although somewhat dated today) is still very useful.

HOWARD G. WILSON
2050 Mandeville Canyon Road
Los Angeles, CA 90049
(310) 471-7197

Former Hughes vice-president, Howard Wilson was the real "make it happen" factor behind GM's Impact and Sunraycer projects.

BOB WING
PO Box 277
Inverness, CA 94937
(415) 669-7402

EV conversions and consulting.

## SUPPLIERS

This category includes those from whom you can obtain complete conversion kits (all the parts you need to build your own EV after you have the chassis); conversion plans; and suppliers specializing in motors, controllers, batteries, chargers, and other components.

### Conversion kits

Companies and individuals listed here are those more likely to provide the parts that go into converting or building an EV once you already have the chassis, such as components, advice, literature, etc.

KEN KOCH
KTA Services Inc.
944 W 21st Street
Upland, CA 91786
(909) 949-7914  Fax: (909) 949-7916

Ken's services to those interested in converting or building EVs are in a class by themselves. His company offers everything from individual components to several levels of kits, and its products are featured throughout this book. You'll find KTA knowledgeable about your needs, prompt in its deliveries, and a pleasure to do business with.

JOHN STOCKBERGER
Electric Auto Crafters
643 Nelson Lake Road, #2S
Batavia, IL 60510
(312) 879-0207

Provides parts, information, and testing for EV builders.

BOB BATSON
Electric Vehicles Of America
PO Box 59
Maynard, MA 01754
(508) 897-9393

Provides kits, components, and literature.

STEVE VAN RONK
Global Light and Power
55 New Montgomery, Suite 424
San Francisco, CA 94105
(415) 495-0494
Kits and components; promotes annual Clean Air Revival.

STEVE DECKARD
King Electric Vehicles
Box 514
East Syracuse, NY 13057
Kits and components.

C. FETZER
Performance Speedway
2810 Algonquin Avenue
Jacksonville, FL 32210
(904) 387-9858
Kits and components.

PAUL SCHUTT AND ASSOCIATES
673 Via Del Monte
Palos Verdes Estates, CA 90274
(310) 373-4063
Represents manufacturers who supply EV components.

UNIQUE MOBILITY
425 Corporate Circle
Golden, CO 80419
(303) 278-2002
Well-known for their prototype vehicles using proprietary technology, Unique offers numerous advanced EV capabilities and components.

## Conversion plans

Listed here are companies and individuals who are more likely to provide vehicle plans, kits, or components rather than completed vehicles.

RICK DORAN
Doran Motor Company
1728 Bluehaven Drive
Sparks, NV 89431
(805) 546-9654, (702) 359-6735
Best known for the Doran three-wheeler and its plans.

DOLPHIN VEHICLES
PO Box 110215
Campbell, CA 95011
(408) 734-2052
Best known for the Vortex three-wheeler and its plans for either internal combustion or electric propulsion.

ROY KAYLOR
Kaylor Energy Products
1918 Menalto Avenue
Menlo Park, CA 94025
(415) 325-6900

Experienced EV player best known for Kaylor Kit Sports Car Volkswagen conversion plans; had early success, but has had ups and downs. Check current status before sending your money.

ROBERT G. BUCY
Lectric Kar
Box 28794
Dallas, TX 75228
(714) 327-7197

Offers useful but 1973-vintage plans for the Renault conversion XE-1 Lectric Car.

## Motors

A considerable number of companies manufacture electric motors. The short list here is only to get you started.

ADVANCED D.C. MOTORS, INC.
219 Lamson St.
Syracuse, NY 13206
(315) 434-9303

Specialists in motors for EV applications, their motor is featured in chapter 8 and used for the conversion in chapter 11.

BALDOR ELECTRIC CO.
5711 South 7th
Ft. Smith, AR 72902
(501) 646-4711

GENERAL ELECTRIC COMPANY
3001 East Lake Road
Erie, PA 16531
(814) 455-5466

PRESTOLITE ELECTRIC
PO Box 904
Toledo, OH 43697
(419) 249-7600

RELIANCE ELECTRIC
PO Box 17438
Cleveland, OH 44117
(216) 266-7000

## Controllers

A considerable number of companies manufacture controllers; again, this short list is only to get you started.

CURTIS PMC
6591 Sierra Lane
Dublin, CA 94568
(510) 828-5001

Specialists in controllers for EV applications, this company's controller is featured in chapter 9 and used in chapter 11's conversion.

AC PROPULSION, INC.
462 Borrego Ct., Unit B
San Dimas, CA 91773
(714) 592-5399

AC Propulsion's ac EV drive systems are simply a better idea whose time will come when economies of scale drive prices down. You'll meet them again in chapter 8.

TECHNICAL PUBLICATIONS DEPARTMENT
Motorola Inc.
Semiconductor Products Sector
PO Box 52073
Phoenix, AZ 85072
(800) 521-6274

WESTINGHOUSE ELECTRONIC SYSTEMS
PO Box 17319
Baltimore, MD 21203
(410) 765-4396

## Batteries

The lead-acid battery culture dominates today's offerings; you are best served by looking in your local Yellow Pages under *Battery* for a distributor in your area. Use the listings here only as a call-in number to find the local distributor or contact.

TROJAN BATTERY CO.
12380 Clark Street
Santa Fe Springs, CA 90670
(800) 423-6569, (213) 946-8381, (714) 521-8215

Trojan has manufactured deep-cycle lead-acid batteries suitable for EV use longer than most companies, and has considerable expertise; its batteries are featured in chapter 9.

ALCO BATTERY CO.
2980 Red Hill Avenue
Costa Mesa, CA 92626
(714) 540-6677

Offers a full line of lead-acid batteries suitable for EVs.

CONCORDE BATTERY CORP.
2009 W. San Bernadino Road
West Covina, CA 91760
(818) 962-4006

Offers lead-acid batteries for aircraft use.

EAGLE-PICHER INDUSTRIES
PO Box 47
Joplin, MO 64802
(417) 623-8000
Offers a full line of lead-acid batteries suitable for EVs.

SAFT
251 Industrial Boulevard
PO Box 7366
Greenville, NC 27835-7366
(919) 830-1600
Manufactures nickel-iron and nickel-cadmium batteries suitable for EVs.

U.S. BATTERY MANUFACTURING CO.
1675 Sampson Avenue
Corona, CA 91719
(800) 695-0945, (714) 371-8090
Manufactures deep-cycle lead-acid batteries suitable for EVs.

VOLTMASTER COMPANY
PO Box 288
Corydon, IA 50060
(515) 872-2044
Manufactures deep-cycle lead-acid batteries suitable for EVs.

YUASSA-EXIDE
9728 Alburtis Avenue
PO Box 3748
Santa Fe Springs, CA 90670
(800) 423-4667, (213) 949-4266
Manufactures deep-cycle lead-acid batteries suitable for EVs.


## Chargers

There are many battery charger manufacturers; this short list is only to get you started.

LESTER ELECTRICAL
625 West A Street
Lincoln, NE 68522
(402) 477-8988
Lester has been manufacturing battery chargers suitable for EV use longer than most companies, and has considerable expertise. Their charger is featured in chapter 10's discussion.

K & W ENGINEERING
3298 Country Home Road
Marion, IA 52302
(319) 378-0866
K & W's lightweight, transformerless chargers designed for onboard use are also featured in chapter 10's discussion.

ROLLS BATTERY ENGINEERING
8 Proctor Street
Salem, MA 01970
(508) 745-3333
Manufactures battery chargers suitable for EVs.

HUGHES POWER CONTROL SYSTEMS
PO Box 2923
Torrance, CA 90509-2923
(310) 517-5855
Provides inductive chargers to infrastructure builders—not set up for public inquiry.

## Other parts

Here you'll find an assortment of goodies designed to assist your EV enjoyment and pleasure; again, it's not an all-inclusive list—just one to get you started.

CRUISING EQUIPMENT CO.
6315 Seaview Avenue
Seattle, WA 98107
(206) 782-8100
Offers the Amp-Hour+ meter for monitoring the state of battery charge.

HYDROCAP CORP.
975 N.W. 95 Street
Miami, FL 33150
(305) 696-2504
Offers lead-acid battery caps that reduce maintenance and improve performance.

INSTANT AUTO HEATER CO.
Box 307
Woodside, NY 11377
(718) 476-1723
Offers EV heaters.

INTEGRAL ENERGY SYSTEMS
109 Argall
Nevada City, CA 95959
(800) 735-6790, (916) 265-8441
Offers photovoltaic panels.

NEW CONCEPTS ENGINEERING
4-B SW Monroe Parkway, #168
Lake Oswego, OR 97035
(503) 781-0270
Offers special EV chargers, heaters, meters, and other components.

HORLACHER AG
PO Box 50, Guterstrasse 9
Mohlin , CH-4313 SWITZERLAND
011-41-61-882118
011-41-61-883200 fax

Specialists in plastics and composite body fiber engineering and manufacturing, they have built numerous record-breaking prototype EVs to demonstrate their awesome capabilities.

## BOOKS, ARTICLES, & PAPERS

There have been a number of books and thousands of articles and papers written about EVs, both technical and nontechnical. Here are some available related books and manuals, and a sampling of a few nontechnical articles that will give you instant expertise in the subject area.

### Books

Michael Hackleman, *Design and Build Your Own Electric Vehicles*, Earthmind, 1977.
William Hamilton, *Electric Automobiles*, McGraw-Hill, 1980.
Ted Lucas and Fred Ries, *How to Convert to an Electric Car*, Crown Publishers, 1980.
Clyde R. Jones, *Convert Your Compact Car to Electric*, Domus Books, 1981.
Douglas F. Marsh, *Electric Vehicles Unplugged*, South Florida EAA, 1991.
Steve McCrea and Richard Minner, *Why Wait for Detroit*, South Florida EAA, 1992.
Sheldon R. Shackett, *The Complete Book of Electric Vehicles*, Domus Books, 1979.
Robert J. Traister, *All About Electric & Hybrid Cars*, Tab Books, 1982.
Ernest H. Wakefield, *The Consumer's Electric Car*, Ann Arbor Science, 1977.
Barbara Whitener, *The Electric Car Book*, Love Street Books, 1981. 96p.

### Manuals

Michael Brown with Shari Prange, *Convert It*, Electro Automotive, 1989.
Derek Chan and Ken Tenure, *Electric Vehicle Purchase Guidelines Manual*, EVAA, 1992.
Clarence Ellers, *Electric Vehicle Conversion Manual*, Self, 1992.
Gary Starr, *Electric Car Conversion Book*, Solar Electric Engineering, 1991.
Bill Williams, *Guide to Electric Auto Conversion*, Williams Enterprises, 1981.

### Articles

"Battery and electric vehicle Update," *Automotive Engineering*, September 1992, p. 17.
Stuart F. Brown, "Chasing Sunraycer Across Australia," *Popular Science*, February 1988, p. 64.
Phillip S. Meyers, "Reducing Transportation Fuel Consumption," *Automotive Engineering*, September 1992, p. 89.
"Propulsion Technology: An Overview," *Automotive Engineering*, July 1992, p. 29.
Paul McCready, "Design, Efficiency and the Peacock," *Automotive Engineering*, October 1992, p. 19.
Gill Andrews Pratt, "EVs: On the road again," *Technology Review*, August 1992, p. 50.
David C. White, et al, "The New Team: Electricity Sources without Carbon Dioxide," *Technology Review*, January 1992, p. 42.
David H. Freedman, "Batteries Included," *Discover*, March, 1992, p. 90.
Reinhardt Krause, "High Energy Batteries," *Popular Science*, February 1993, p. 64.
Len Frank and Dan McCosh, "Power to the People," *Popular Science*, August 1992, p. 103.

Len Frank and Dan McCosh, "Alternate Fuel Follies," *Popular Science*, July 1992, p. 54.

Len Frank and Dan McCosh, "Electric Vehicles Only," *Popular Science*, May 1991, p. 76.

Ron Cogan, "Electric Cars; the Silence of the Cams," *Motor Trend*, September 1991, p. 71.

## Publishers

Here are a few companies that specialize in publications of interest to EV converters.

BATTERY COUNCIL INTERNATIONAL
401 N. Michigan Avenue
Chicago, IL 60611
(312) 644-6610

Publishes battery-related books and articles.

INSTITUTE FOR ELECTRICAL AND ELECTRONIC ENGINEERS (IEEE)
IEEE Technical Center
Piscataway, NJ 08855

Publishes numerous articles, papers, and proceedings. Expensive, but one of the best sources for recent published technical information on EVs.

LEAD INDUSTRIES ASSOCIATION
292 Madison Avenue
New York, NY 10017

Publishes information on lead recycling.

SAE (SOCIETY OF AUTOMOTIVE ENGINEERS) INTERNATIONAL
400 Commonwealth Drive
Warrendale, PA 15096-0001
(412) 772-7129

Publishes numerous articles, papers, and proceedings. Also expensive, but the other best source for recent published technical information on EVs.

## Newsletters

Here are a few companies that specialize in newsletter-type publications of interest to EV converters.

ELECTRIC GRAND PRIX CORP.
6 Gateway Circle
Rochester, NY 14624
(716) 889-1229

ELECTRIC VEHICLE CONSULTANTS
327 Central Park W.
New York, NY 10025
(212) 222-0160

ELECTRIC VEHICLE NEWS
1911 N. Fort Meyer Drive, Suite 703
Arlington, VA 22209
(703) 276-9093

ELECTRIC VEHICLE PROGRESS
215 Park Avenue S., Suite 1301
New York, NY 10003
(212) 228-0246

ELECTRIFYING TIMES
63600 Deschutes Market Road
Bend, OR 97701
(503) 388-1908

SOLAR MIND
759 S. State Street, #81
Ukiah, CA 95482
(707) 468-0878

## Electric vehicle directories

Here are a few companies that specialize in directory-type publications of interest to
EV converters.

RANDALL ASHERLAN & ASSOCIATES
P.O. Box 166
Honokaa, HI 96727
(808) 775-0301

SPIRIT PUBLICATIONS
PO Box 23417
Tucson, AZ 85734
(602) 822-2030

# 6

# Chassis & design

*A 20-HP electric motor will easily push your*
*4000-pound vehicle at 50 mph.*

The chassis is the foundation of your electric vehicle conversion. While you might never build your own chassis from scratch, there are fundamental chassis principles that can help you with any EV conversion or purchase—things that never come up when using internal combustion engine vehicles—such as the influence of weight, aerodynamic drag, rolling resistance, and drivetrains.

This chapter will step you through the process of optimizing, designing, and buying your own electric vehicle. You'll become familiar with the chassis trade-offs involved in optimizing your EV conversion. Then you'll design your EV conversion to be sure the components you've selected accomplish what you want to do. When you have figured out what's important to you and verified your design will do what you want, you'll look at the process of buying your chassis.

Knowledge of all these steps will help you immediately (when reading about chapter 11's pickup truck conversion), and eventually (when picking the best EV chassis for yourself). The principles are universal, and you can apply them whether buying, building, or converting.

## CHOOSE THE BEST CHASSIS FOR YOUR EV

The chassis you pick is the foundation for your EV—choose it wisely. That's the message of this chapter in a nutshell. Since you're likely to be converting rather than building from scratch, there's not a lot you can do after you've made your chassis selection. The secret is to ask yourself the right questions and be clear about what you want to accomplish before you make your selection.

Like a youngster's soapbox derby racer, you want a chassis with an aerodynamic shape and thin wheels, so that you can just give a shove and it runs almost forever. But its frame must also be big enough and strong enough to carry you and your passengers along with the motor, drivetrain/controller, and batteries. In addition, if you want to drive it on the highway, federal and state laws require it be roadworthy and adhere to certain safety standards.

The first step is to know your options. Your EV should be as light in weight as possible; streamlined, with its body optimized for minimum drag; optimized for minimum rolling resistance from its tires, brakes, and steering; and optimized for minimum drivetrain losses.

The motor-drivetrain-battery combination must match the body style you've se-lected. It must also be capable of accomplishing the task most important to you: high speed, long range, or a utility commuter vehicle midway between the two. So step two is to design for the capability that you want. Your EV's weight, motor and battery placement, aerodynamics, rolling resistance, handling, gearing, and safety features must also meet your needs. You now have a plan.

Step three is to execute your plan—to buy the chassis that meets your needs. At its heart this is a process no different from any other vehicle purchase you've ever made, except that the best solution to your needs might be a vehicle that the owner or dealer can't wait to get rid of soon enough—one with a diesel engine, a bad engine, or no engine—so the tables are completely turned from a normal buying situation. Used is usually the least expensive—but not too used. You want to feel confident and good about converting the vehicle you choose before you leave the lot. If it's too small or cramped to fit all the electrical parts—let alone the batteries—you know you have a problem. Or if it's particularly dirty, greasy, or rusty, you need to think twice.

Figure 6-1 gives you the quick picture. The rest of the chapter covers the details. Let's get started.

## OPTIMIZE YOUR EV

Optimizing is always step number one. Even if you go out to buy your electric vehi-cle ready-made, you still want to know what kind of a job has been done—so you can decide if you're getting the best model for you. In all other cases, you'll be doing the optimizing—either by the choices you make in chassis selection up front or by your conscious optimizing decisions later on. In this section, you'll be looking to minimize the following resistance factors:

- Weight and climbing.
- Aerodynamic drag and wind.
- Rolling and cornering resistance.
- Drivetrain system.

You'll look at equations that define each of these factors, and construct a table of real values normalized for a 1000-pound vehicle and nine specific vehicle speeds. These values should be handy regardless of what you do later—just multiply by your own EV's weight ratio and use directly, or interpolate between the speed values.

You'll immediately see a number of values reassembled in the design section of this chapter, when a real vehicle's torque requirements are calculated to see if the torque available from the electric motor and drivetrain selected is up to the task. This design process can be infinitely adapted and applied to whatever EV you have.

### Conventions & formulas

This book uses the U.S. automotive convention of miles, miles per hour, feet per sec-ond, pounds, pound-feet, etc., rather than the kilometers, newton-meters, etc., in com-mon use overseas. Any formulas borrowed from the Bosch handbook[1] have been converted to U.S. units. Speaking of formulas, you will find the following twelve use-ful; they have been grouped in one section for your convenience:

1. Power (ft-lb/sec) = Torque (ft-lb) × Speed (radians/sec) = FV
2. 1 Horsepower (HP) = 550 ft-lb/sec

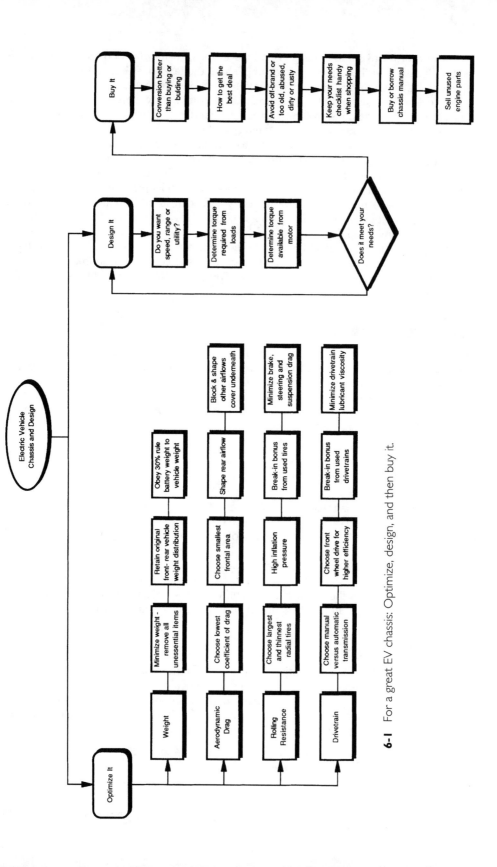

**6-I** For a great EV chassis: Optimize, design, and then buy it.

Applying this to equation (1) gives you:

3. Horsepower (HP) = FV/550 where V is speed expressed in feet/sec
4. 88 feet/sec = 60 mph
Multiply feet/sec by (60 × 60)/5280 to get mph.
5. Horsepower (HP) = FV/375
where V is speed expressed in mph.
6. Horsepower (HP) = (Torque × RPM)/5252 = 2$\pi$/60 × FV/550
7. Wheel RPM = (mph × Revolutions/mile)/60
8. Power (kW) = 0.7457 × HP
9. g = the gravitational constant = 32.2 ft/sec$^2$
10. Mass (M) = Weight (W) / g = W/g
11. Torque = (F(5280/2$\pi$))/(Revolutions/mile) = 840.34 × F/(Revolutions/mile)
12. Torque$_{wheel}$ = Torque$_{motor}$/(overall gear ratio × overall drivetrain efficiency)
13. Speed$_{vehicle}$ (in mph) = (RPM$_{motor}$ × 60)/(overall gear ratio × Revolutions/mile)

## IT AIN'T HEAVY, IT'S MY EV

In real estate they say the three most important things are location, location, and location. In electric vehicle conversion the three most important things are weight, weight, and weight. In this section, you'll be taking a closer look at the items in the *Weight* row of FIG. 6-1 and the supreme importance of minimizing weight in your EV.

### Remove all unessential weight

You don't want to carry around a lot of unnecessary weight. But, unless you're starting with a build-from-scratch design, you're inheriting the end result of someone else's weight trade-offs. This means you need to carefully go over everything with regard to its weight versus its value at three different times:

*Before you purchase the conversion vehicle.* Think about its weight-reduction potential before you buy it. Is it going to be easy (pickup) or difficult (van) to get the extra weight out? What about hidden-agenda items? Has a previous accident resulted in a lead-filled fender on your prospective purchase (take along a magnet during your exam)? Does its construction lead to ease of weight removal or substitution of lighter weight parts later? Think about these factors as you look.

*During conversion.* As you remove the internal combustion engine parts, it's likely you'll discover additional parts that you hadn't seen or thought of taking out before— parts snuggled up against the firewall or mounted low on the fenders are sometimes nearly invisible in a crowded and/or dirty engine compartment. Get rid of all unnecessary weight, but do exercise logic and common sense in your weight reduction quest. Substituting a lighter weight cosmetic body part is a great idea; drilling holes in a load-bearing structural frame member is not.

*After conversion.* Break your nasty internal combustion engine vehicle habits. Toss out all extras that you might have continued to carry, including spare tire and tools.

After all your work, give yourself a pat on the back. You've probably removed from 400 to 800 pounds or more from a freshly-cleaned-up former internal combustion engine vehicle chassis that's soon to become a lean and mean EV machine. The reason for all your work is simple—weight affects every aspect of an EV's performance: acceleration, climbing, speed, and range. Let's see exactly how.

## Weight affects acceleration

When Sir Issac Newton was bonked in the head with an apple, he was allegedly pondering one of the basic relationships of nature—his Second Law:

$$F = Ma; \text{ or force (F) equals mass (M) times acceleration (a)}$$

For EV purposes, it can be rewritten as:

$$F_a = C_i Wa$$

where $F_a$ is acceleration force, W is vehicle weight in pounds, a is acceleration in mph/second, and $C_i$ is a units conversion factor that also accounts for the added inertia of the vehicle's rotating parts. The force required to get the vehicle going varies directly with the vehicle's weight; twice the weight means twice as much force is required.

$C_i$, the mass factor that represents the inertia of the vehicle's rotating masses (wheels, drivetrain, flywheel, clutch, motor armature, and other rotating parts) is given by,

$$C_i = 1 + 0.04 + 0.0025(N_c)^2$$

where $N_c$ represents the combined ratio of the transmission and final drive.[2] The mass factor depends upon the gear in which you are operating. For internal combustion engine vehicles, the mass factor is typically: high gear—1.1; 3rd gear—1.2; 2nd gear—1.5; and 1st gear—2.4. For EVs, where a portion of the drivetrain and weight has typically been removed or lightened, it is typically 1.06 to 1.2.

TABLE 6-1 shows the acceleration force $F_a$, for three different values of $C_i$, for ten different values of acceleration a, for a vehicle weight of 1000 pounds. The factor a' is the acceleration expressed in ft/sec[2] rather than in mph/second—21.95 = 32.2 × (3600/5280)—used only in the formula (because acceleration expressed in mph/second is a much more convenient and familiar figure to work with). Notice that an acceleration of 10 mph/sec, an amount that takes you from zero to 60 mph in 6 seconds nominally requires extra force of 500 pounds; 5 mph/sec, moving from zero to 50 mph in 10 seconds, requires 250 pounds.

### Table 6-1 Acceleration force, $F_a$ (in pounds), for different values of $C_i$

| a (in mph/sec) | a'= a/21.95 | $F_a$ (in pounds) $C_i = 1.06$ | $F_a$ (in pounds) $C_i = 1.1$ | $F_a$ (in pounds) $C_i = 1.2$ |
|---|---|---|---|---|
| 1 | 0.046 | 48.3 | 50.1 | 54.7 |
| 2 | 0.091 | 96.6 | 100.2 | 109.3 |
| 3 | 0.137 | 144.8 | 150.3 | 164.0 |
| 4 | 0.182 | 193.1 | 200.4 | 218.6 |
| 5 | 0.228 | 241.4 | 250.5 | 273.3 |
| 6 | 0.273 | 289.7 | 300.6 | 328.0 |
| 7 | 0.319 | 338.0 | 350.7 | 382.6 |
| 8 | 0.364 | 386.3 | 400.8 | 437.3 |
| 9 | 0.410 | 434.5 | 450.9 | 491.9 |
| 10 | 0.455 | 482.8 | 501.0 | 546.6 |

To use TABLE 6-1 with your EV, multiply by the ratio of your vehicle weight and use the $C_i = 1.06$ column for lighter vehicles and $C_i = 1.2$ column for heavier ones. For example, the 3800-pound Ford Ranger pickup truck of chapter 11 would require 5 mph/sec = 3.8 × 273.3 = 1038.5 pounds.

## Weight affects climbing

When you go hill climbing, you add another force:

$$F_h = W\sin\phi$$

where $F_h$ is hill-climbing force, W is vehicle weight in pounds, and $\phi$ is angle of incline as shown in FIG. 6-2. The *degree* of the incline—the way hills or inclines are commonly referred to—is different from the *angle* of the incline, but FIG. 6-2 should clear up any confusion for you. Notice that $\sin\phi$ varies from 0 at no incline (no effect) to 1 at 90 degrees; in other words, the full weight of the vehicle is trying to pull it back down the incline. Again, weight is directly involved—acted upon this time by the steepness of the hill.

$$\text{Degree of incline} = 1\% = \frac{1\text{ foot}}{100\text{ feet}} = \frac{\text{Rise}}{\text{Run}}$$

$$\text{Angle of incline, } \emptyset = \text{Arc tan}\frac{\text{Rise}}{\text{Run}} = \text{Arc tan } 0.01 = 0 \text{ degrees 34 minutes}$$

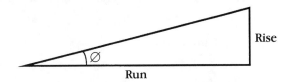

**6-2** Angle of incline defined.

TABLE 6-2 shows the hill-climbing force $F_h$, for fifteen different incline values for a vehicle weight of 1000 pounds. Notice that the tractive force required for acceleration of 1 mph/sec equals that required for hill-climbing of a 5-percent incline, 2 mph/sec for 10-percent incline, etc., on up through a 30-percent incline. This handy relationship will be used later in the design section.

To use TABLE 6-2 with your EV, multiply by the ratio of your vehicle weight. For example, the 3800-pound Ford Ranger pickup truck of chapter 11 going up a 10 percent incline would require = 3.8 × 99.6 = 378.5 pounds.

## Weight affects speed

Although speed also involves other factors, it's definitely related to weight. As horsepower and torque are related to speed per equation (3):

$$HP = FV/550$$

where HP is motor horsepower, F is force or torque in foot-pounds and V is speed in feet/sec. Armed with this information, Newton's Second Law equation can be rearranged as:

$$a = (1/M) \times F$$

## Table 6-2  Hill-climbing force $F_h$ for 15 different values of incline

| Degree of incline | Incline angle ø | sin ø | $F_h$ (in pounds) | a (in mph/sec) |
|---|---|---|---|---|
| 1% | 0° 34' | 0.00989 | 9.9 | |
| 2% | 1° 9' | 0.02007 | 20.1 | |
| 3% | 1° 43' | 0.02996 | 29.6 | |
| 4% | 2° 17' | 0.04013 | 40.1 | |
| 5% | 2° 52' | 0.05001 | 50.0 | 1 |
| 6% | 3° 26' | 0.05989 | 59.9 | |
| 8% | 4° 34' | 0.07062 | 79.6 | |
| 10% | 5° 43' | 0.09961 | 99.6 | 2 |
| 15% | 8° 32' | 0.14838 | 148.4 | 3 |
| 20% | 11° 19' | 0.19623 | 196.2 | 4 |
| 25% | 14° 2' | 0.24249 | 242.5 | 5 |
| 30% | 16° 42' | 0.28736 | 287.4 | 6 |
| 35% | 19° 17' | 0.33024 | 330.2 | |
| 40% | 21° 48' | 0.37137 | 371.4 | |
| 45% | 24° 14' | 0.41045 | 410.5 | |

and because M = W/g (10) and F = (550 × HP)/V, they can be substituted to yield:

$$a = 550(g/V)(HP/W)$$

Finally, a and V can be interchanged to give:

$$V = 550(g/a)(HP/W)$$

where V is the vehicle speed in ft/sec, W is the vehicle weight in pounds, g is the gravitational constant 32.2 ft/sec² and the other factors you've already met. For any given acceleration, as weight goes up, speed goes down because they are inversely proportional.

### Weight affects range

Distance is simply speed multiplied by time:

$$D = Vt; \text{ therefore } D = 550(g/a)(HP/W)t$$

So weight again enters the picture. For any fixed amount of energy you are carrying onboard your vehicle, you will go farther if you take longer (drive at a slower speed) or carry less weight. You already encountered the practical results of this trade-off in FIG. 4-22 of chapter 4.

If you want to know still more about how weight affects acceleration, climbing, speed, and range, consult Gillespie's book or the Bosch handbook. Besides the primary task of eliminating all unnecessary weight, there are two other important weight-related factors to keep in mind when doing EV conversions: front-to-rear weight distribution, and the 30 percent rule.

### Remove the weight but keep your balance

Always focus on keeping your vehicle's front-to-rear weight distribution intact and not exceeding its total chassis and front/rear axle weight loading specifications. Figure 6-3

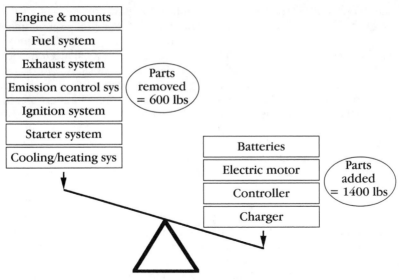

**6-3** Front to rear weight distribution trade-off.

shows the magnitude of your problem for a Ford Ranger pickup truck similar to the one used in chapter 11's conversion. You have pulled out 600 pounds in engine, fuel, exhaust, emission, ignition starter and heating/cooling systems. But you're going to be putting 1400 pounds back in, including 1200 pounds of batteries (20 at about 60 pounds each). How do you handle it?

TABLE 6-3 provides the answers. Notice the first row shows the 3000-pound curb weight nominally distributed 60 percent front (1800 pounds) and 40 percent rear

### Table 6-3 Electric vehicle conversion weights compared

| Item | Curb weight (lbs) | Front axle weight (lbs) | Rear axle weight (lbs) | Payload weight (lbs) |
|---|---|---|---|---|
| Ford Ranger pickup before conversion | 3000 | 1800 | 1200 | 1200 |
| Less IC engine & system parts | <600> | <500> | <100> | |
| Subtotal before conversion | 2400 | 1300 | 1100 | |
| Plus electric vehicle batteries, motor, etc. | 1400 | 400 | 1000 | |
| Ford Ranger pickup after conversion | 3800 | 1700 | 2100 | 400 |
| Battery weight 20 ea 6-volt @ 60 lbs | 1200 | | | |
| Ratio battery weight to vehicle weight | 32% | | | |

(1200 pounds) with a 1200-pound payload capacity. The second row shows that most of the weight you took out came from in and around the engine compartment—the 600 pounds you removed took 500 pounds off the front axle and 100 pounds off the rear axle. The secret is to put the weight back in a reasonably balanced fashion. This is accomplished by mounting four of the batteries—approximately 240 pounds—up front in the engine compartment along with the motor, controller, and charger. This puts about 400 pounds up front and about 1000 pounds worth of batteries in the rear. The fifth row shows the results. You're up to a curb weight of 3800 pounds with a 1700-pound to 2100-pound front-to-rear weight distribution, but you're still inside the GVWR, front/rear GAWR, weight distribution, and payload specifications. Furthermore, when you go out and drive the vehicle, it exhibits the same steering, braking, and handling capability that it had before the conversion.

As long as you keep within your vehicle's original internal combustion engine weight loading and distribution specifications, suspension and other support systems will never notice that you've changed what's under the hood. Overloading your EV chassis makes no more sense than overloading your internal combustion engine vehicle chassis. The best and safest solution is to get another larger or more heavy duty chassis. A postscript—some owners actually prefer to adjust the shocks and springs of their conversion vehicle at this point to give a slightly firmer ride, or (in the case of pickup truck owners) to return it to its previous firmer ride.

### Remember the 30-percent rule

The "30 percent or greater" rule-of-thumb (battery weight should be at least 30 percent of gross vehicle weight when using lead-acid batteries) is a very useful target to shoot for in an EV conversion. You'll want to do even better if you're optimizing for either high-speed or long-range performance goals. TABLE 6-3 shows that battery weight was 32 percent of gross vehicle weight for this conversion. Notice that if you opt for a 144-volt battery system (adding four more batteries and 240 pounds more weight), the ratio goes up to 1440/4040 or 36 percent. Going the other way (taking out four batteries and 240 pounds), the ratio drops to 960/3560 or 30 percent for a 96-volt system. Taking out four more batteries and going to a 72-volt system, the ratio drops to 720/3320 or 22 percent. The rule of thumb proves to be correct in this case, because you'd be unhappy with the performance of a 72-volt system in this vehicle; even 96 volts is marginal.

## STREAMLINE YOUR EV THINKING

Until fairly recently, most of the automobile industry's wedge-front designs—while attractive—are actually 180 degrees away from aerodynamic streamlining. Look at nature's finest and most common example, the falling raindrop: rounded and bulbous in front, it tapers to a point at its rear—the optimum aerodynamic shape. The cover of the January, 1993 *Technology Review* shows that new bicycle-racing helmets adhere perfectly to this principle.

While airplanes, submarines, and bullet trains have for decades incorporated the raindrop's example into their designs, automakers' design shops have eschewed this idea as unappealing to the public's taste. With plenty of internal combustion engine horsepower at their disposal, they didn't need aerodynamics, they needed style. Because batteries provide only one percent as much power per weight as gasoline, you and your EV do need aerodynamic awareness.

In this section, you'll look at the *Aerodynamic drag* row of FIG. 6-1 and learn about the factors that come with the turf when you select your conversion vehicle, and the items that you can change to help any EV conversion slip through the air more efficiently.

## Aerodynamic drag force defined

Mike Kimbell, an EV consultant, said it best: "Below 30 mph you could put an electric motor on a brick and never notice the difference." The reason is simple: aerodynamic drag force varies with the square of the speed. If you're not moving, there's no drag at all. Once you get rolling it builds up rapidly and soon swamps all other factors. Let's see exactly how.

The aerodynamic drag force can be expressed as:

$$F_d = (C_d A V^2)/391$$

where $F_d$ is the aerodynamic drag force in pounds, $C_d$ is coefficient of drag of your vehicle, A is its frontal area in square feet, and V is the vehicle speed in mph. To minimize drag for any given speed you must minimize $C_d$, the coefficient of drag, and A, its frontal area.

## Choose the lowest coefficient of drag

The coefficient of drag $C_d$, has to do with streamlining and air turbulence flows around your vehicle—characteristics that are inherent in the shape and design of the conversion vehicle you choose. $C_d$ is not easily affected or changed later, so if you're optimizing for either high-speed or long-range performance goals, it's important that you keep this critical performance factor foremost in your mind when selecting your conversion vehicle. Figure 6-4 shows the value of $C_d$ for different shapes and types of vehicles. Notice that $C_d$ has declined significantly with the passage of time—the 1920s Ford sedan had a $C_d$ around 0.85; today's Ford Taurus has a $C_d$ of 0.32. The values of $C_d$ for typical late 1980s cars, trucks and vans is:

- Cars—0.30 to 0.35
- Vans—0.33 to 0.35
- Pickup trucks—0.42 to 0.46

While car $C_d$ values have typically declined from 0.5 in the 1950s, to 0.42 in the 1970s, to 0.32 today—don't be misled. That snazzy contemporary open cockpit roadster can still have a $C_d$ of 0.6. That's because $C_d$ has everything to do with air turbulence caused by open windows, cockpits, and pickup box areas—not with streamlining alone.

TABLE 6-4 shows how different areas—in this case taken for a 1970s-vintage car—contribute to the $C_d$ of a vehicle. Contrary to what you might think, the body's rear area contributes more than 33 percent of $C_d$ by itself, followed by the wheel wells at 21 percent, the underbody area at 14 percent, the front body area at 12 percent, projections (mirrors, drip rails, window recesses) at 7 percent, and engine compartment and skin friction at 6 percent each. That's why General Motor's highly-optimized Impact electric vehicle $C_d$ is 0.19—it has a finely sculptured rear, sculptured or covered wheel wells with thin tires, an enclosed underbody, a low nose with highly sloping windshield and low ground clearance, no projections, and only two small openings to the engine compartment.

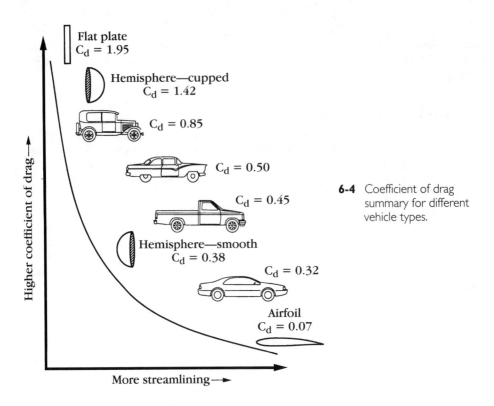

Flat plate
$C_d = 1.95$

Hemisphere—cupped
$C_d = 1.42$

$C_d = 0.85$

$C_d = 0.50$

$C_d = 0.45$

Hemisphere—smooth
$C_d = 0.38$

$C_d = 0.32$

Airfoil
$C_d = 0.07$

Higher coefficient of drag→

More streamlining→

**6-4** Coefficient of drag summary for different vehicle types.

### Table 6-4  Contribution of different car areas to overall $C_d$ for a typical 1970's-vintage car

| Car Area | $C_d$ Value | Percentage of total |
|---|---|---|
| Body-Rear | 0.14 | 33.3 |
| Wheel wells | 0.09 | 21.4 |
| Body-Under | 0.06 | 14.3 |
| Body-Front | 0.05 | 11.9 |
| Projections & Indentations | 0.03 | 7.1 |
| Engine compartment | 0.025 | 6.0 |
| Body-Skin friction | 0.025 | 6.0 |
| Total | 0.42 | 100.0 |

### Choose the smallest frontal area

The A (frontal area) for typical late-model cars, trucks, and vans is in the 18- to 24-square-foot range. A 4-foot by 8-foot sheet of plywood held up vertically in front of your vehicle would have a frontal area of 32 square feet. It has to do with the effective area your vehicle presents to the onrushing air stream. Frontal area is also not easily affected or changed later, and there's not too much you can do to significantly minimize it except choose a different vehicle body type. If you're optimizing for either high-speed or long-range performance, keep this critical performance factor in your mind when selecting your conversion vehicle.

## Relative wind contributes to aerodynamic drag

Drag force is measured nominally at 60 degrees Fahrenheit and a barometric pressure of 30 inches of Mercury in still air. Normally those are adequate assumptions for most calculations. But very few locations have still air, so an additional drag component due to relative wind velocity has to be added to your aerodynamic drag force calculation. This is the additional wind drag pushing against the vehicle from the random local winds. The equation defining the relative wind factor, $C_w$ is

$$C_w = (0.98(w/V)^2 + 0.63(w/V))C_{rw} - 0.40(w/V)$$

where w is the average wind speed of the area in mph, V is the vehicle speed, and $C_{rw}$ is a relative wind coefficient that is approximately 1.4 for typical sedan shapes, 1.2 for more streamlined vehicles, and 1.6 for vehicles displaying more turbulence or sedans driven with their windows open.[3]

TABLE 6-5 shows $C_w$ calculated for seven different vehicle speeds—assuming the U.S average value of 7.5 mph for the average wind speed—for the three different $C_r$ values.

To get your total aerodynamic drag force (still plus relative wind) multiply and add,

$$F_{td} = F_d + C_w F_d \text{ or } F_d(1 + C_w)$$

## Table 6-5  Relative wind factor $C_w$ at different vehicle speeds for three $C_{rw}$ values

| $C_{rw}$ at average wind=7.5mph | $C_w$ factor at V=5 mph | $C_w$ factor at V=10 mph | $C_w$ factor at V=20 mph | $C_w$ factor at V=30 mph | $C_w$ factor at V=45 mph | $C_w$ factor at V=60 mph | $C_w$ factor at V=75 mph |
|---|---|---|---|---|---|---|---|
| 1.2 | 3.180 | 0.929 | 0.299 | 0.163 | 0.159 | 0.063 | 0.047 |
| 1.4 avg sedan | 3.810 | 1.133 | 0.374 | 0.206 | 0.185 | 0.082 | 0.062 |
| 1.6 | 4.440 | 1.338 | 0.449 | 0.250 | 0.212 | 0.101 | 0.076 |

## Aerodynamic drag force data you can use

TABLE 6-6 puts the $C_d$ and A values for actual vehicles together and calculates their drag force for seven different vehicle speeds. Notice that drag force is least on a small car and greatest on the small pickup—but the small car might not have the room to mount the batteries to deliver the performance that you need. Notice also that an open cockpit roadster—even though it has a small frontal area A—has drag force identical to the pickup truck.

To use TABLE 6-5 and TABLE 6-6 with your EV, pick out your vehicle type in TABLE 6-6, then multiply its drag force number by the relative wind factor at the identical vehicle speed using the appropriate $C_{rw}$ row for your vehicle type. For example, the 3800-pound Ford Ranger pickup truck of chapter 11 has a drag force of 24.86 pounds at 30 mph using TABLE 6-6. Multiplying by the relative wind factor of 0.250 from the

## Table 6-6  Aerodynamic drag force $F_d$ at different vehicle speeds for typical vehicle $C_d$ and A values

| Vehicle | $C_d$ | A | V=5 mph | V=10 mph | V=20 mph | V=30 mph | V=45 mph | V=60 mph | V=75 mph |
|---|---|---|---|---|---|---|---|---|---|
| Small car | 0.3 | 18 | 0.35 | 1.38 | 5.52 | 12.43 | 27.97 | 49.72 | 77.69 |
| Larger car | 0.32 | 22 | 0.45 | 1.80 | 7.20 | 16.20 | 36.46 | 64.82 | 101.28 |
| Van | 0.34 | 26 | 0.57 | 2.26 | 9.04 | 20.35 | 45.78 | 81.39 | 127.17 |
| Sm pickup | 0.45 | 24 | 0.69 | 2.76 | 11.05 | 24.86 | 55.93 | 99.44 | 155.37 |
| Roadster | 0.6 | 18 | 0.69 | 2.76 | 11.05 | 24.86 | 55.93 | 99.44 | 155.37 |

bottom row ($C_{rw}$ = 1.6) of TABLE 6-5 gives you 6.22 pounds. Your total aerodynamic drag forced is then 24.86 + 6.22 or 31.08 pounds.

## Shape rear airflow

If you've seen a movie of a wind tunnel test—with smoke added to make the air currents visible—you've noticed a vortex or turbulence area at the rear of most vehicles tested. Those without access to wind tunnels notice the same effect when a semitrailer truck blows past you on the highway.

Recalling the falling raindrop shape, a boat tail or rocket ship nose shape is the ideal. While this is difficult to achieve, and no production chassis designs are available to help, you *can* benefit from rounding your vehicle's rear corners and eliminating all sharp edges. As you saw in TABLE 6-4, the body rear by itself makes up approximately 33 percent of the $C_d$—so any change you can make here should pay big dividends.

In the practical domain of our pickup truck example, putting on a lightweight and streamlined (rounded edges) camper shell should be your first choice. Station wagon shapes actually have better $C_d$ figures than fastback sedan shapes (though the difference in weight negates the advantage). A pickup truck with a streamlined camper shell is very close to a station wagon shape. The next choice is to put a cover over the pickup box. If even this is not feasible, leave the tailgate down and use a cargo net or wire screen instead, or simply remove the tailgate if this presents you with no functional problems.

## Shape wheel well & underbody airflow

Next we pay attention to the wheels and wheel well area. TABLE 6-4 showed that the tire and wheel well area by itself contributes approximately 21 percent of the $C_d$, so small streamlining changes here can have large benefits. Smooth wheel covers, thinner tires, rear wheel well covers, no mud flaps, even lowering the vehicle height all come to mind as immediate beneficial steps. Every little bit helps; anything you can do to reduce drag or turbulence is good.

The next obvious area is underneath the vehicle—TABLE 6-4 showed this contributes approximately 14 percent of your $C_d$. Automobile designers have traditionally ignored this area because the public doesn't see it. But the onrushing air does. So the built-from-the-ground-up General Motor's Impact EV designers didn't ignore this area. You shouldn't either.

The immediate solution is simple: close the bottom of the engine compartment. There are no longer any bulky internal combustion engine components in the engine compartment, so cover the entire open area with a lightweight sheet of material. You probably still have transmission—and definitely have steering/suspension—components to deal with, so it might take several sheets of material. Whatever material you use, you don't want it rumpling when the onrushing air strikes it, so choose a thickness that eliminates this possibility and fasten the sheet (or sheets) securely to the chassis.

Make the area from behind the underside of your vehicle's nose to underneath the firewall region (or just beyond it) as smooth as possible, using as lightweight a material as you can. If you can go all the way to the rear of your vehicle, so much the better. A fully streamlined underside reduces drag and turbulence, and can only help you.

### Block and/or shape front airflow

As you saw in TABLE 6-4, the body front and the engine compartment combined comprise approximately 18 percent of your $C_d$. While you cannot make significant changes to the $C_d$ and A values you inherit when you purchase your conversion chassis, you can replace or cover its sharp-cornered front air entry grill and block off the flow of air to the engine compartment. An EV doesn't need the massive air intake required in internal combustion vehicles to feed cooling air to the radiator and engine compartment. A 3-inch-diameter duct directing air to your electric motor is perfectly adequate. So anything you can do to round or streamline your vehicle's nose area (the smooth Ford Taurus or Thunderbird shapes of the 1990s as opposed to the shining and sharp grillwork shapes of the 1980s and earlier) is fair game. What you want is the maximum streamlined effect with minimum weight, so use modern kit car plastic and composite materials and techniques—no lead weight auto-body-filler noses if you please!

Air entering your EV conversion's now somewhat vacant engine compartment has the negative effect of creating under-the-hood turbulence, so blocking the incoming airflow with a sheet of lightweight material (such as aluminum) placed behind the grill works wonders. Whatever material you choose, just make sure it's heavy enough and fastened securely enough not to buckle, rumple, or vibrate when the air strikes it. Remember to leave a small opening for your electric motor's cooling air duct.

## ROLL WITH THE ROAD

As they said in the movie *Days of Thunder*, "tires is what wins the race." Today tires are fat, have wide tread, and are without low rolling resistance characteristics—they've been optimized for good adhesion instead. As an electric vehicle owner, you need to go against the grain of current tire thinking and learn to roll with the road to win the performance race.

In this section, you'll look at the *Rolling resistance* row of FIG. 6-1 and learn how to maximize the benefit from those four (or three) tire-road contact patches that are no bigger than your hand.

### Rolling resistance defined

The rolling resistance force is defined as:

$$F_r = C_r W \cos \phi$$

where $C_r$ is the rolling resistance factor, W is your vehicle weight in pounds and $\phi$ is the angle of incline as shown in FIG. 6-2. Notice that cos $\phi$ varies from 1 at no incline (maximum effect) to 0 at 90 degrees (no effect). Again vehicle weight is a factor—this time modulated by the vehicle's tire friction. The rolling factor $C_r$ might at its most elementary level be estimated as a constant. For a typical under-5000-pound EV, it is approximately:

- 0.015 on a hard surface (concrete).
- 0.08 on a medium-hard surface.
- 0.30 on a soft surface (sand).

If your calculations require more accuracy, $C_r$ varies linearly with speed at lower speeds and can be represented by:

$$C_r = 0.012 \; (1 + V/100)$$

where V is vehicle speed in mph.

## Pay attention to your tires

Tires are important to an EV owner. They support the vehicle and battery weight while cushioning against shocks; develop front-to-rear forces for acceleration and braking; and develop side-to-side forces for cornering.

Tires are almost universally of radial-ply construction today. Typically one or more steel-belted plies run around the circumference (hence, radial) of tire. These deliver vastly superior performance to the bias ply types (several plies woven crosswise around the tire carcass, hence, bias or on an angle) of earlier years that were replaced by radials as the standard in the 1960s. A tire is characterized by its rim width, the size wheel rim it fits on, section width (maximum width across the bulge of the tire), section height (distance from the bead to the outer edge of the tread), aspect ratio (ratio of height to width), overall diameter, and load. In addition, the Tire and Rim Association[4] defines the standard tire naming conventions:

- 5.60 × 15 (typical VW Bug) bias tire size—*5.60* denotes section width in inches and *15* denotes the rim size in inches.
- 155R13 (typical Honda Civic) radial tire size—*155* denotes section width in millimeters and *13* denotes the rim size in inches.
- P185/75R14 (typical Ford Ranger pickup) more recent P-metric radial tire size—*P* denotes passenger car, *185* denotes section width in millimeters, *75* denotes aspect ratio, *R* denotes radial, *B* belted, or *D* bias, and *14* denotes the rim size in inches.

Although Goodyear has taken a leadership role with their GFE (Greater Fuel Efficiency) tires, Firestone has their Concept EVT series, and other tire-makers are also developing designs that are ideal for EVs, doing a conversion today means using what's readily available. TABLE 6-7 gives you a comparison of the published characteristics for the Goodyear Decathalon tire family. This is an economical class of all-weather radials (other tire-makers have similar families) that should be more than adequate for most EV owner's needs.

Everything in TABLE 6-7 is from Goodyear's published spec sheets except the *Revolutions per mile* column, which is a nominal value calculated directly from the *Overall diameter* column rather than using actual measured data. The calculated value is slightly lower than a measured value when new and, as tread wears down, you are looking at a difference of 0.4 to 0.8 inches less in the tire's diameter, which translates

## Table 6-7 Comparison of Goodyear's decathalon tire family[1]

| Tire size | Wheel rim width | Tire rim width | Section width | Tread width | Overall diameter | Revolutions per mile | Maximum load (lbs) |
|---|---|---|---|---|---|---|---|
| P155/80R13 | 4.5–5.5 | 4.5 | 6.18 | 3.8 | 22.76 | 886 | 959 |
| P165/80R13 | 4.5–6.0 | 4.5 | 6.5 | 4.2 | 23.39 | 862 | 1069 |
| P175/80R13 | 4.5–6.0 | 5.0 | 6.97 | 4.1 | 24.02 | 840 | 1179 |
| P185/80R13 | 5.0–6.5 | 5.0 | 7.24 | 4.4 | 24.65 | 818 | 1301 |
| P185/75R14 | 5.0–6.5 | 5.0 | 7.24 | 4.6 | 24.96 | 808 | 1290 |
| P195/75R14 | 5.0–7.0 | 5.5 | 7.72 | 4.9 | 25.51 | 791 | 1400 |
| P205/75R14 | 5.5–7.5 | 5.5 | 7.99 | 5.1 | 26.14 | 772 | 1532 |
| P205/75R15 | 5.5–7.5 | 5.5 | 7.99 | 5.0 | 27.13 | 743 | 1598 |
| P215/75R15 | 5.5–7.5 | 6.0 | 8.5 | 5.3 | 27.68 | 729 | 1742 |
| P225/75R15 | 6.0–8.0 | 6.0 | 8.79 | 5.4 | 28.31 | 712 | 1874 |
| P235/75R15 | 6.0–8.0 | 6.5 | 9.25 | 5.8 | 28.86 | 699 | 2028 |

[1] All dimensions in inches

into even more revolutions per mile. The difference might be 30 revolutions out of 900—a difference of 3 percent—but if this figure is important to your calculations, measure your tire's actual circumference in your driveway. A chalk mark on the sidewall and a tape measure and one full turn of your tire is all you need to tell if you're in the ballpark.

Ideally, you want soapbox derby or at least motorcycle tires on your EV: thin (little contact area with road), hard (little friction), and large diameter (fewer revolutions per mile, and thus higher mileage, longer wear). From engineering studies on the rolling loss characteristics of solid rubber tires,

$$F_t = C_t(W/d)(t_h/t_w)^{1/2}$$

where $F_t$ is the rolling resistance force, $C_t$ is a constant reflecting the tire material's elastic and loss characteristics, W is the weight on the tire, d is the outside diameter of the tire, and $t_h$ and $t_w$ are the tire section height and width, respectively. This is the last you'll see or hear of this equation in the book, but the point is, the rolling resistance force is affected by the material (harder is better for EV owners), the loading (less weight is better), the size (bigger is better) and the aspect ratio (a lower $t_h/t_w$ ratio is better).

The variables in more conventional tire rolling resistance equations are usually tire inflation pressure (decreases with increasing inflation pressure—harder is better), vehicle speed (increases with increasing speed), tire warmup (warmer is better), and load (less weight is better).

## Use radial tires

Radial tires are nearly universal today, so tire construction is no longer a factor. But you might buy an older chassis that doesn't have radials on it, so check to be sure because bias-ply or bias-belted tires deliver far inferior performance to radials in terms of rolling resistance versus speed, warmup, and inflation.

## Use high tire inflation pressures

While you don't want to overinflate and balloon out your tires so that they pop off their rims, there is no reason not to inflate your EV's tires to their limit (and then some) to suit your purpose. The upper limit is established by your discomfort level from the road vibration transmitted to your body. Rock hard tires are fine; the only real caveat is not to overload your tires while simultaneously overinflating them.

## Use used tires

On the practical side, you're limited to using the tires that are manufactured today. But nothing stops you from using used tires. Not only are they less expensive, but broken-in tires with less tread depth have less rolling resistance than new tires. Go for a tire whose tread has uniformly worn down to roughly half its original tread depth. Go up to a larger diameter tire so that you get the minimum number of revolutions per mile. This might decrease your 0-to-60-mph elapsed times slightly, but will increase your range and your tire lifetimes. Go to a thinner profile tire that gives you the minimum footprint with the road. Without ignoring tire and condition, you would do well to purchase a matched set of broken-in radials for your EV, save money up front, and enjoy their money-saving, reduced-friction benefits during your ownership period.

## Brake drag & steering scuff add to rolling resistance

In addition to tires, rolling resistance comes from brake drag and steering, and suspension alignment "scuff." Brake drag is another reason used vehicles are superior to new ones in the rolling resistance department—brake drag usually goes away as the vehicle is broken in. Alignment is another story. At worst it's like dragging your other foot behind you turned 90 degrees to the direction you're walking in. You want to check and make sure front wheel alignment is at manufacturing spec levels, and you haven't accidently bought a chassis whose rear wheels are tracking down the highway in sideways fashion. But neither of these contributes an earth-shattering amount to rolling resistance: the brake drag coefficient can be estimated as a constant 0.002 factor, and steering/suspension drag as a constant 0.001 factor. Taken together, they add only 0.003—or an additional 3 pounds of force required by a vehicle weighing 1000 pounds travelling on a level surface.

## Rolling resistance force data you can use

For most purposes, the nominal $C_r$ of 0.015 (for concrete) with the nominal brake and steering drag of 0.003 added to it—total 0.018—is all you need. This generates 18.0 pounds of rolling resistance force for a 1000-pound vehicle. The 3800-pound Ford Ranger pickup truck of chapter 11 would have a rolling resistance of 68.4 pounds (3800 pounds of pickup weight $\times$ 0.018, or 3.8 $\times$ 18 pounds). As you can see, Mike's comment in the aerodynamic section was right. At 30 mph, the aerodynamic drag force on the Ford Ranger pickup truck of chapter 11 is 31.08 pounds—less than half the contribution of its 68.4 pounds of rolling resistance drag.

Figure 6-5 shows the aerodynamic drag force and the rolling resistance force on the Ford Ranger pickup truck of chapter 11 plotted for several vehicle speeds. These two forces—along with the acceleration or hill-climbing forces—constitute the *propul-*

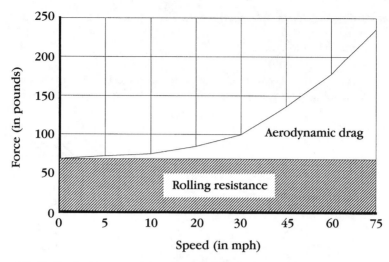

**6-5** Rolling resistance and aerodynamic drag versus speed.

*sion* or *road load*. Notice the 68.4 pounds of rolling resistance force is the main component of drag until the aerodynamic drag force takes over above 45 mph. Adding the force required to accelerate at a 1 mph/sec rate—nominally equivalent to that required to climb a 5 percent incline—merely shifts the combined aerodynamic drag-rolling resistance force curve upwards by 207.9 pounds (3.8 × 54.7 pounds) for the pickup. We'll look at these forces once again in the design section.

## LESS IS MORE WITH DRIVETRAINS

In this section, you'll look at the *Drivetrain* row of FIG. 6-1 and see how to get the most out of the internal combustion engine vehicle drivetrain components you adopt for your EV conversion.

The drivetrain in any vehicle comprises those components that transfer its motive power to the wheels and tires. The problem is, two separate vocabularies are used when talking about drivetrains for electric motors as opposed to those for internal combustion engines. This section will discuss the basic components, cover differences in motor-versus-engine performance specifications, discuss transmission gear selection, and look at the trade-offs of automatic versus manual transmission; new versus used; and heavy versus light fluids on drivetrain efficiencies.

### Drivetrains

Let's start with what the drivetrain in a conventional internal combustion engine vehicle must accomplish. In practical terms, the power available from the engine must be equal to the job of overcoming the tractive resistances discussed earlier for any given speed. The obvious mission of the drivetrain is to apply the engine's power to driving the wheels and tires with the least loss (or highest efficiency). But overall, the drivetrain must perform a number of tasks:

- Convert torque and speed of the engine to vehicle motion—traction.
- Change directions—enable forward and backward vehicle motion.

- Permit different rotational speeds of the drive wheels when cornering—differential.
- Overcome hills and grades.
- Maximize fuel economy.

The drivetrain layout shown in simplified form in FIG. 6-6 is most widely used to accomplish these objectives today. The function of each component is listed below.

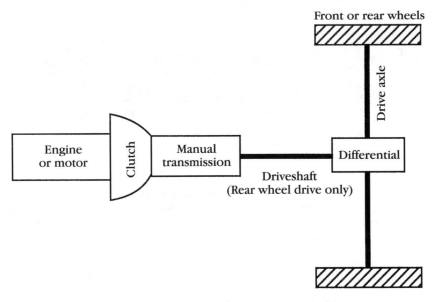

**6-6** Simplified EV drivetrain layout.

*Engine (or electric motor).* Provides the raw power or torque to propel the vehicle.

*Clutch.* For internal combustion engines, separates or interrupts the power flow from the engine so that transmission gears can be shifted and, once engaged, the vehicle can be driven from standstill to top speed.

*Manual transmission.* Provides a number of alternative gear ratios to the engine so that vehicle needs—maximum torque for hill-climbing or minimum speed for economical cruising—can be accommodated.

*Driveshaft.* Connects the drive wheels to the transmission in rear-wheel-drive vehicles—not needed in front wheel drive vehicles.

*Differential.* Accommodates the fact that outer wheels must cover a greater distance than inner wheels when a vehicle is cornering, and translates drive force 90 degrees in rear-wheel-drive vehicles (might or might not in front wheel drive vehicles, depending on how engine is mounted).

*Drive axles.* Transfer power from the differential to the drive wheels.

TABLE 6-8 shows that you can typically expect 90 percent or greater efficiencies—slightly better for front wheel drive vehicles—from today's drivetrains.

Internal combustion engine vehicle drivetrains provide everything necessary to allow an electric motor to be used in place of the removed engine and its related components to propel the vehicle. But the drivetrain components are usually complete

## Table 6-8  Comparison of front and rear wheel drivetrain efficiencies

| Drivetrain type | Manual transmission | Driveshaft | Differential drive | Drive axle | Overall efficiency |
|---|---|---|---|---|---|
| Front wheel drive | 0.96 | not required | 0.97 | 0.98 | 0.91 |
| Rear wheel drive | 0.96 | 0.99 | 0.97 | 0.98 | 0.90 |

overkill for the EV owner. The reason has to do with the different characteristics of internal combustion engines versus electric motors, and the way they are specified.

### Difference in motor versus engine specifications

Comparing electric motors and internal combustion engines is not an "apples to apples" comparison. If someone offers you either an electric motor and an internal combustion engine with the same rated horsepower, take the electric motor—it's far more powerful. Also, an electric motor delivers peak torque upon startup (zero RPM), whereas an internal combustion engine delivers nothing until you wind up its RPMs. An electric motor is so different from an internal combustion engine that a brief discussion of terms is necessary before going further.

There is a substantial difference in the way an electric motor and an internal combustion engine are rated in horsepower. Figure 6-7's purpose is to show at a glance that an electric motor is more powerful than an internal combustion engine of the same rated horsepower. All internal combustion engines are rated at specific RPM levels for maximum torque and maximum horsepower. Internal combustion engine max-

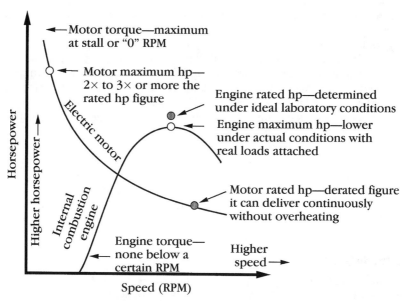

**6-7**  Comparison of electric motor vs. internal combustion engine characteristics.

imum horsepower ratings are typically derived under idealized laboratory conditions (for the bare engine without accessories attached), which is why the rated HP point appears above the maximum peak of the internal combustion engine horsepower curve in FIG. 6-7. Electric motors, on the other hand, are typically "derated" (i.e., used at less than their maximum) by codes and/or conventions. Horsepower rating reflects either the one hour (or more) or continuous output level it can maintain without overheating. As you can see from FIG. 6-7, the rated HP point for an electric motor is far down from its peak output, which is typically two to three times higher or more.

There is another substantial difference. While an electric motor can produce a high torque at zero speed, an internal combustion engine produces negative torque until some speed is reached. An electric motor can therefore be attached directly to the drive wheels and accelerate the vehicle from a standstill—without the need for the clutch, transmission, or torque converter required by the internal combustion engine. Everything can be accomplished by controlling the drive current to the electric motor. While an internal combustion engine can only deliver peak torque in a relatively narrow speed range, and requires a transmission and different gear ratios to deliver high torque over a wide vehicle speed range, electric motors can be designed to deliver high torque over a broad speed range with no need for transmission at all.

All these factors mean that current EV conversions put a lighter load on their borrowed-from-an-internal-combustion-engine-vehicle drivetrains, and future EV conversions eliminate the need for several drivetrain components altogether. Let's briefly summarize:

*Clutch.* Although basically unused, a clutch is handy to have in today's EV conversions because its front end gives you an easy place to attach the electric motor, and its back end is already conveniently mated to the transmission. In short, it saves the work of building adapters, etc. In the future, when widespread adoption of ac motors and controllers eliminates the need for a complicated mechanical transmission, the electric motor can be directly coupled to a simplified, lightweight, one-direction, one- or two-gear ratio transmission—eliminating the need for a clutch.

*Transmission.* Another handy item in today's EV conversions, the transmission's gears not only match the vehicle you are converting to a variety of off-the-shelf electrical motors, but also give you a mechanical reversing control that eliminates the need for a two-direction motor and controller—again simplifying your work. In the future, when widespread adoption of ac motors and controllers provides directional control and eliminates the need for a large number of mechanical gears to get the torques and speeds you need, today's transmission will be able to be replaced by a greatly simplified (and even more reliable) mechanical device.

*Driveshaft, differential, drive axles.* These components are all used intact in today's EV conversions. Because contemporary, built-from-the-ground-up electric vehicles like General Motor's Impact use two ac motors and place them next to the drive wheels, it's not too difficult to envision even simpler solutions for future EVs, because electric motors (with only one moving part) are so easily designed to accommodate different roles.

## Going through the gears

The transmission gear ratios, combined with the ratio available from the differential (or *rear end* as it's sometimes called in automotive jargon), adapt the internal combustion engine's power and torque characteristics to maximum torque needs for hill-climbing or minimum economy needs for cruising. Figure 6-8 shows these at a glance for

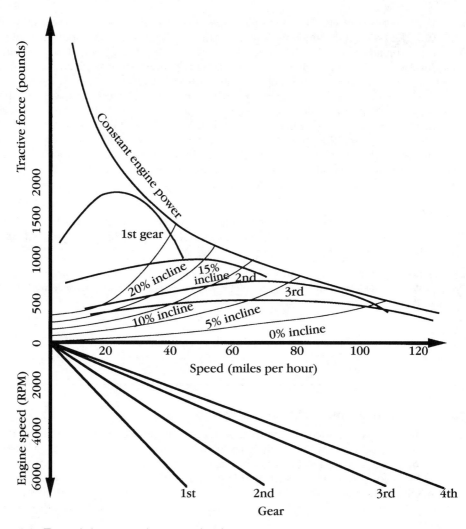

**6-8** Transmission gear ratio vs. speed and power summary.

a typical internal combustion engine with four manual transmission gears—horse-power/torque characteristics versus vehicle speed appear above the line and RPM versus vehicle speed appear below. The constant engine power line is simply equation (5), HP = FV/375 (V in mph), less any drivetrain losses. The tractive force line for each gear is simply the characteristic internal combustion engine torque curve (similar to the one shown in FIG. 6-7) multiplied by the ratios for that gear. The superimposed incline force lines are the typical propulsion or road-load force components added by acceleration or hill-climbing forces (recall the shape of this curve in FIG. 6-5). The intersection of the incline or road-load curves and the tractive-force curves are the maximum speed that can be sustained in that gear. The upper half of FIG. 6-8 illustrates how low first gearing for startup and high fourth gearing for high speed driving apply to engine torque capabilities.

The lower part of FIG. 6-8 shows road speed versus engine speed—for each gear appears. The point of this drawing is to illustrate how gear selection applies to engine speed capabilities. Normally, the overall gear ratios are selected to fall in a geometric progression: 1st/2nd = 2nd/3rd, etc. Then individual gears are optimized for starting (1st), passing (2nd or 3rd), and fuel economy (4th or 5th).

TABLE 6-9 shows how these ratios turn out in an actual production car—in this case a Ford 1989 Taurus SHO. Notice the first two gear pairs are in a 1.5 ratio, whereas the next two move to 1.35. TABLE 6-9 also calculates the actual transmission, differential, and overall gear ratios (overall equals transmission times differential) for the 1987 Ford Ranger pickup truck that will be later used in the design section. Notice that the Ranger is optimized at both ends of the range but lower in the middle versus the Taurus—reflecting the difference in car versus truck design.

### Table 6-9  Comparison of overall transmission gear ratios for Ford 1989 Taurus SHO and 1987 Ranger Pickup

| Transmission Gear | 1989 Taurus SHO Overall | Ratio to next gear | 1987 Ranger Pickup Transmission | 1987 Ranger Pickup Differential | 1987 Ranger Pickup Overall |
|---|---|---|---|---|---|
| 1st | 12.01 | 1.5 | 3.96 | 3.45 | 13.66 |
| 2nd | 7.82 | 1.5 | 2.08 | 3.45 | 7.18 |
| 3rd | 5.16 | 1.35 | 1.37 | 3.45 | 4.73 |
| 4th | 3.81 | 1.35 | 1.00 | 3.45 | 3.45 |
| 5th | 2.79 | N/A | 0.84 | 3.45 | 2.90 |

## Automatic versus manual transmission

The early transmission discussion purposely avoided automatic transmissions. The reason is simple: EV owners need efficiency, and automatic transmissions are tremendously inefficient. There's another good reason—even with off-the-shelf components you are not going to shift gears as much with an EV. If you're driving around town, you might only use one or two gears—you just put your EV in the gear you need and go. There's far less need for the clutch too. Remember, when you're sitting at an intersection your electric motor is not even turning.

Two ordinary household fans can demonstrate the torque converter principle that automatic transmissions use. Set them up facing each other about two feet apart and turn one on; the other starts to rotate with it. An automatic transmission uses transmission fluid rather than air as the coupling medium, but the principle is the same. An automatic transmission has a maximum efficiency of 80 percent, which drops dramatically at lower engine speed or vehicle torque levels. Automatic transmissions are also typically matched in characteristics to a given family of limited peak torque range internal combustion engines—exactly opposite in behavior to electric motors. In short, choose a model with a clutch and a manual transmission for your internal combustion engine conversion vehicle.

## Use a used transmission

There's a bonus here for the efficiency-seeking EV converter. Drivetrains take thousands of miles to wear into their minimum loss condition. When added to the fact that used tires have less rolling resistance, brake drag reduces as drums and linings wear smooth, and oil and grease seals have less drag after a period of break-in wear, you can expect about 20 percent less rolling and drivetrain resistance from a used vehicle than from a brand new one (assuming the used vehicle has not been badly worn or abused). And a used vehicle costs less, so it's a real bonus!

## Heavy versus light drivetrains and fluids

Efficiency-seeking EV converters should not only avoid automatic transmissions, but also oversize axles, transmission, clutches, or anything that adds weight and reduces efficiency. Even a manual-transmission-based drivetrain will exhibit higher losses when operating at a low fraction of the gear's design maximum torque, which is the normal EV mode. The lower EV load will result in a lower efficiency out of a heavy duty part than out of a regular or economy part. Look for a vehicle with a lighter engine and manual transmission (e.g., four cylinders rather than six, etc.) in your purchasing search. Earlier vintage models in a series are preferable because manufacturers tend to introduce higher performance options in successive model years.

The corollary of all this is the lubricant you choose. Using a lighter viscosity fluid in your differential lets things turn a lot easier. You're not breaking any rules here. Instead of shoving 500 horsepower through your drivetrain, you're at the opposite extreme—you're putting in an electric motor that lets you cruise at 10 percent of the peak torque load used by the internal combustion engine you just replaced. You're shifting less, using a lower peak torque, and probably using it less often. As a result your electric motor is putting only the lightest of loads on your internal combustion vehicle drivetrain, and you're probably using 50 percent (or less) of your drivetrain's designed capability. So less wear and tear on the gears means you can use a lighter viscosity lubricant and recover the additional benefit of further increased efficiency.

## DESIGN YOUR EV

This is step number two. Look at your big picture first. Before you buy, build, or convert, decide what the main mission of your EV will be: a high speed dragster to quietly blow away unsuspecting opponents at a stoplight; a long-range flyer to be a winning candidate at Electric Auto Association meetings; or a utility commuter vehicle to take you to work or grocery shopping, with capabilities midway between the former two. Your EV's weight is of primary importance to any design, but high acceleration off the line will dictate one type of design approach and gear ratios, while a long-range design will push you in a different direction. If it's a utility commuter EV you seek, then you'll want to preserve a little of both while optimizing your chassis flexibility toward either highway commuting or neighborhood hauling and pickup needs. In this section, you'll learn how to match your motor-drivetrain combination to the body style you've selected by going through the following steps:

- Learn when to use horsepower, torque, or current units and why.
- Look at a calculation overview.
- Determine the required torque needs of your selected vehicle's chassis.
- Determine the available capabilities of your selected electric motor and drivetrain torque.

The design process described here can be infinitely adapted to any EV you want to buy, build, or convert.

## Horsepower, torque, and current

Let's start with some basic formulas. Earlier in the chapter, equation (2) casually introduced you to the fact that

$$(2)\ 1\ \text{Horsepower (HP)} = 550\ \text{ft-lb/sec}$$

This was then conveniently bundled into equation (5):

$$(5)\ \text{Horsepower (HP)} = FV/375$$

where V is speed expressed in mph.

Horsepower is a rate of doing work. It takes 1 HP to raise 550 pounds one foot in one second. But the second equation, which relates force and speed, brings horsepower to you in more familiar terms. It takes 1 horsepower to move 37.5 pounds at 10 mph. Great—but you can also move 50 pounds at 7.5 mph with 1 horsepower. The first instance might describe the force required to push a vehicle forward on a level slope, the second describes the force required to push this same vehicle up an incline. Horsepower is equal to force or torque times speed, but you need to specify the force and speed you are talking about. For example, since we already know that 146.19 pounds is the total drag force on the 3800-pound Ranger pickup at 50 mph, and equation (5) relates the actual power required at a vehicle's wheels as a function of its speed and the required tractive force:

$$HP = (146.19 \times 50)/375 = 19.49$$

or approximately 20 horsepower.

Only about 20 horsepower is necessary—at the wheels—to propel this pickup truck along at 50 mph. In fact, a rated 20-HP electric motor will easily propel a 4000-pound vehicle at 50 mph—a fact that might amaze those who think in terms of the typical rated 90 HP or 120 HP internal combustion engine that might just have been removed from the pickup.

The point here is to condition yourself to think in terms of force or torque values, which are relatively easy to determine, rather than in terms of a horsepower figure that is arrived at differently for engines versus electric motors, and that means little until tied to specific force/torque and speed values anyway.

Another point (covered in more detail in chapter 7's discussion of electric motors and chapter 10's discussion of the electrical system) is to think in terms of current when working with electric motors. The current that is directly related to motor torque on the one hand is the same that flows through the controller, system wiring, and connectors, and from the batteries on the other. Through the torque-current relationship, you can directly link the mechanical and electrical worlds.

## Calculation overview

Notice that the starting point in the calculations was the ending point of the force value required. Once you know the forces acting on your vehicle chassis at a given speed, the rest is easy. For your calculation approach, first determine these values, then plug in your motor and drivetrain values for its *design center* operating point—be it a 100-mph speedster, a 20-mph economy flyer, or a 50-mph utility vehicle. A 50-mph speed will be the design center for our pickup truck utility vehicle example.

In short, you need to select a speed, select an electric motor for that speed, choose the RPM at which the motor delivers that horsepower, choose the target gear ratio based on that RPM, and see if the motor provides the torque over the range of level and hill-climbing conditions you need. Once you go through the equations, worksheets, and graphed results covered in this section, and repeat them with your own values, you'll find the process quite simple.

The entire process is designed to give you graphic results you can quickly use to see how the torque available from your selected motor and drivetrain meets your vehicle's torque requirements at different vehicle speeds. If you have a microcomputer with a spreadsheet program, you can set it up once, and afterwards graph the results of any changed input parameter in seconds. In equation form, what we are saying is:

$$\text{Available engine power = Tractive resistance demand}$$
$$\text{Power = (Acceleration + Climbing + Rolling + Drag + Wind) Resistance}$$

Plugging into the force equations gives you:

$$\text{Force} = F_a + F_h + F_r + F_d + F_w$$
$$\text{Force} = C_i Wa + W\sin\phi + C_r W\cos\phi + C_d AV^2 + C_w F_d$$

You've determined every one of these earlier in the chapter. Under steady-speed conditions, acceleration is zero, so there is no acceleration force. If you are on a level surface, $\sin\phi = 0$, $\cos\phi = 1$ and the force equation can be rewritten,

$$\text{Force} = C_r W\cos\phi + C_d AV^2 + C_w F_d$$

This is the propulsion or road-load force you met at the end of the rolling resistance section and graphed for the Ford pickup in FIG. 6-5. You need to determine this force for your vehicle at several candidate vehicle speeds, and add back in the acceleration and hill-climbing forces. This is easy if you recall that the acceleration force equals the hill-climbing force over the range from 1 mph/sec to 6 mph/sec.

You can now calculate your electric motor's required horsepower for your EV's design center. Recalling equations (6) and (7),

$$(6) \text{ Horsepower (HP)} = (\text{Torque} \times \text{RPM})/5252 = 2\pi/60 \times FV/550$$
$$(7) \text{ Wheel RPM} = (\text{mph} \times \text{Revolutions/mile})/60$$

Equation (7) can be substituted in equation (6) to give,

$$\text{HP}_{wheel} = (\text{Torque}_{wheel} \times \text{mph} \times \text{Revolutions/mile})/(5252 \times 60)$$

but,

$$(14) \text{ HP}_{motor} = \text{HP}_{wheel}/n_o$$

where $n_o$ is the overall drivetrain efficiency. Substituting the previous equation into this gives you,

$$(15) \text{ HP}_{motor} = (\text{Torque}_{wheel} \times \text{mph} \times \text{Revolutions/mile})/(315120 \times n_o)$$

Plugging the values for torque, speed, and Revolutions/mile (based on your vehicle's tire diameter) into equation (15) will give you the required horsepower for your electric motor.

After you have chosen your candidate electric motor, the manufacturer will usually provide you with a graph or table showing its torque and current versus speed performance based on a constant voltage applied to the motor terminals. From these figures or curves, you can derive the RPM at which your electric motor delivers closest to its rated horsepower. Using this motor RPM figure and the wheel RPM figure from equation (7), which gives you the wheel RPM from your target speed and RPM, you can determine your best gear or gear ratio from:

$$(16) \text{ Overall gear ratio} = RPM_{motor}/RPM_{wheel}$$

This—or the one closest to it—is the best gear for the transmission in your selected vehicle to use; if you were setting up a one-gear-only EV, you would pick this ratio. With all the other motor torque and RPM values you can then calculate wheel torque and vehicle speed using equations (12) and (13) for the different overall gear ratios in your drivetrain:

$$(12) \text{ Torque}_{wheel} = \text{Torque}_{motor}/(\text{overall gear ratio} \times n_o)$$
$$(13) \text{ Speed}_{vehicle} \text{ (in mph)} = (RPM_{motor} \times 60)/(\text{overall gear ratio} \times \text{Revolutions/mile})$$

You now have the family of torque available curves versus vehicle speed for the different gear ratios in your drivetrain. All that remains is to graph the torque required data and the torque available data on the same grid. A quick look at the graph tells you if you have what you need or need to go back to the drawing board.

### Torque required worksheet

TABLE 6-10 computes the torque required data for a 1987 Ranger pickup—the vehicle to be converted in chapter 11. You've met all the values going into the level drag force before, but not in one worksheet. Now they are converted to torque values using equation (11), and new values of force and torque are calculated for incline values from 5 percent to 25 percent. Conveniently, these correspond rather closely to the acceleration values for 1 mph/sec to 5 mph/sec, respectively, and the two can be used interchangeably. The vehicle assumptions all appear in TABLE 6-10. If you were preparing a microcomputer spreadsheet, all of this type information would be grouped in one section so that you could see the effects of changing chassis weight, $C_dA$, $C_r$ and other parameters. You might also want to graph speed values at 5 mph intervals to present a more accurate picture.

### Torque available worksheet

There are a few preliminaries to go through before you can prepare the torque available worksheet. First, you have to determine the horsepower of electric motor using equation (15):

$$(15) \text{ HP}_{motor} = (\text{Torque}_{wheel} \times \text{mph} \times \text{Revolutions/mile})/(315120 \times n_o)$$

Plugging in real values for the Ranger pickup with P185/75R14 tires and using the torque value from the torque required worksheet at the 50 mph design center speed (recall that overall efficiency for rear wheel drive manual transmission was 0.90 from TABLE 6-8),

$$\text{HP}_{motor} = (152.04 \times 50 \times 808)/(315120 \times 0.9) = 21.66$$

This corresponds quite nicely to the capabilities of the Advanced DC Motors model FB1-4001 rated at 22 HP. From the manufacturer's torque versus speed curves for this motor driven at a constant 120 volts and using equation (6),

$$(6) \text{ HP} = (\text{Torque} \times RPM)/5252 = (25 \times 4600)/5252 = 21.89$$

## Table 6-10  Torque required worksheet for 1987 Ford Ranger pickup at different vehicle speeds and inclines[1]

| Vehicle speed (mph) | 10 | 20 | 30 | 40 | 50 | 60 | 70 | 80 | 90 |
|---|---|---|---|---|---|---|---|---|---|
| Tire rolling resistance $C_r$ | 0.015 | 0.015 | 0.015 | 0.015 | 0.015 | 0.015 | 0.015 | 0.015 | 0.015 |
| Brake and steering resistance $C_r$ | 0.003 | 0.003 | 0.003 | 0.003 | 0.003 | 0.003 | 0.003 | 0.003 | 0.003 |
| Total rolling force (lbs) | 68.4 | 68.4 | 68.4 | 68.4 | 68.4 | 68.4 | 68.4 | 68.4 | 68.4 |
| Still air drag force (lbs) | 2.76 | 11.05 | 24.60 | 44.19 | 69.05 | 99.43 | 135.35 | 176.78 | 223.73 |
| Relative wind factor $C_w$ | 1.338 | 0.449 | 0.25 | 0.169 | 0.126 | 0.101 | 0.083 | 0.071 | 0.062 |
| Relative wind drag force (lbs) | 3.70 | 4.96 | 6.21 | 7.47 | 8.73 | 9.99 | 11.25 | 12.51 | 13.77 |
| Total drag force, level (lbs) | 74.86 | 84.40 | 99.47 | 120.07 | 146.19 | 177.83 | 215.00 | 257.69 | 305.91 |
| Total drag torque, level (ft-lbs) | 77.85 | 87.78 | 103.45 | 124.87 | 152.04 | 184.95 | 223.60 | 268.00 | 318.15 |
| Sin ø, ø = Arc tan 5% incline | 0.0500 | 0.0500 | 0.0500 | 0.0500 | 0.0500 | 0.0500 | 0.0500 | 0.0500 | 0.0500 |
| Cos ø, ø = Arc tan 5% incline | 0.9988 | 0.9988 | 0.9988 | 0.9988 | 0.9988 | 0.9988 | 0.9988 | 0.9988 | 0.9988 |
| Incline force WSin ø (lbs) | 190.04 | 190.04 | 190.04 | 190.04 | 190.04 | 190.04 | 190.04 | 190.04 | 190.04 |
| Rolling drag force $C_r$WCos ø (lbs) | 68.31 | 68.31 | 68.31 | 68.31 | 68.31 | 68.31 | 68.31 | 68.31 | 68.31 |
| Total drag force, 5% (lbs) | 264.81 | 274.36 | 289.43 | 310.02 | 336.14 | 367.78 | 404.95 | 447.64 | 495.86 |
| Total drag torque, 5% (ft-lbs) | 275.41 | 285.34 | 301.01 | 322.43 | 349.59 | 382.50 | 421.16 | 465.56 | 515.71 |
| Sin (Arc tan 10% incline) | 0.0996 | 0.0996 | 0.0996 | 0.0996 | 0.0996 | 0.0996 | 0.0996 | 0.0996 | 0.0996 |
| Cos (Arc tan 10% incline) | 0.9950 | 0.9950 | 0.9950 | 0.9950 | 0.9950 | 0.9950 | 0.9950 | 0.9950 | 0.9950 |
| Incline force WSin ø (lbs) | 378.52 | 378.52 | 378.52 | 378.52 | 378.52 | 378.52 | 378.52 | 378.52 | 378.52 |
| Rolling drag force $C_r$WCos ø (lbs) | 68.06 | 68.06 | 68.06 | 68.06 | 68.06 | 68.06 | 68.06 | 68.06 | 68.06 |
| Total drag force, 10% (lbs) | 453.04 | 462.58 | 477.65 | 498.25 | 524.37 | 556.01 | 593.18 | 635.87 | 684.08 |
| Total drag torque, 10% (ft-lbs) | 471.17 | 481.10 | 496.77 | 518.19 | 545.35 | 578.26 | 616.92 | 661.32 | 711.46 |
| Sin ø, ø = Arc tan 15% incline | 0.1484 | 0.1484 | 0.1484 | 0.1484 | 0.1484 | 0.1484 | 0.1484 | 0.1484 | 0.1484 |
| Cos ø, ø = Arc tan 15% incline | 0.9889 | 0.9889 | 0.9889 | 0.9889 | 0.9889 | 0.9889 | 0.9889 | 0.9889 | 0.9889 |
| Incline force WSin ø (lbs) | 563.84 | 563.84 | 563.84 | 563.84 | 563.84 | 563.84 | 563.84 | 563.84 | 563.84 |
| Rolling drag force $C_r$WCos ø (lbs) | 67.64 | 67.64 | 67.64 | 67.64 | 67.64 | 67.64 | 67.64 | 67.64 | 67.64 |
| Total drag force, 15% (lbs) | 637.94 | 647.49 | 662.56 | 683.16 | 709.27 | 740.92 | 778.09 | 820.78 | 868.99 |
| Total drag torque, 15% (ft-lbs) | 663.48 | 673.41 | 689.08 | 710.50 | 737.66 | 770.57 | 809.23 | 853.63 | 903.77 |
| Sin ø, ø = Arc tan 20% incline | 0.1962 | 0.1962 | 0.1962 | 0.1962 | 0.1962 | 0.1962 | 0.1962 | 0.1962 | 0.1962 |
| Cos ø, ø = Arc tan 20% incline | 0.9806 | 0.9806 | 0.9806 | 0.9806 | 0.9806 | 0.9806 | 0.9806 | 0.9806 | 0.9806 |
| Incline force WSin ø (lbs) | 745.67 | 745.67 | 745.67 | 745.67 | 745.67 | 745.67 | 745.67 | 745.67 | 745.67 |
| Rolling drag force $C_r$WCos ø (lbs) | 67.07 | 67.07 | 67.07 | 67.07 | 67.07 | 67.07 | 67.07 | 67.07 | 67.07 |
| Total drag force, 20% (lbs) | 819.20 | 828.75 | 843.82 | 864.41 | 890.53 | 922.18 | 959.34 | 1002.0 | 1050.3 |
| Total drag torque, 20% (ft-lbs) | 851.99 | 861.92 | 877.59 | 899.01 | 926.18 | 959.08 | 997.74 | 1042.1 | 1092.3 |
| Sin ø, ø = Arc tan 25% incline | 0.2425 | 0.2425 | 0.2425 | 0.2425 | 0.2425 | 0.2425 | 0.2425 | 0.2425 | 0.2425 |
| Cos ø, ø = Arc tan 25% incline | 0.9702 | 0.9702 | 0.9702 | 0.9702 | 0.9702 | 0.9702 | 0.9702 | 0.9702 | 0.9702 |
| Incline force WSin ø (lbs) | 921.46 | 921.46 | 921.46 | 921.46 | 921.46 | 921.46 | 921.46 | 921.46 | 921.46 |
| Rolling drag force $C_r$WCos ø (lbs) | 66.36 | 66.36 | 66.36 | 66.36 | 66.36 | 66.36 | 66.36 | 66.36 | 66.36 |
| Total drag force, 25% (lbs) | 994.28 | 1003.8 | 1018.9 | 1039.5 | 1065.6 | 1097.3 | 1134.4 | 1177.1 | 1225.3 |
| Total drag torque, 25% (ft-lbs) | 1034.1 | 1044.0 | 1059.7 | 1081.1 | 1108.3 | 1141.2 | 1179.8 | 1224.2 | 1274.4 |

[1]Values computed for: 1987 Ford Ranger pickup; weight = 3800 lbs; coefficient of drag, $C_d$ = 0.45; frontal area, A = 24 square feet; relative wind factor, $C_{rw}$ = 1.6; relative wind, w = 7.5 mph; tires = P185/75R14; revolutions/mile = 808; torque multiplier = 840.34/(revolutions/mile) = 1.04.

This motor produces approximately 22 HP at 4600 RPM at 25 ft-lbs of torque and 170 amps. Next calculate Wheel RPM using equation (7),

(7) Wheel RPM = (mph × Revolutions/mile)/60 = (50 × 808)/60 = 673.33

You can then calculate the best gear using equation (16):

(16) Overall gear ratio = $RPM_{motor}/RPM_{wheel}$ = 4600/673.33 = 6.83

From TABLE 6-9, the 2nd gear overall ratio of 7.18 for the Ford Ranger pickup comes quite close to this figure, which means it should deliver the best overall performance.

Now you are ready to prepare the torque available worksheet shown in TABLE 6-11. This worksheet sets up motor values on the far left. Wheel torque and vehicle speed values for the 1st through 5th gears go from left to right across the worksheet, and at higher values of torque and current from top to bottom of the worksheet within each gear. The net result is you now have the wheel torque available at a given vehicle speed for each one of the vehicle's gear ratios. A microcomputer spreadsheet would make quick work of this; you could quickly see the effects of varying motor voltage, tires sizes, etc.

## Torque required & available graph

All that remains is to graph the torque required data from TABLE 6-10 and the torque available data from TABLE 6-11 on the same grid versus vehicle speed. This is done in FIG. 6-9. Notice its similarity to the curve in FIG. 6-8 drawn for the internal combustion engine (except that the torque available curves resemble the electric motor characteristics—a comparison made earlier in FIG. 6-7).

How do you read FIG. 6-9 and what does it tell you? The usable area of each gear is the area to the left and below it—bounded at the bottom by the torque required at the level condition curve. You want to work as far down the torque available curve for each gear as possible for minimum current draw and maximum economy and range. The graph confirms that 2nd gear is probably the best overall selection. You could put the EV into second gear and leave it there, because it gives you 2 mph/sec acceleration at startup, hill-climbing ability up to 15 percent inclines, and provides you with enough torque to take you up to about 52.5 mph. For mountain climbing or quick pops off the line, 1st gear gives you everything you could hope for at the expense of really sucking down the amps, current-wise. But at the other end of 1st gear, if you drive like there's an egg between your foot and the accelerator pedal, it actually draws only 100 amps at 45 mph versus the 210 amps required by 2nd gear. At higher speeds, 3rd gear lets you cruise at 60 mph at 270 amps, and 4th gear lets you cruise at 70 mph at 370 amps. At any speed, 5th gear appears marginal in this particular vehicle; though it can possibly hold 78 mph, it requires 440 amps to do so. At any other speed, other gears do it better with less current draw. While current draw is your first priority, too much for too long overheats your motor. You don't want to exceed your motor's speed rating either—as you would do if you drove much above 45 mph in 1st gear. Is this a usable motor and drivetrain combination for this vehicle? Definitely. If you want to make minor adjustments, just raise or lower the battery voltage. This will shift the torque available curve for each gear to the right (higher voltage) or to the left (lower voltage). A larger motor in this particular vehicle will give you better acceleration and upper end speed performance; the torque available curves for each gear would be

## Table 6-11 Torque available worksheet for 120-volt dc series motor powered 1987 Ford Ranger Pickup at different motor speeds and gear ratios[1]

| Vehicle gear | 1st | 2nd | 3rd | 4th | 5th |
|---|---|---|---|---|---|
| Overall gear ratio | 13.66 | 7.18 | 4.73 | 3.45 | 2.9 |
| Motor torque multiplier, equation (12) | 12.294 | 6.462 | 4.257 | 3.105 | 2.61 |
| RPM multiplier, equation (13) | 165.56 | 87.02 | 57.33 | 41.81 | 35.15 |

| Motor Current | Motor Torque | Motor RPM | 1st Wheel Torque | 1st Vehicle Speed | 2nd Wheel Torque | 2nd Vehicle Speed | 3rd Wheel Torque | 3rd Vehicle Speed | 4th Wheel Torque | 4th Vehicle Speed | 5th Wheel Torque | 5th Vehicle Speed |
|---|---|---|---|---|---|---|---|---|---|---|---|---|
| 100 | 10 | 7750 | 122.94 | 46.81 | 64.62 | 89.06 | 42.57 | 135.19 | 31.05 | 185.34 | 26.10 | 220.50 |
| 125 | 15 | 6400 | 184.41 | 38.66 | 96.93 | 73.54 | 63.86 | 111.64 | 46.58 | 153.06 | 39.15 | 182.09 |
| 150 | 20 | 5000 | 245.88 | 30.20 | 129.24 | 57.46 | 85.14 | 87.22 | 62.10 | 119.58 | 52.20 | 142.26 |
| 170 | 25 | 4600 | 307.35 | 27.78 | 161.55 | 52.86 | 106.43 | 80.24 | 77.63 | 110.01 | 65.25 | 130.88 |
| 190 | 30 | 4100 | 368.82 | 24.76 | 193.86 | 47.11 | 127.71 | 71.52 | 93.15 | 98.05 | 78.30 | 116.65 |
| 210 | 35 | 3900 | 430.29 | 23.56 | 226.17 | 44.82 | 149.00 | 68.03 | 108.68 | 93.27 | 91.35 | 110.96 |
| 230 | 40 | 3700 | 491.76 | 22.35 | 258.48 | 42.52 | 170.28 | 64.54 | 124.20 | 88.49 | 104.40 | 105.27 |
| 250 | 45 | 3500 | 553.23 | 21.14 | 290.79 | 40.22 | 191.57 | 61.05 | 139.73 | 83.70 | 117.45 | 99.58 |
| 270 | 50 | 3400 | 614.70 | 20.54 | 323.10 | 39.07 | 212.85 | 59.31 | 155.25 | 81.31 | 130.50 | 96.73 |
| 290 | 55 | 3350 | 676.17 | 20.23 | 355.41 | 38.50 | 234.14 | 58.44 | 170.78 | 80.12 | 143.55 | 95.31 |
| 305 | 60 | 3250 | 737.64 | 19.63 | 387.72 | 37.35 | 255.42 | 56.69 | 186.30 | 77.73 | 156.60 | 92.47 |
| 320 | 65 | 3150 | 799.11 | 19.03 | 420.03 | 36.20 | 276.71 | 54.95 | 201.83 | 75.33 | 169.65 | 89.62 |
| 335 | 70 | 3050 | 860.58 | 18.42 | 452.34 | 35.05 | 297.99 | 53.20 | 217.35 | 72.94 | 182.70 | 86.78 |
| 355 | 75 | 3000 | 922.05 | 18.12 | 484.65 | 34.47 | 319.28 | 52.33 | 232.88 | 71.75 | 195.75 | 85.35 |
| 370 | 80 | 2950 | 983.52 | 17.82 | 516.96 | 33.90 | 340.56 | 51.46 | 248.40 | 70.55 | 208.80 | 83.93 |
| 390 | 85 | 2900 | 1045.0 | 17.52 | 549.27 | 33.33 | 361.85 | 50.59 | 263.93 | 69.35 | 221.85 | 82.51 |
| 405 | 90 | 2850 | 1106.5 | 17.21 | 581.58 | 32.75 | 383.13 | 49.71 | 279.45 | 68.16 | 234.90 | 81.09 |
| 420 | 95 | 2800 | 1167.9 | 16.91 | 613.89 | 32.18 | 404.42 | 48.84 | 294.98 | 66.96 | 247.95 | 79.66 |
| 440 | 100 | 2750 | 1229.4 | 16.61 | 646.20 | 31.60 | 425.70 | 47.97 | 310.50 | 65.77 | 261.00 | 78.24 |

Current in amps, torque in ft-lbs, vehicle speed in mph

[1]Values computed for: 1987 Ford Ranger pickup; tires = P185/75R14; revolutions/mile = 808; overall drivetrain efficiency = 0.90; dc series traction motor is Advanced DC Motors Model FB1-4001; battery pack is 120 volts; equation (12) is $T_{wheel} = T_{motor}/$(overall gear ratio x overall drivetrain efficiency); equation (13) is $Speed_{vehicle} = (RPM_{motor} \times 60)/$(overall gear ratio x revolutions/mile).

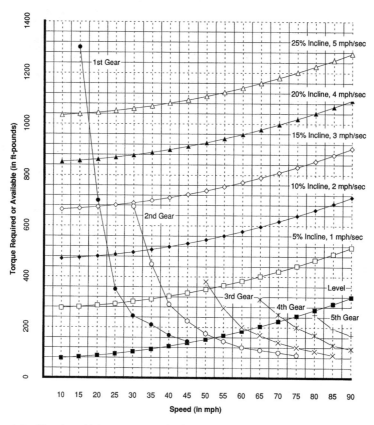

**6-9** Electric vehicle torque required vs. torque available graph.

shifted higher. But the penalty would be higher weight and increased current draw—shorter range. A smaller motor would shift the torque available curves lower while returning a small weight and current draw advantage. But beware of underpowering your vehicle. If given the choice, always go for slightly more rather than slightly less horsepower than you need. The result will almost always be higher satisfaction with your finished EV conversion.

## BUY YOUR EV CHASSIS

This is step number three. Even if you go out to buy your EV ready-made, you still want to know what kind of a job they've done, so you can decide if you're getting the best model for you. In all other cases, you'll be doing the optimizing—either by the choices you've made up front in chassis selection or by other decisions you make later on during the conversion. In this section, you'll be looking at key points that contribute to buying smart:

- Review why conversion is best—the pro side
- Why conversion might not be for you—the con side
- How to get the best deal
- Avoid off-brand, too old, abused, dirty or rusty

**158    Chassis & design**

- Keep your needs list handy
- Buy or borrow the chassis manual
- Sell unused engine parts

## Why conversion is best

In the real world—where time is money—converting an existing internal combustion engine vehicle saves money in terms of large capital investment and a large amount of labor. By starting with an existing late-model vehicle, the EV converter's bonus is a structure that comes complete with body, chassis, suspension, steering and braking systems—all designed, developed, tested, and safety-proven to work together. Provided the converted electric vehicle does not greatly exceed the original vehicle's GVWR overall weight or GAWR weight per axle specifications, all systems will continue to deliver their previous performance, stability and handling characteristics. And the EV converter inherits another body bonus—its bumpers, lights, safety-glazed windows, etc. are already preapproved and tested to meet all safety requirements.

There's still another benefit—you save more money. Automobile junkyards make money by buying the whole car (truck, van, etc.) and selling off its pieces for more than they paid for the car. When you build—rather than convert—an EV, you are on the other side of the fence. Unless you bought a complete kit, building from scratch means buying chassis tubing, angle braces and sheet stock plus axles/suspension, brakes, steering, bearings/wheels/tires, body/trim/paint, windshield/glass/wipers, lights/electrical, gauges, instruments, dashboard/interior trim/upholstery, etc., parts that are bound to cost you more ala carte than buying them already manufactured and installed in a completed vehicle.

## The other side of conversion

What's the downside? It's likely that any conversion vehicle you choose will not be streamlined like a soapbox derby racer. It will be a lot heavier than you'd like it, and have tires designed for traction rather than low rolling resistance. You do the best you can in these departments depending on your end-use goals: EV dragster, commuter, or highway flyer.

It's equally likely your conversion vehicle comes with a lot of parts you no longer need: internal combustion engine and mounts, and its fuel, exhaust, emission control, ignition, starter, and cooling/heating systems. These you remove and, if possible, sell.

Then you have additional conversion vehicle components that you might wish to change or upgrade for better performance, such as drivetrain, wheels/tires, brakes, steering, and battery/low voltage accessory electrical system. On these just do what makes sense.

## How to get the best deal

There are a number of trends that help you to get into your EV conversion at the best possible price, but you still have to do the shopping. Shopping for your EV chassis is no different from buying any vehicle in general—put your boots on. Grab a good book on buying new or used cars to help you.[5] Just remember not to divulge your true intentions while bargaining, so you can entertain scenarios like this: When the salesperson says, "Well, to be honest, only two of its four cylinders are working," you say, "No problem. How much will you knock off the price?" Or the ideal situation:

"Frankly, that's the cleanest model on the lot but its engine doesn't work," and you say, "No problem. Let me take it off your hands for $100." Your best deal will be when finding exactly the nonworking "lemon" that someone wants to sell—specifically, a lemon in the engine department. If you find a $5000 vehicle that doesn't run because of an engine problem (but everything else works okay), you can pick it up for a fraction of its value and save nearly the entire cost of conversion. Seek these deals out; they can save you a substantial amount of money.

Specific vehicle characteristics aside, what's the best-vintage vehicle for conversion in terms of cost? As a rule of thumb, used is better than new, but not too used. While older is better in terms of lower cost, you lose a lot of the more-recent vehicle safety features if you go too far back. If you go back farther than 10 to 15 years, you begin getting into body deterioration and mechanical high mileage problem territory. And if you go back to the classics—the 1960s and earlier-vintage models—the price starts going up again. But several important trends are working in the EV converter's benefit.

*Late-model used vehicles (late 1980s onward)*. These late-model vehicles make ideal EV conversions. Not only are they available at a lower price than new vehicles with all the depreciation worked out, but vehicles with 20,000 to 50,000 miles on the odometer are better for EV converters—because vehicle drivetrains, brakes and wheels/tires generate less friction (burrs and ridges are worn down, shoes no longer drag and seals are seated, etc.) and roll/turn more easily.

*Stripped-down, late-model used vehicles*. These are an even better deal. Almost everyone wants the deluxe, V-8, automatic transmission, power-everything model. You, on the other hand, are interested in a straight stick, 4-cylinder, no-frills model that nobody wants. The salesperson will fall all over himself/herself trying to help you—try to keep composed. And don't say what it's for until you've finished converting.

*Lightweight, early 1980s, 4-cylinder cars/trucks*. These represent problem sales for used-auto dealers and are frequently discounted just to move them. Surprise the salesperson who sold you that 4-cylinder lemon—visit next month in your fire-breathing 120-volt EV conversion.

*Older lightweight diesel or rotary cars/trucks*. These also represent problem sales for used-auto dealers because potential buyers are unsure of engine repairability and parts. With current owners, its more likely they have just gone out of favor. In either case, these represent a buying opportunity for you.

*Cars/trucks with blown engines or no engines at all*. These are a real problem to move for any owner but a gift for you. It's a marriage made in heaven and you can usually call the terms. Scan the newspapers for these deals.


## Avoid the real junk

While late-model, nonrunning bargains are great, avoid the problem situations. Avoid buying off-brand, too old, abused, dirty, or rusty chassis. The parts and labor that you have to add to them to bring them up to the level of normal used models makes them no bargain. The salesperson might offer you that used 1957 Mingus for only $50 but where do you get the parts for it?

Dirt and rust are okay in moderation but too much of anything is too much. If you can't tell what kind of engine it is because you can't see it in the engine compartment, pass on this choice. You don't want to spend as much cleaning as you do converting—or pay someone else to do it—only to find that essential parts you thought were present under the dirt are either in poor condition or not present at all.

Rust is rust, and what you can't see is the worst of it. How can you put your best

EV foot forward in a rust bucket for a chassis? Don't do it; pass on this choice in favor of a rust-free vehicle or one with minimal rust.

## Keep your needs list handy

Regardless of which vehicle you choose for conversion, you want to feel good about your ability to convert it before you leave the lot. If it's too small and/or cramped to fit all the electrical parts—let alone the batteries—you know you have a problem. Or if it's very dirty, greasy, or rusty, you might want to think twice. Here's a short checklist to keep in mind when buying:

*Weight.* With 120 volts and a 22 HP series dc motor, 4000 to 5000 pounds is about the upper limit. On the other hand, the same components will give you blistering performance and substantially more range in a 2 000- to 3000-pound vehicle. Weight is everything in EVs—decide carefully.

*Aerodynamic drag.* You can tweak the nose and tail of your vehicle to produce less drag and/or turbulence, but what you see before you buy it is basically what you've got. Choose wisely and aerodynamically.

*Rolling resistance.* Special EV tires are still expensive, so look for a nice set of used radials and pump them up hard.

*Drivetrain.* You don't want an automatic; a 4- or 5-speed manual will do nicely, and front wheel drive typically gives you more room for mounting batteries. Avoid 8- and 6-cylinders in favor of 4-cylinders, and choose the smallest, lightest engine/drivetrain combinations. Avoid heavy duty anything or 4-wheel drive.

*Electrical system.* Pass on air conditioning, electric windows, and any power accessories.

*Size.* Will everything you want to put in (batteries, motor, controller, and charger) have room to fit? How easy will it be to do the wiring?

*Age & condition.* These determine whether you can get parts for it, and how easy it is to restore it to a condition fit to serve as your car.

## Buy or borrow the manuals

Manuals are invaluable. If possible, seek them out to read about any hidden problems before you buy the vehicle. After you own it, don't spend hours finding out if the red-striped or the green-striped wire goes to dashboard terminal block number 3; just flip to the appropriate schematic in the manual and locate it in minutes. Component disassembly is easy when you learn that you must always disengage bolt number 2 in a clockwise direction before turning bolt number 1 in a counterclockwise direction, etc. Believe me, these are labor savers.

## Sell unused engine parts

Somebody somewhere wants that 4-cylinder engine you are removing from your EV chassis. Emblazon this above your workbench area before the removed parts begin to accumulate dust or crowd out all other items from your garage. First make a few phone calls to place want ads; then call dealers and junk yards. If no cash consideration is offered, see if you can trade the parts for something of value. Do all of this early, before you are in the middle of your conversion. Nothing worse than stubbing your toe on an engine block while nonchalantly going for your voltmeter.

# 7
# Electric motors

*The superior ac system will replace the entrenched but inferior dc one.*
George Westinghouse (from *Tesla: Man Out of Time*)

The heart of every electric vehicle is its electric motor. Electric motors come in all sizes, shapes, and types and are the most efficient mechanical devices on the planet. Unlike an internal combustion engine, an electric motor emits zero pollutants. With only one moving part, an electric motor will outlive an internal combustion engine several times over. Widespread adoption of EVs is a planet-saving proposition.

What this means to you is ownership of a fun-to-drive, high-performance EV that will deliver years of low-maintenance, hassle-free driving at minimum cost, mostly because of the inherent characteristics of its electric motor—power and economy. Depending on your design and component choices, the electric motor in your EV conversion can smoke its tires and routinely offer 60-mile range on a dollar's worth of electricity.

The objective of this chapter is to guide you towards the best candidate motor for your EV conversion today, and suggest the best electric motor type for your future EV conversion. To accomplish these goals this chapter will: review electric motor basics and provide you with useful equations; introduce you to the different types of electric motors and their advantages and disadvantages for EVs; introduce the best electric motor for your EV conversion today and its characteristics; and introduce you to the electric motor type that you should closely follow and investigate for future EV conversions.

## WHY AN ELECTRIC MOTOR?

In chapter 1 you learned that the electric motor is ubiquitous because of its simplicity. All motors by definition have a fixed *stator* or stationary part, and a *rotor* or moveable part. While these can be interchanged in some actual designs, all electric motors have only one moving part. This simplicity is the secret of their dependability, and why—in direct contrast to the internal combustion engine with its hundreds of moving parts—electric motors are a far superior source of propulsion.

Electric motors are inherently powerful. By selecting a design that delivers peak torque at or near stall, you can literally move a mountain. That's why electric traction motors have powered our trolley cars, subways, and diesel-electric railroad locomotives for so many years. There is no waiting, as with an internal combustion engine, while it winds up to its peak torque RPM range. Apply electric current to it and you've instantly got torque to spare. If any EV's performance is wimpy, it's due to a poor design or electric motor selection—not the electric motor itself.

Electric motors are inherently efficient. You can expect to get 90 percent or more of the electrical energy you put into an electric motor out of it in the form of mechanical torque. Few other mechanical devices even come close to this efficiency.

Electric motors are inherently flexible. That translates directly to convenience. They can be designed to accommodate any job you need done, can be powered from dc or ac, can be large or small in any shape, weight, or configuration. With only one moving part, an electric motor that is correctly designed, manufactured, and used is virtually impervious to failure and indestructible in use.

## DC ELECTRIC MOTORS

An electric motor is a mechanical device that converts electrical energy into motion, and that can be further adapted to do useful work such as pulling, pushing, lifting, stirring, or oscillating. It uses the fundamental properties of magnetism and electricity. Before looking at dc motors and their properties, let's review some fundamentals.

### Magnetism & electricity

Magnetism and electricity are opposite sides of the same coin. Electrical and electronics design engineers who regularly utilize Maxwell's four laws of electromagnetism—based on Faraday's and Ampere's earlier discoveries—in their daily work might tell you, "magnetism and electricity are inextricably intertwined in nature." In simple fact, you don't have one without the other. But usually you only look at one or the other—unless you are discussing electric motors or other devices that involve both.

If you set a gallon jug of water on the edge of your sink and poke a small pencil-point-sized hole in the bottom, water squirts out into the sink. When the jug is filled to the top with water, it squirts out all the way across the sink. When only a little water remains in the jug, it squirts out fairly close to the jug. If you refill the jug with water and enlarge the hole, water flows out of it faster, i.e., there is less resistance to the flow.

You hook up a light bulb to a battery by completing the wire connections from the battery's positive and negative terminals to the light bulb, and the bulb lights. Hook up two batteries in series to double the voltage, and the bulb shines even brighter. The water's force or potential—the height of water in the jug—corresponds to the battery's force or potential or *voltage*; the size of the hole in the bottom of the jug corresponds to the resistance to the flow or the *resistance* of the light bulb in this case; and the flow of water from the jug corresponds to the flow of electricity or *current*. The electrical equation, commonly known as Ohm's law, is:

$$(1) \quad V = IR$$

where V is voltage in volts, I is current in amps, and R is resistance in ohms. When you double the voltage, you send twice as much current through the wire and the light bulb becomes brighter.

The simple bar magnet, which you probably encountered in your school science class, has two ends or poles—north and south. Either end attracts magnetizable objects to it, usually objects containing iron in some form, such as iron filings, paper clips, etc. When two bar magnets are used together, opposite poles (north-south) attract one another and identical poles (north-north or south-south) repel one another.

The compass needle is a magnetized object with its own north and south pole. Lightweight and delicately balanced, it aligns itself with the earth's magnetic field, and tells you which direction is north. But bring the bar-style magnet near it and it will point toward the magnet instead, demonstrating an important fact: Magnetism varies with distance. By moving one end of the magnet around the compass, you move the compass needle along with it. And you can move the needle either by attraction (opposite poles) or repulsion (identical poles). The compass demonstrates the way a motor works: through an attractive or repulsive force created by a magnetic field.

You can create a magnetic field with electricity. Take an iron nail from your toolbox, wrap a few turns of insulated copper wire around it, and hook the ends up to a battery. The plain nail is transformed into a bar-style magnet that can perform the identical feat with the compass.

If you were to write a magnetic equation, it would look like:

$$\text{mmf} = \phi R = \text{flux} \times \text{reluctance}$$

In this equation, commonly known as Rowland's law, mmf is the magneto motive force, $\phi$ is the flux (the flow of magnetism), and $R$ is reluctance, the opposition to the flow. Because designers of electrical transformers and motors work in terms of magnetic field strength H, flux density B, and permeability (u) of a given transmission medium, and because

$$\text{mmf} = H \times l; \ \phi = B \times A; \text{ and } R = l/(u \times A)$$

(where l is the length and A is the area), Rowland's law usually appears in its more common form (for a linear range) as

$$(2) \ B = Hu$$

In this equation, B is the flux density or induction force in lines per square inch, H is the magnetic field strength or flow of magnetism in the material in ampere-turns per inch, and u is the permeability or resistance of the material to magnetizing force in henrys per inch.

## Conductors & magnetic fields

If you had a horseshoe-shaped magnet with the ends close together and you moved a wire through its pole area at a right angle to the flux as shown in FIG. 7-1, the equation describing it would be

$$(3) \ V = Blv \times 10^{-8}$$

In this equation, a simplified form of Faraday's law, V is the induced voltage in volts, B is the flux density in lines per square inch, l is the length of that part of the conductor actually cutting the flux, and v is the velocity in inches per second. For example, if you moved a 1-inch wire or conductor at a right angle to a field of 50,000 lines per square inch at 50 feet per second,

$$V = 50000 \times 1 \times 50 \times 12 \times 10^{-8} = 0.3 \text{ volts}$$

This relationship holds true whether the field is stationary and the wire is moving or vice versa. The faster you cut the lines of flux, the greater the voltage, but you must do so at right angles (at any angle $\phi$ other than the 90-degree right angle the equation becomes $V = Blv \times \sin \phi \times 10^{-8}$). What you have just demonstrated is Faraday's law or the *generator rule*: Motion of a conductor at a right angle through a magnetic field induces a voltage across the conductor. If the circuit is closed, the induced voltage will cause a current to flow. A handy way to remember the relationships is the right-hand rule: the thumb of your right hand points upward in the direction of the motion of the

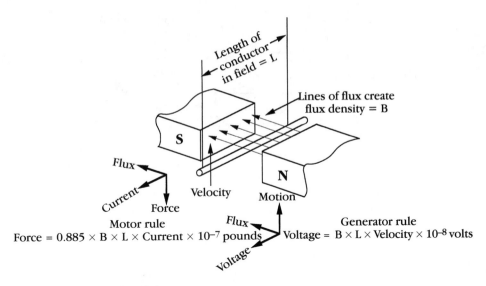

Length of conductor in field = L

Lines of flux create flux density = B

Flux

Current

Force

Velocity

Motion

S

N

**Motor rule**

Force = $0.885 \times B \times L \times Current \times 10^{-7}$ pounds

Flux

Voltage

**Generator rule**

Voltage = $B \times L \times Velocity \times 10^{-8}$ volts

**7-1** dc motor basics—the motor and generator rules.

conductor, your index finger extends at right angles to it in the direction of the flux (from north to south pole), and your third finger extends in a direction at right angles to the other two, indicating the polarity of the induced voltage or the direction in which the current will flow as shown in FIG. 7-1.

The flip-side or corollary to the generator rule is the *motor rule* whose equation is

$$(3)\ V = 0.885 \times Bl \times I \times 10^{-7}$$

where F is the force generated on the conductor in pounds, B is the flux density in lines per square inch, l is the length of that part of the conductor actually cutting the flux, and I is the current in amperes. For example, if you had a 1-inch wire at right angles to a field of 50,000 lines per square inch with a current of 100 amperes flowing through it:

$$V = 0.885 \times 50000 \times 1 \times 100 \times 10^{-7} = 0.44\ \text{pounds}$$

What we have just demonstrated here is Ampere's law or the motor rule: Current flowing through a conductor at a right angle to a magnetic field produces a force upon the conductor. The right-hand rule is again a handy way to remember the relationships, but this time the thumb of your right hand points toward you in the direction of the current through the conductor, your extended index finger at right angles to it points in the direction of the flux (from north to south pole), and your extended third finger points downwards in a direction at right angles to the other two, indicating the direction of the generated force as shown in FIG. 7-1. Now you're ready to talk seriously about motors.

## dc motors in general

If you could support the conductor shown in FIG. 7-1 so that it could rotate in the magnetic field, you would create the condition shown in the upper part of FIG. 7-2. Now

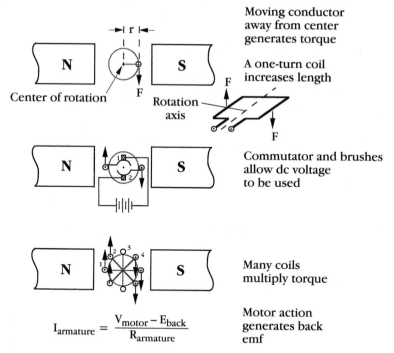

**7-2** dc motor basics—obtaining torque from moving conductors.

the current through the conductor (which flows into the page—you're seeing the back of the arrow) would exert a force that would tend to rotate it in the clockwise direction. The magnitude of the torque would be given by

$$(4)\ T = Fr$$

where T is torque in foot-pounds, F is force in pounds and r is the distance measured perpendicularly from the direction of F to the center of rotation in feet. Now you have a motor design—on paper. You need to take some steps to make it real.

First you need to make it more "force-full." Since the force varies with the length of the conductor, if you make a coil of wire, as shown in FIG. 7-2 upper right, twice as much length is cutting the lines of flux. The force generated on the right-hand wire is downward and the force generated on the left-hand wire is upward; they would assist one another in producing rotation and result in twice the torque.

To further assist rotation, you add the commutator and brushes, shown in the middle of FIG. 7-2. This arrangement allows you to power your motor from a constant supply of direct current (dc) voltage. Switching the polarity of the coils when they reach the 12 o'clock or 6 o'clock position (minimum flux point) guarantees that current will always flow in through brush number 2 and out through brush number 1, thus always producing upward force on the left-hand conductor and downward force on the right-hand conductor, and creating a constant rotation.

To further increase the motor's torque abilities, you add additional coils as shown at the bottom of FIG. 7-2. In reality, each coil can have many windings, and you can arrange commutator segments to match the number of coils so that you have the force on each of these coils acting in unison with the force on all the other coils. Notice that the force on the conductor in position 1 is the same as the force on the conductor in position 2. But the torque is less in position 2 because r, the distance measured perpendicularly from the direction of F to the center of rotation, is less. In position 3, r = 0, so there is no torque generated at all. But if the inertia of a moving coil allows it to move to position 4, a torque is again generated. Now if you substitute equation (3) (the motor equation) into equation (4) (the torque equation), you get

$$T = V = 0.885 \times Bl \times I \times 10^{-7} \times r$$

Because the flux $\phi$ can be expressed as BA, where B is the flux density and A the area it is acting over, and the dimensions, parallel paths, and number of conductors (or number of poles, magnetic material, dimensions of the airgap, etc.) are constant for any given motor, this equation can be rewritten as

$$(5) \ Torque = K \times \phi \times I_a$$

where K is a constant incorporating all the motor fixed characteristics, $\phi$ is the flux in lines, and $I_a$ is the current flowing through the conductors in the armature of the motor in amperes or amps. This is a handy equation, because if you know the value of torque at one value of current (e.g., 10 foot-pounds), and you want to find it when the current level is twice this value, just substitute

$$T_2 \times (K \times \phi \times 2I_1)/(K \times \phi \times I_1) = 20 \times 2 = 20 \ foot\text{-}pounds$$

Now recall that, while a voltage would be induced in the conductor when the conductor cut the lines of flux, the resulting current that flows in the conductor itself generates a force that tends to oppose the applied voltage. So using equation (1) the real armature current can be described as

$$(6) \ I_a = (V_t - E)/R_a$$

where $V_t$ is the motor terminal voltage, E is the counter electro-motive-force (emf) voltage, and $R_a$ is the motor armature resistance. Rewriting this in its alternate form gives you

$$(7) \ V_t = E + I_a R_a$$

Another useful equation involves the speed of the motor. With a constant voltage applied to the motor's terminals, the speed of the motor depends on the flux:

$$E = K \times \phi \times S$$

This can be rewritten as

$$(8) \ S = (V_t - I_a R_a)/K\phi$$

where S is the motor speed in revolutions per minute (you've seen all the other components before).

### dc motors in the real world

Now it's time you met real world dc motors—their construction, definitions, and efficiency. Let's start by looking at their components.

**Armature**  The armature is the main current-carrying part of a motor that normally rotates (brushless motors tend to blur this distinction) and produces torque via the action of current flow in its coils. It also holds the coils in place, and provides a low reluctance path to the flux. (Reluctance is defined as $(H \times l)/\phi$ and measured in ampere-turns per lines of flux.) The armature usually consists of a shaft surrounded by laminated sheet steel pieces called the *armature core*. The laminations reduce eddy current losses; steel is replaced by more efficient metals in newer designs. There are grooves or slots parallel to the shaft around the outside of the core; the sides of the coils are placed into these slots. The coils (each with many turns of wire) are placed so that one side is under the north pole and the other is under the south pole; adjacent coils are placed in adjacent slots, as shown at the bottom of FIG. 7-2. The end of one coil is connected to the beginning of the next coil so that the total force then becomes the sum of the forces generated on each coil.

**Commutator** The smart part of the motor that permits constant rotation by reversing the direction of current in the windings each time they reach the minimum flux point. Physically, it's a part of the armature (typically located near one end of the shaft) that appears as a ring split into segments surrounding the shaft. These segments are insulated from one another and the shaft.

**Field poles** In the real world, electromagnets (recall your toolbox nail with a few turns of insulated copper wire wrapped around it) are customarily used instead of the permanent magnets you saw in FIG. 7-1 and FIG. 7-2. (Permanent magnet motors are, in fact, used, and you'll be formally introduced to them and their advantages later in this section.) In a real motor the lines of flux are produced by an electromagnet created by winding turns of wire around its poles or pole pieces. A pole is normally built up of laminated sheet steel pieces, which reduce eddy current losses; as with armatures, steel has been replaced by more efficient metals in the newer models. The pole pieces are usually curved where they surround the armature to produce a more uniform magnetic field. The turns of copper wire around the poles are called the field windings. How these windings are made and connected determines the motor type. A coil of a few turns of heavy wire connected in series with the armature is called a *series motor*. A coil of many turns of fine wire connected in parallel with the armature is called a *shunt motor*.

**Brushes** Typically consisting of rectangular-shaped pieces of graphite or carbon, brushes are held in place by springs whose tension can be adjusted. The brush holder is an insulated material that electrically isolates the brush itself from the motor frame. A small flexible copper wire embedded in the brush (called a *pigtail*) provides current to the brush. Smaller brushes can be connected together internally to support greater current flows.

**Motor case, frame, or yoke** Whatever you wish to call it, the function of this part is not only to provide support for the mechanical elements, but also to provide a magnetic path for the lines of flux to complete their circuit—just like the lines of flux around a bar magnet. In the motor's case, the magnetic path goes from the north pole, through the air gap, the magnetic material of the armature, and the second air gap, to the south pole and back to the north pole again via the case, frame, or yoke.

Motors operating in the real world are subject to losses from three sources:

*Mechanical.* All torque available inside the motor is not available outside because torque is consumed in overcoming friction of the bearings, moving air inside the motor (known as windage), and because of brush drag.

*Electrical.* Power is consumed as current flows through the resistance of the armature, field windings, and brushes.

*Magnetic.* Additional losses are caused by eddy current and hysteresis losses in the armature and the field pole cores.

In equation form the total internal motor losses could be expressed as

$$P_m = P_t - P_f - P_a - \text{stray power losses}$$

where $P_m$ is the power output of the motor, $P_t$ is the power input to the motor (the power delivered to its terminals), $P_f$ is the power or copper loss of its field windings, $P_a$ is the power or copper loss of its armature, and stray power losses are the other mechanical and magnetic losses. This is marginally useful because it only covers internal power (not power available at the output of the motor) because of losses due to the rotation of the armature. What you really want is efficiency.

To determine the efficiency of a motor, you measure its output mechanical torque relative to the input electrical power you are providing to it. Let's back up a few steps. In electrical terms, power is defined as

$$(9) \quad \text{Power} = \text{Voltage} \times \text{Current or } P = V \times I$$

Substituting equation (1) into (9):

$$(10) \quad P = (IR) \times I = I^2R$$

This equation gives the power lost in an electrical component in terms of the resistance and current flowing through that component, where power is measured in watts, current in amps, and resistance in ohms.

Returning to efficiency, it could be expressed as

$$\text{Efficiency} = \text{Torque}_{out}/\text{Torque}_{in}$$

Using equation (5) you could express this as

$$\text{Efficiency} = \text{Torque}_{out}/K \times \phi \times I_{in}$$

Using equation (8) you could rewrite this as

$$\text{Efficiency} = (\text{Torque}_{out} \times \text{RPM})/((V_t - I_tR_t)/ \times I_t)$$

From equation (6) in chapter 6

$$\text{Horsepower} = (\text{Torque} \times \text{RPM})/5252 \text{ or } (2\pi \times \text{Torque} \times \text{RPM})/33000$$

Your equation can now be rewritten in horsepower terms if you multiply by the conversion factor (equation (8) in chapter 6) of 745.7 watts per HP to put the denominator into the right horsepower units. Now you have

$$\text{Efficiency} = ((2\pi \times 745.7)/33000) \times \text{Torque}_{out} \times \text{RPM})/((V_t - I_tR_t)/ \times I_t)$$

This can simply be rewritten as

$$(11) \quad \text{Efficiency} = ((0.142) \times \text{Torque}_{out} \times \text{RPM})/(V_t I_t - R_t \times (I_t)^2)$$

where $\text{Torque}_{out}$ is the torque from the motor in foot-pounds, V is the open circuit voltage impressed across the motor's terminals (before you impose a load), $I_t$ is the total current flowing into the motor's terminals, and $R_t$ is the total of all the resistance in the circuit (battery plus circuit external to motor plus field plus armature). The ratio you get from equation (11) can simply be converted to efficiency in percent. Notice the numerator of this equation is horsepower expressed in torque; also notice that the denominator of the equation gives you the input power minus the $I^2R$ losses of the motor and circuit power—both factors converted to horsepower units via the conversion factor.

## dc motor types

Now that you've been introduced to dc motors in theory and in the real world, it's time to compare the different motor types. dc motors appear in the following forms:

- Series
- Shunt
- Compound
- Permanent magnet
- Brushless
- Universal

The last three motor types are just variations of the first three, fabricated with different construction techniques. The compound motor is a combination of the series and shunt motors. For the first three motor types we'll look at the circuit showing how the motor circuit field windings are connected, and then look at the characteristics of the torque and speed versus armature current and shaft horsepower curves that describe their operation (FIG. 7-3).

Each of the motor types will be examined for its torque, speed, reversal, and regenerative braking capabilities—the factors important to EV users. The motor types will all be compared at full load shaft horsepower—the only way to compare different motor types of equal rating. Efficiency is a little harder to determine because—as equation (11) showed—it also depends on the external resistance of the circuit to which the motor is connected. So efficiency has to be calculated for each individual case.

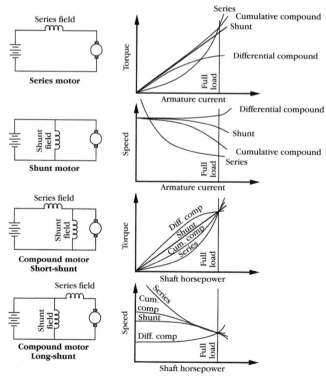

**7-3** Summary of dc motor types—windings and characteristics.

## Series dc motors

The most well-known of the dc motors, and the one which comes to mind for traction applications (like propelling EVs) is the series dc motor. It's so named because its field winding is connected in series with the armature (FIG. 7-3). Because the same current must flow through this field winding as through the motor armature itself, it is wound with a few turns of heavy gauge wire.

**Torque** As equation (5) showed, Torque = $K \times \phi \times I_a$, and the current in the series field is $I_a$. $I_a$ can be substituted for $\phi$ and the series motor torque equation can be rewritten as,

$$(12)\ \text{Torque} = K' \times I_a^2$$

This shows that, in a series motor, torque varies with the square of the current—a fact substantiated by FIG. 7-3's actual torque versus armature current graph. The graph is actually a little misleading because at startup there is no counter emf to impede the flow of current in the armature of a series motor, and startup torque can be enormous; with no current limiting, you can start up at torque values far to the right of the full load line for a series motor. In actual use, the armature reaction and magnetic saturation of the series motor at high currents sets upper limits on both torque and current, although you might prefer to limit your circuit and components to far lower values. High starting torque makes series motors highly desirable for traction applications.

**Speed** As equation (8) showed,

$$\text{Speed} = (V_t - I_a R_a)/K\phi$$

Once again, $I_a$ can be substituted for $\phi$ and the series motor speed equation can be rewritten as

$$(13)\ \text{Speed} = (V_t - I_a(R_a + R_f))/KI_a$$

where $R_a$ and $R_f$ are the resistances of the armature and field, respectively. This shows that speed becomes very large as the current becomes small in a series motor—a fact substantiated by FIG. 7-3's actual speed versus armature current graph. High RPMs at no load is the series motor's Achilles heel. You need to make sure you are always in gear, have the clutch in, have a load attached, etc., because the series motor's tendency is to run away at no load. Just be aware of this and back off immediately if you hear a series motor rev up too fast.

**Field weakening** This technique is an interesting way to control series motor speed. You place an external resistor in parallel with the series motor field winding, in effect diverting part of its current through the resistor. Keep it to 50 percent or less of the total current (resistor values equal to or greater than 1.5 times the motor's field resistance). The byproduct is a speed increase of 20 to 25 percent at moderate torques without any unstability in operation (hunting in RPM, etc.). Used in moderation, it's like getting something for nothing.

**Reversing** The same current that flows through the armature flows through the field in a series motor, so reversing the applied voltage polarity does not reverse the motor direction. To reverse motor direction you have to reverse or transpose the direction of the field winding with respect to the armature. This characteristic also makes it possible to run series dc motors from ac (more on this later in the section).

**Regenerative braking** All motors simultaneously exhibit generator action—motors generate counter emf—as you read earlier in this section. The reverse also holds true—generators produce counter torque. Regenerative braking allows you to slow down the speed of your EV (and save its brakes) and put power back into its battery (and extend its range) by harnessing its motor to work as a generator after it is up and running at speed. While all motors can be used as generators—the series motor has rarely been used as a generator in practice because of its unique and relatively unstable generator properties. A series generator's torque versus armature current graph is the flip side of a series motor's—with a load attached, any current flowing through the armature creates increased flux buildup in the series field, which in turn creates higher induced voltage that further increases current, etc., until magnetic saturation is reached. In the saturation region above full load current, the series generator looks likes a constant-current source—a nearly straight line returning back down to zero torque. A series motor's generator characteristics made manual regenerative braking control difficult; many workable solutions were found, but they operated in a "Rube Goldberg" style (a lot of switches, relays, contactors, etc.) because manual regeneration had to be applied intermittently to overcome the instability factor. The arrival of modern, solid-state chopper control electronic circuits made series motor regenerative braking relatively straightforward to accomplish—although still not as simple or effective as using other motor types.

## Shunt dc motors

The second most well-known of the dc motors is the shunt dc motor, so named because its field winding is connected in parallel with the armature (FIG. 7-3). Because it doesn't have to handle the high motor armature currents, a shunt motor field coil is typically wound with many turns of fine gauge wire and has a much higher resistance than the armature.

**Torque** Because the shunt field is connected directly across the voltage source, the flux in the shunt motor remains relatively constant. Its torque is directly dependent upon the armature current, as described by torque equation (5):

$$(14) \text{ Torque} = K \times \phi \times I_a$$

This shows that, in a shunt motor, torque varies directly with the current, and the straight-line relationship shown in FIG. 7-3 results. Although there is initially no counter emf to impede the flow of startup current in the armature of a shunt motor, the shunt motor's linear relationship is quickly established; as a result, the shunt motor does not produce nearly as much startup torque as the series motor. This translates to reduced acceleration performance for owners of shunt motor-powered EVs.

**Speed** When a load is applied to a shunt motor, it will tend to slow down and, in turn, reduce the counter emf produced in accordance with the earlier-introduced equation

$$E = K \times \phi \times S$$

This produces a corresponding rise in the shunt motor's armature current governed by equation (6):

$$I_a = (V_t - E)/R_a$$

The increase in armature current results in an increase in torque to take care of the added load. So the shunt motor might be viewed as a constant-speed motor over a wide range of armature current (as FIG. 7-3 attests), and its speed equation can be written

$$(15) \text{ Speed} = (V_t - I_aR_a)/K\phi \text{ or } K'E/\phi_f$$

where $\phi_f$ is the shunt field flux. This equation shows that as $\phi_f$ is constant, counter emf E tends to remain constant over a wide range of armature current values, and produces a fixed speed curve that only droops slightly at high armature currents. The shunt motor's linear torque and fixed speed versus armature current characteristics have two undesirable side effects for traction applications when controlled manually. First, when a heavy load (hill climbing, extended acceleration) is applied, a shunt motor does not slow down appreciably as does a series motor, and the excessive current drawn through its armature by the continuous high torque requirement makes it more susceptible to damage by overheating. Next, in contrast to the "knee" bend in the series motor torque-speed equations (FIG. 6-9), shunt motor torque-speed curves are nearly straight lines. This means more speed control or shifting is necessary to achieve any given operating point. Series motors are therefore used where there is a wide variation in both torque and speed and/or heavy starting loads. But the arrival of modern, solid-state electronics has made possible controllers that compensate for the shunt motor's deficiencies; other than having much lower startup torque, shunt motors can perform as well as series motors in EVs when electronically controlled.

**Field weakening**  You can also achieve a higher-than-rated shunt motor speed by reducing the shunt coil current—in this case you place an external control resistance in series with the shunt motor field winding. But here, unlike the series motor's runaway RPM region at no load, you are playing with fire, because the loaded shunt motor armature has an inertia that does not permit it to respond instantly to field control changes. If you do this while your shunt motor is accelerating, you might cook your motor or have motor parts all over the highway by the time you adjust the resistance back down to where you started. Be careful with field weakening in shunt motors.

**Reversing**  Current flows through the field in a shunt motor in the same direction as it flows through the armature, so reversing the applied voltage polarity reverses both the direction of the current in the armature and the direction of the field-generated flux, and does not reverse the motor rotation direction. To reverse motor direction you have to reverse or transpose the direction of the shunt field winding with respect to the armature.

**Regenerative braking**  A shunt motor is instantly adaptable as a shunt generator. Most generators are in fact shunt wound, or variations on this theme. The linear or nearly linear torque and speed versus current characteristics of the shunt motor manifest as nearly linear voltage versus current characteristics when used as a generator. This also translates to a high degree of stability that makes a shunt motor both useful for and adaptable to regenerative braking applications—either manually or electronically controlled.

## Compound dc motors

A compound dc motor is a combination of the series and shunt dc motors. The way its windings are connected, and whether they are connected to boost (assist) or buck (oppose) one another in action determines its type. Its basic characterization comes from whether current flowing into the motor first encounters a series field coil—*short-shunt compound motor*—or a parallel shunt field coil—*long-shunt compound motor*—as shown in FIG. 7-3. If, in either one of these configurations, the coil windings are hooked up to oppose one another in action, you have a *differential compound motor*. If the coil windings are hooked up to assist one another in action, you have a *cumulative compound motor*. The beauty of the compound motor is its ability to bring the best of both the series and the shunt dc motors to the user.

**Torque**  The torque in a compound motor has to reflect the actions of both the series and the shunt field coils—so it will look similar to equation (14) but reflect dual variables:

$$(16)\ \text{Torque} = K \times (\phi_f + \phi_s) \times I_a$$

or

$$\text{Torque} = K \times (\phi_f - \phi_s) \times I_a$$

Depending on whether you are hooked up in the *differential* or *cumulative* position, the shunt torque $\phi_f$ and the series torque $\phi_s$ either subtract to a difference figure or add together. The effect of these hookup arrangements on torque is illustrated in FIG. 7-3, where the differential compound motor builds more slowly to a lower torque value than the shunt curve and the cumulative compound motor builds to a slightly higher torque value than the shunt curve at a slightly higher rate.

**Speed**  Similar to torque, the speed action in a compound motor will also reflect the dual variables of series and shunt field coil action:

$$(17)\ \text{Speed} = (V_t - I_a(R_f + R_s))/K(\phi_f + \phi_s) = K'E/(\phi_f + \phi_s) \text{ or } K'E/(\phi_f - \phi_s)$$

In this equation, the speed of the compound motor increases or decreases versus a shunt motor depending on whether the shunt and series windings are hooked up in the cumulative or differential configuration; FIG. 7-3 shows the speed curves. One of the initial benefits of the compound configuration is that runaway conditions at low field current levels for the shunt motor and at lightly loaded levels for the series motor can be eliminated. While the differential compound configuration is of questionable value—its curve shows a tendency to runaway speeds at high armature current values—the cumulative compound motor appears to offer benefits to EV operation. You can tailor a cumulative compound motor to your EV needs by picking one whose series winding delivers good starting torque and whose shunt winding delivers lower current draw and regenerative braking capabilities once up to speed. When you look, you might find these characteristics already exist in an off-the-shelf model. Obviously your life is made even easier if you have modern solid-state electronics controlling your cumulative compound motor. A cumulative compound motor's downside is its increased complexity, which means it will cost more than a series or shunt motor.

**Reversing**  A short-shunt compound motor resembles a series motor, so reversing its applied voltage polarity normally does not reverse the motor direction. A long-shunt compound motor resembles a shunt motor so reversing the voltage supply leads normally reverses the motor. As with speed and torque, compound motors can be tailored to do whatever you want to do in the reversing department.

**Regenerative braking**  A compound motor is as easily adaptable as a shunt motor to regenerative braking. Its series winding gives it additional starting torque, but can be bypassed during regenerative braking, and its shunt winding allowed to give it the more desirable shunt generator characteristic. Manual or solid-state electronics controlled, compound motors are usable for and adaptable to regenerative braking applications.

## Permanent magnet dc motors

When you were first introduced to the dc motor topic, permanent magnets were used as an example because of their simplicity. Permanent magnet motors are, in fact, being increasingly used today because new technology—various alloys of Alnico magnet material, ferrite-ceramic magnets, etc.—enables them to be made smaller and lighter in weight than equivalent wound field coil motors of the same horsepower rating. While commutator and brushes are still required, you save the complexity and expense of fabricating a field winding, and gain in efficiency because no current is needed for the field.

Permanent magnet motors approximately resemble the shunt motor in their torque, speed, reversing, and regenerative braking characteristics; either motor type can usually be substituted for the other in control circuit designs. But because modern materials can support higher levels of magnetizing force—the H factor in equation (2)—the much smaller armature reaction of permanent magnet motors greatly extends the linear characteristics of conventional shunt motor speed/torque curves down to zero speed. This means that permanent magnet motors have starting torques several times that of shunt motors, and their speed-versus-load characteristics are more linear and easier to predict.

## Brushless dc motors

The next step in improving efficiency is to replace the dc motor's commutator and brushes. This thought—radical to earlier motor designers—is now a reality with today's power semiconductors that can handle high currents at high voltages and high frequencies. Common current designs build on the permanent magnet motor principles, but transpose the fixed field function and the rotating armature parts. Its rotating part (the rotor) now consists of two or more permanent magnet poles, and its fixed part (the stator) carries the main windings that you formerly found in the armature. Position sensors on the rotor shaft send signals to semiconductor switches, which control the winding currents in relation to the rotating magnetic rotor position. With no brushes to replace or commutator parts to maintain, brushless motors promise to be the most long-lived and maintenance-free of all motors. You can now custom-tailor the motor's characteristics with electronics (because electronics now represent half the motor), and the distinction between dc motor types blurs. In fact, the brushless motor more closely resembles an ac motor (which you'll meet in the next section) in construction. Assume that brushless dc motors resemble their permanent magnet dc motor cousins in characteristics—shunt motor plus high starting torque plus linear speed/torque—with the added kicker of even higher efficiency due to no commutator or brushes.

## Universal dc motors

Although any dc motor can be operated on ac, not all dc motor types run as well on ac and some might not start at all (but will run once started). If you want to run a dc motor on ac you have to design for it. A series dc motor type is usually chosen as the starting point for universal motors that are to be run on either dc or ac. dc motors designed to run on ac typically have improved lamination field and armature cores to minimize hysteresis and eddy current losses; a larger field pole area wound with fewer turns to lower the flux density and reduce the iron losses and reactive voltage drop; an increased number of conductors in the armature coil windings to compensate for the decreased flux density; and an increased number of commutator segments to reduce commutator sparking. Additional compensating or interpole windings can be added to the armature to further reduce commutation problems by reducing the flux at commutator segment transitions. In general, series dc motors operating on ac perform almost the same (high starting torque, etc.), but are less efficient at any given voltage point.

## AC ELECTRIC MOTORS

Now that you've met dc electric motors, it's time to meet the motor you encounter most often in your everyday life—the ac electric motor. The great majority of our homes, offices, and factories are fed by alternating current (ac). Because it can easily be transformed from high voltage for transmission into low voltage for use, more ac motors are in use than all the other motor types put together. Before looking at ac motors and their properties, let's look at transformers.

### Transformers

Transformers are the true unsung heros of the modern age—our entire technology depends on them: homes, businesses, factories, radios, TVs, telephones, computers, and vehicles—all have one or more transformers in them. They are even more ubiquitous than electric motors, are more efficient (typically 98 to 99 percent), and they rarely wear out because they have no moving parts. While on the one hand Michael Faraday's basic induction discoveries of the 1830s are just as valid today, the increasing subtleties of modern transformer design (even when well-armed with a microcomputer) frequently approach art rather than pure science or engineering. The importance of transformers to you is twofold: first, understanding transformer action is a prelude to understanding ac motor operation; and second, transformer size is related to frequency (increasing the frequency means a reduction in size), a fact that holds equally true for motors. This fact is or will be of vital importance to you as an EV converter—if not now, then certainly at some point in the future.

In its simplest and most familiar form, a transformer consists of two copper coils wound on a ferromagnetic core (FIG. 7-4). The primary is normally connected to a source of alternating electric current of the form

$$(18) \quad E_{instant} = E_{max}\sin \omega t = E_{max}\sin 2\pi f$$

The secondary is normally connected to the load. When a changing current is applied to the primary coil a changing magnetic field common to both coils results in the transfer of electrical energy to the second coil. An alternate form of Faraday's law shows the relationship:

$$E_{average} = N \times (\phi/t) \times 10^{-8}$$

where $\phi$ is the flux in lines and $N$ is the number of turns. When this equation is used to express sine waves of the form shown in equation (18), it is written

$$(19) \quad E_{rms} = 4.44Nf\phi_{max} \times 10^{-8}$$

or

$$E_{square} = 4Nf\phi_{max} \times 10^{-8}$$

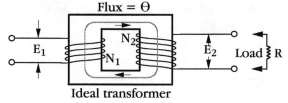

Ideal transformer

$$E_{instant} = E_{max}\sin 2\pi f$$

$$E_{average} = N\left(\frac{\Theta}{t}\right)10^{-8}$$

$$\frac{E_1}{E_2} = \frac{N_1}{N_2}$$

Power in = Power out
$$E_1I_1 = E_2I_2$$

$$E_{rms} = 4.44fnB_{max}A10^{-8} \qquad E_{sq\ wave} = 4fnB_{max}A10^{-8}$$

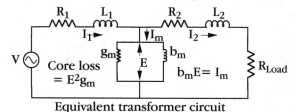

Equivalent transformer circuit

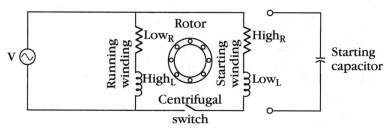

Single phase induction motor circuit

Ideal transformer turns ratio "transforms" voltage, current, or impedance

Equations prove the higher the frequency, the smaller the size and weight of a transformer or motor

Transformer model allows analysis at any frequency

Induction motor resembles transformer with rotor as load

Starting capacitor

Additional "split phase" winding neccesary to start— either inductor or capacitor

**7-4** ac motors resemble transformers.

The difference in the factors is due to the rms (root-mean-square) value of a sine wave being 1.11 times the average value, while for a square wave the maximum is the average value. Because of this relationship, in an ideal transformer with no losses you can readily establish that

$$(20)\ E_1/E_2 = N_1/N_2 = n = \text{turns ratio}$$

It also follows that

$$N_1I_1 = N_2I_2$$

or

$$(21)\ I_2/I_1 = N_1/N_2 = n = \text{turns ratio}$$

This equation says that the primary and secondary currents vary in inverse ratio to the voltage or turns. All that's left to define is power and impedance:

$$Power\ In = Power\ Out$$
$$(22)\ E_1I_1 = E_2I_2$$
$$Impedance = Z = E/I$$
$$(23)\ (E_1/I_1)/(E_2/I_2) = Z_1/Z_2 = n^2$$

This equation says that a transformer is also capable of transforming or matching impedances via its turns ratio relationship.

If the relationship used earlier in deriving equation (2), $\phi = B \times A$, is substituted into equation (19), you have

$$(24)\ E_{rms} = 4.44NAfB_{max} \times 10^{-8}$$

In this equation, you tie together the five principal parameters governing transformers: voltage E, turns N, cross-sectional area A, frequency f, and flux density B. As the flux density varies inversely with frequency (decreasing with increasing frequency), and the power lost in the core of a transformer also decreases with frequency, increasing the frequency decreases the size and weight of the transformer and motor (generator, alternator, etc.). Modern magnetic materials extend this advantage even further. In today's electronics, smaller is better because less power consumption and less weight ultimately means less cost. Why all the formulas? Easy—if 60 hertz transformers and motors are great, 400 hertz has to be even better. If you, as an EV converter or owner, can coax the same power output from a motor and controller combination that weighs half as much and occupies half the space in the engine compartment, it translates directly to improved performance and increased range. Research higher frequency components—it will pay you dividends.

The other aspect of transformers useful to you (highlighted by FIG. 7-4) is that an equivalent circuit of a transformer can be drawn for any frequency, and you can study what is going on. This is useful and directly applicable to ac induction motors.

## ac induction motors

The ac induction motor, patented by Nikola Tesla back in 1888, is basically a rotating transformer. Think of it as a transformer whose secondary load has been replaced by a rotating part. In simplest form, this rotating part (rotor) only requires its conductors be rigidly held in place by some conducting end plates attached to the motor's shaft. When a changing current is applied to the primary coil (the stationary part or stator), the changing magnetic field results in the transfer of electrical energy to the rotor via induction. This is called the "squirrel-cage" ac induction motor—now you know why you studied transformers! As energy is received by the rotor via induction without any direct connection, there is no longer a need for any commutator or brushes. Because the rotor itself is simple to make yet extremely rugged in construction (typically a copper bar or conductor embedded in an iron frame), induction motors are far more economical than their equally rated dc motor counterparts in both initial cost and ongoing maintenance.

While ac motors come in all shapes and varieties, the ac induction motor—the most widely-used variety—holds the greatest promise for EV owners because of its significant advantages over dc motors:

- No brushes or commutators required
- Lower maintenance requirements

- Ability to operate in dirty and explosive environments
- Better reliability
- Higher efficiency
- More rugged
- Lower cost
- Lighter weight
- Smaller volume
- Less inertia

This is an impressive list. But until the arrival of modern, solid-state components, ac induction motors were primarily used in applications requiring constant speed—dc motors dominated variable speed applications. These new solid-state components have resulted in ac induction motors appearing in variable speed drives that meet or exceed dc motor performance—a trend that will surely accelerate in the future as more-efficient solid-state components are introduced at ever-lower costs. This section will examine the ac induction motor for its torque, speed, reversal, and regenerative braking capabilities—the factors important to EV users. Other ac motor types not as suitable for EV propulsion will not be covered.

## Single-phase ac induction motors

Although the squirrel-cage ac induction motor works well, when you first connect it to single-phase ac—the kind you would normally find in your house—it's not likely to start turning by itself.

Recall the universal dc motor discussed earlier. When you connect it to a single-phase ac source, you have little difference in its motor action because changing the polarity of the line voltage reverses both the current in the armature and the direction of the flux, and the motor starts up normally and continues to rotate in the same direction. Not so in an induction motor driven from a single-phase ac source. At startup you have no net torque (or more correctly, balanced opposing torques) operating on its motionless rotor conductors. Once you manually twist or spin the shaft, however, the rotating flux created by the stator currents now cuts past the moving conductors of the rotor, creates a voltage in them via induction, and builds up current in the rotor that follows the rotating flux of the stator.

How do you overcome the problem? The bottom of FIG. 7-4 shows the key. If you introduce a second winding that is physically at right angles to the main stator winding, you induce a rotor current *out of phase* with the main rotor current that is sufficient to start the motor. This *split-phase* induction motor design—or some variation of it—is the one you are most likely to encounter on typical smaller motors that power fans, pumps, shop motors, etc. To maximize the electrical phase difference between the two windings, the resistance of the starting winding is much higher and its inductance much lower than the running winding. To minimize excessive power dissipation and possible temperature rise after the motor is up and running, a shaft-mounted centrifugal switch is connected in series with the starting winding that opens at about three-fourths of synchronous speed. Figure 7-4 looks like a representation of a shunt motor: the smaller split-phase induction motor speed characteristics look like those of dc shunt motors, but their starting torque is much greater.

The most common split-phase induction motor is one that uses a *capacitor-start*, also shown in FIG. 7-4. The capacitor automatically provides a greater electrical phase difference than inductive windings. This greater phase difference—nearly 90 electrical degrees—also gives capacitor-start split-phase induction motors a much higher starting torque (three to five times rated torque is common). The principle was discovered quite early by Charles Steinmetz and others, but capacitor technology had to catch up before it could be widely introduced on production motors. Capacitor-start design variations include two types: separate starting and running capacitors; and permanent capacitor with no centrifugal cutout switch. The two-capacitor approach brings you the best of both the starting and the running worlds; the permanent capacitor type gives you superior speed control during operation at the expense of lower starting torque. The other common split-phase induction motor design, called *shaded pole* applies mostly to smaller motors; you are more likely to find it in your electric alarm clock than in your EV, so we'll skip it here.

## Polyphase ac induction motors

*Polyphase* means more than one phase. ac is the prevailing mode of electrical distribution. Single-phase 110-volt to 120-volt ac is the most prevalent form found in your home and office. Three-phase 220-volt, 440-volt, and higher ac is the most prevalent form for home washer/dryer/heaters, as well as offices, factories, and much, much higher voltages for transmission. A polyphase ac induction motor usually means a three-phase ac induction motor. These are widely available in nearly every city in the industrialized world. If one phase is good, then three phases are better, right? Well, usually. Stationary three-phase electric induction motors are inherently self-starting and highly efficient, and electricity is conveniently available. Moving induction motors located in EVs are likewise inherently self-starting and highly efficient, but you would need to carry around three times the amount of controller components—six times if you are using digital control—to generate the ac drive and control currents from the dc batteries.

Three-phase ac connected to the stator windings of a three-phase ac induction motor produces currents that look like those shown at the top of FIG. 7-5—they are of the same amplitude, but 120 degrees out of phase with one another. Physically, the stator looks like the bottom of FIG. 7-6. Six stator poles are spaced 60 degrees apart, and the three-phase windings each go from one pole to its companion pole located directly across the stator from it: A1 to A1, B1 to B1, and C1 to C1. Current flow in any of the three phases sets up a flux field in the air gap that looks like that shown for the B phase in FIG. 7-5. The middle of FIG. 7-5 (above the stator) shows how these fields act on the rotor. The equivalent flux vectors derived from the input currents are shown at three time instants: $t_1$, $t_2$ and $t_3$. For convenience, the maximum current is assumed to have a value of 10. At time instant $t_1$, the A phase is at maximum positive value of +10, the B and C phases have values of −5. The flux components of these three currents act to produce a single flux in the direction shown—because the flux at any instant of time is the vector sum of the flux fields generated by the individual currents. The same is true of times $t_2$ and $t_3$.

Three-phase ac connected to a three-phase ac induction motor produces a rotating flux in the air gap that can be varied only by changing the number of poles on the stator, or by changing the frequency of the supply. This speed, known as the *synchronous speed* of the motor, is given by the relationship

$$(25) \quad RPM_{synchronous} = (120 \times frequency)/(\text{\# of poles})$$

For any given load, the difference in speed between the synchronous speed and the actual operating speed of the motor is called the *slip*, and is given by

$$(26) \quad Slip = (RPM_{synchronous} - RPM_{rotor})/(RPM_{synchronous})$$

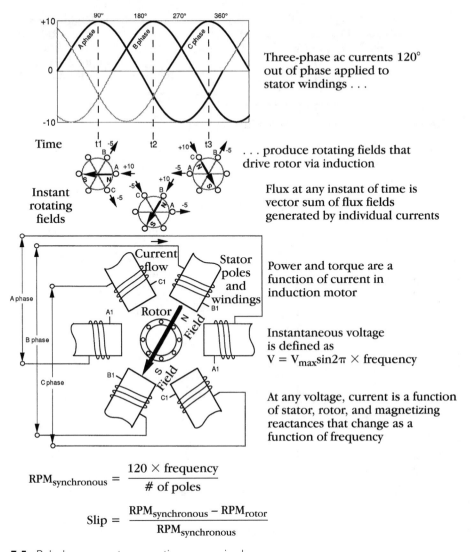

Three-phase ac currents 120° out of phase applied to stator windings . . .

. . . produce rotating fields that drive rotor via induction

Flux at any instant of time is vector sum of flux fields generated by individual currents

Power and torque are a function of current in induction motor

Instantaneous voltage is defined as
$V = V_{max}\sin 2\pi \times$ frequency

At any voltage, current is a function of stator, rotor, and magnetizing reactances that change as a function of frequency

$$RPM_{synchronous} = \frac{120 \times \text{frequency}}{\# \text{ of poles}}$$

$$Slip = \frac{RPM_{synchronous} - RPM_{rotor}}{RPM_{synchronous}}$$

**7-5** Polyphase ac motor operation summarized.

As in a dc motor, power and torque are also a function of current in an induction motor. Because the current is equal to the voltage divided by the motor reactance, at any given voltage, current is a function of stator, rotor, and magnetizing reactances that change as a function of frequency. The top of FIG. 7-6 shows this at a glance. Here the same transformer equivalent circuit model has been adapted to induction motor use. Notice that the resistance of the rotor is governed by the slip factor in the model. Looking into the motor's terminals, its input impedance is governed by

$$(27) \quad Z_{in} = (R_1 + (R_2/s)) + j2\pi f(L_1 + L_2)$$

This is an important relationship, because for maximum power transfer, the controller output impedance should match the motor's input impedance.

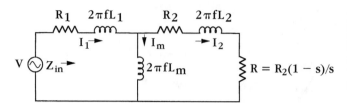

Induction motor must lag a few RPM behind rotating field, even at no load, to overcome retarding effect of motor losses

Equivalent circuit—one phase of induction motor

$$Z_{in} = \left(R_1 + \frac{R_2}{s}\right) + j2\pi f(L_1 + L_2)$$

Controller output impedance must match motor input impedance for maximum power transfer

$$Torque_{out} = \frac{3R_2 \times \# \text{ of poles}}{4\pi fs} \cdot \frac{V^2}{\left(R_1 + \frac{R_2}{s}\right)^2 \times (2\pi f)^2(L_1 + L_2)^2}$$

Torque varies with slip at any given voltage and frequency

$$Torque_{max} = \frac{3 \times \# \text{ of poles} \times V^2}{4(2\pi f)2 \times (L_1 + L_2)}$$

Maximum torque can be maintained if voltage to frequency ratio is held constant

**7-6** Polyphase ac motor's unique speed, torque, and slip characteristics vs. voltage and frequency.

The output torque of the induction motor is given by

$$(28)\quad T_{out} = (3R_2 \times \# \text{ of poles} \times V^2)/(4\pi fs((R_1 + (R_2/s))^2 + (2\pi f)^2(L_1 + L_2)^2))$$

This impressive-looking equation, also derived from the model of FIG. 7-6, says that at any given voltage and frequency, torque varies with the slip. Substituting for the value of s at maximum torque, and using the fact that $2\pi f(L_1 + L_2) \gg R_1$ at higher frequencies,

$$(29)\quad T_{max} = (3 \times \# \text{ of poles} \times V^2)/(4(2\pi f)^2(L_1 + L_2))$$

This important equation means that if the voltage-to-frequency ratio (V/f) can be held constant, then maximum torque can be maintained throughout the range. Solid-state induction motor controllers make use of this vital fact, as you will learn in chapter 8.

The characteristic induction motor torque to slip graph, shown in FIG. 7-6 for both its motor and generator operating regions, offers insight into induction motor operation. If an induction motor is started at no load, it quickly comes up to a speed that might only be a fraction of one percent less than its synchronous speed. When a load is applied, speed decreases, thereby increasing slip; an increased torque is generated to meet the load up to the area of full load torque, and far beyond it up to the maximum torque point (a maximum torque of 350 percent rated torque is typical). When an induction motor is first started, slip = 1.0 and if operating at a higher frequency such that $2\pi f(L_1 + L_2) \gg (R_1 + R_2)$, the starting torque is

$$(30) \quad T_{start} = (3R_2 \times \# \text{ of poles} \times V^2)/(2(2\pi f)^3(L_1 + L_2))$$

This value is also higher than full load torque, and is actually specified by NEMA (National Electrical Manufacturers Association) at values from 150 percent to 105 percent for 60-Hz squirrel cage induction motors of 2 to 16 poles, respectively. NEMA also categorizes squirrel cage induction motors into Classes A through F, based on their slip, starting and running torque, and current characteristics.

Speed and torque are relatively easy to handle and determine in an induction motor. So are reversing and regenerative braking. If you reverse the phase sequence of its stator supply (i.e., reverse one of the windings), the rotating magnetic field of the stator is reversed, and the motor develops negative torque and goes into generator action, quickly bringing the motor to a stop and reversing direction. Regenerative braking action—pumping power back into the source—is readily accomplished with induction motors. How much regenerative braking you apply creates braking (moves the steady state induction motor operating point down the torque-slip curve), and generates power (instantaneously runs with negative slip in the generator region, supplying power back to the source).

## Wound-rotor induction motors

You might also encounter the wound-rotor induction motor in your EV power source quest. This is an induction motor with rotor windings rather than squirrel cage construction. A wound rotor's windings are brought out through slip rings—conductive rings on the rotor's shaft—through brushes (analogous to dc motor construction) to an external resistance in series with each winding. The difference is that slip rings are continuous with no commutator slots. The advantage of the wound-rotor induction motor over the squirrel cage induction motor is that resistance control can be used to vary both the motor's speed and its torque characteristics. Increasing the resistance causes maximum torque to be developed at successively higher values of slip until the point slip = 1.0 (commonly called breakdown torque) is reached. Along the way the wound rotor has better starting characteristics and more flexible speed control than its squirrel cage counterpart. What you give up is efficiency at an increased complexity and cost. New solid-state induction motor controllers can be viewed as either augmenting the wound-rotor induction motor's advantage (your control can be made even more precise), or making it obsolete (you can get all the control you need with a less expensive squirrel cage induction motor).

## TODAY'S BEST EV MOTOR SOLUTION

Different motor types in their numerous variations give you an almost unlimited bag of tricks in terms of solutions, but you have to figure out which one to use. If someone tells you there is only one motor solution for a given application—ask another person. Their not-so-surprising and different answer ought to convince you that there are no black and white motor solutions—only shades of gray. You can probably find three or four good motors for any need—and you should look for and ask the questions that uncover these alternate solutions. On any given motor solutions list, you might choose the motor with the lowest price because that is your most important criteria. Another person might choose the motor with the lightest weight. Still another might choose the motor with the best startup torque. Regenerative braking might be most important to another person, and so on. Just remember shades of gray. Following along with this theme, the solution recommended here is not *the* solution, it is just *a* solution that happened to work best in this case.

## Series dc motors win the race

Series dc motors are available from many sources, they work well, controllers are readily available, adapters to different vehicles are easily made or purchased for them, and the price is right. A series dc motor might not be the ultimate or even the best current solution, but it's one that most EV converters will have no trouble implementing today. After you do your first conversion and know what works (and what doesn't) in your particular case, you can get fancy and exotic.

The series dc motor used for the example here is from Advance D.C. Motors, Inc. of Syracuse, New York. Don't read anything into the tea leaves about its appearance here. They are only one out of a large number of motor manufacturers—the list to help you get started appeared in chapter 5—and the motor recommended is only one out of a number of motors they manufacture.

## Today's winner

The Advance FB1-4001 model series dc motor was already discussed in arriving at chapter 6's chassis trade-offs for the Ford Ranger pickup conversion of chapter 11. Figure 7-7 gives you a cutaway view of one of the smaller motors offered in the line; you can see its rugged construction, with dual brushes, heavy duty terminals, and quality material used throughout in the windings, etc. This motor is a 9.1-inch-diameter model that weighs 143 pounds and is rated at 20 HP with a 70-HP peak.

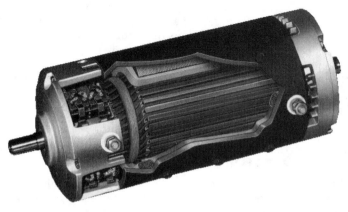

**7-7** Advance series dc motor cutaway view.

The Advance FB1-4001 was designed to propel EVs—one of its big advantages. Another advantage that should not be underestimated is that you can acquire it new, from a reputable vendor. A used compound wound dc aircraft starter motor for $200 might sound great, but powering EVs was never intended to be its true mission in life, and how much use has it seen already? With the FB1-4001 you get a motor that you can return if it doesn't work. In a reputable vendor you have someone to turn to for answers to questions, technical data, and more. A surplus dealer is rarely able to offer this capability.

Figure 7-8 shows the performance curves used in deriving chapter 6's data, this time shown for values of 72 through 120 volts. TABLE 7-1 gives you the data from its S-2 DIN & ISO thermal tests. Another advantage to buying a new motor from a reputable dealer is that you have the curves and data you need to help you optimize your EV conversion.

There are better solutions. But regenerative braking was not important for our conversion, and there was already a matching controller available. We wanted good middle-of-the-road performance at a good price, as well as a product that any potential EV converter reading this book could use and get working the first time up to bat. This motor delivers all that and more.

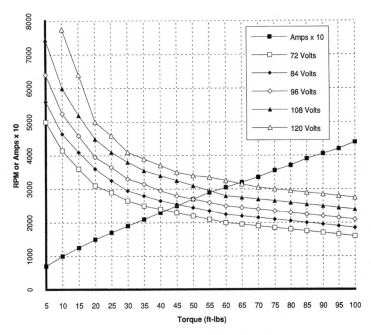

**7-8** Advance FB1-4001 series dc motor's speed and current vs. torque curves.

## TOMORROW'S BEST EV MOTOR SOLUTION

While the series dc motor is unquestionably the best for today's first-time EV converters, the bias of this chapter toward ac induction motors was not accidental. Improvements in solid-state ac controller technology clearly put ac motors on the fast track for EV conversions of the future. ac motors are inherently more efficient, more rugged, and less expensive than their dc counterparts; the reason why they are not in more widespread use today has to do with controllers, as you'll learn in chapter 8.

### Table 7-1  Data for advance model FB1-4001 series dc motor S-2 DIN & ISO thermal tests

| Test Voltage | Time - On | Volts | Amps | RPM | HP | Peak HP | KW |
|---|---|---|---|---|---|---|---|
| 75 volts - .03I | 5 minutes | 63.5 | 380 | 1900 | 27.0 | | 20.3 |
| 75 volts - .03I | 1 hour | 68.0 | 240 | 2550 | 19.0 | | 14.3 |
| 75 volts - .03I | continuous | 69.0 | 210 | 2800 | 17.0 | | 12.8 |
| 75 volts - .03I | | | | | | 42.0 | |
| 96 volts - .03I | 5 minutes | 88.0 | 360 | 3300 | 35.0 | | 26.5 |
| 96 volts - .03I | 1 hour | 89.0 | 210 | 3600 | 23.0 | | 17.3 |
| 96 volts - .03I | continuous | 90.0 | 190 | 3900 | 20.0 | | 15.0 |
| 96 volts - .03I | | | | | | 70 | |

These underlying facts made my surprise all the greater when I asked Darwin Gross (the "engineer's engineer" referenced in the acknowledgments) the question, "What's the best electric vehicle motor solution for tomorrow?" His answer was brief, to the point, and thought-provoking. It forms the basis for this section. Right or wrong, the answer should get your creative juices flowing. Figure 7-9 tells the story at a glance. The ideal EV drive for tomorrow has:

- Low weight.
- Streamlined design.
- Thin tires.
- Simple drivetrain (one or two speeds).
- dc to get started.
- ac to run above 30 mph.
- High frequency components (≥)400 Hz).
- dc motor that gets 96 volts.
- Single-phase ac motor that gets 192 volts.
- Matching controller and motor impedance.

Let's look more closely at some of the pieces.

**dc and ac working together**  The series dc motor provides the best starting torque, and the ac induction motor is most efficient at speed. By using them together with some mechanism to switch back and forth (e.g., at around 30 mph on the level with a kickdown for hills), you get the best of both worlds. Only about 96 volts are needed to get the dc motor started in your average utility commuter EV. When the ac motor takes over, you can add the rest of the battery pack toward powering it—120, 144, 192 volts, or whatever. Darwin emphasized, "Pick a good core material with some nickel in it to achieve 96 percent efficiency, and remember size and weight go down as the frequency goes up."

**Less is more**  Less means less weight, less obstacles to streamlined airflow, less tire width and cross-section, and less drivetrain. It means fewer phases in the ac motor (avoid extra components by using a single phase), and less iron in motors and inverter transformer cores (a higher design frequency—like 400 Hz—makes this possible). It

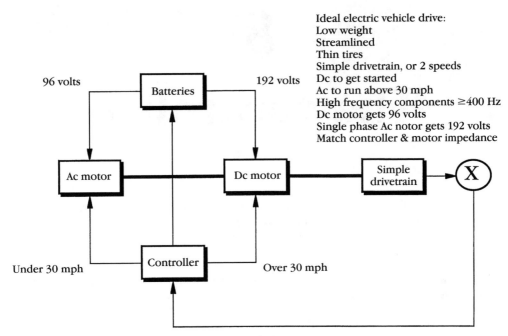

**7-9** Tomorrow's best electric vehicle motor solution.

also means less weight in batteries (although Darwin would not elaborate further, he left the clear message that we're not even close with today's battery technology). In short, less of everything that would make an EV less efficient.

**Tuning** Electrical and mechanical systems deliver better performance when balanced and tuned up. Race car mechanics set up their cars for the average case, then tune and adjust them for maximum performance. Should you do anything less with your EV? Match the impedance of the motor to the controller for maximum power transfer, and go through everything else with a fine-tooth comb.

**Keep it simple** What's the simplest controller that can implement this approach? Do you even need to use solid-state designs? How few batteries can you get by with? Which are the smallest motors you can use? Can you do something innovative with their placement? How can you simplify the drivetrain? Can you do something better with the tires? You get the idea.

# 8

# The controller

*If automobiles had improved as much as electronics in the past few decades, they would go a million miles per hour, cost only pennies, and last for decades.*

The controller is another pillar of every electric vehicle. If one area could take credit for renewed interest in EVs, the controller would be it. Today's solid-state electronics developments have routinely made dc motor controllers available at a good price. You can buy a controller, plug it in, and be up and running in no time—something earlier EV enthusiasts could only dream about. Future electronics development can be expected to deliver the same in the ac motor arena.

Your controller decision follows logically from your initial EV conversion choice: tire-smoker, high-range flyer, or around town commuter. But, unlike the plethora of choices with electric motors, your controller choices are rather simple, and are dictated by the electric motor you use, your desire to make or buy, and the size of your wallet or purse.

In this chapter you'll learn what the different types of controllers are, how they work, and their advantages and disadvantages. Then you'll encounter the best type of controller to choose for your EV conversion today (the type used in chapter 11's conversion), and the electric motor controller you're likely to be seeing a lot more of in the future.

## CONTROLLER OVERVIEW

You can relax now. The useful but largely theoretical equations of chapter 7 are behind you. Controllers can generate even more equations, but we'll save those for the engineers who are building them. The objective here is to give you a brief controller background and introduce you to a working controller for your EV with minimum fuss. (We'll drop some hints along the way for those who are into building controllers.)[1] Although you can read about all sorts of promising developments going on in the labs, the actual controller choices available for your EV conversion today are relatively straightforward: dc motor battery switch with resistive start, dc motor solid-state, and ac motor solid-state.

dc controllers of the pulse width modulation variety (PWM) have been used in forklifts and EVs for decades, and are available off-the-shelf now. Improved versions, capable of powering the largest EV dc motors, are being tested in the labs today. They take advantage of the faster and heavier-duty solid-state components that have recently been introduced. Off-the-shelf sources were mentioned in chapter 5; this chapter looks at the dc controller technology,[2] and simple controller you can build yourself.

Although ac induction motors are superior to their dc counterparts, they're not quite ready for use in your EV conversion today because ac controllers have not caught up with their dc counterparts. Generally, ac motor controllers will be four to six times more complex than dc motor controllers. While you can buy a reliable dc controller from a number of sources today at a good price, ac controllers are still in the prototype stage, and to buy one can easily cost more than was budgeted for the entire EV conversion in chapter 4. This is one area in which, if you are an electronics expert, your "sweat equity" can pay real dividends. You can be the first on your block to own an ac-powered EV with ac controller that you built yourself. This chapter also looks at ac technology[3] and how one company is building a foundation in future controllers.

The best news might be that you don't need an expensive controller at all. If most of your driving is confined to stop-and-go urban situations, you might be perfectly happy with a simple battery switch and resistor setup that you can assemble yourself for only a fraction of the cost of a controller. While it's not the recommended approach of this book, it might be well-suited to some EV converters, and this chapter will give you a look at how you might put such a system together.

To understand how such a radically simple thing as a switch and resistor controller could be included in a modern book on EVs, we need to step back and take a look at how EVs handle when driven, and how they differ dramatically from internal combustion vehicles—particularly in urban driving.

## Urban driving review

The Environmental Protection Agency (EPA) has its Federal Urban Driving Schedule (FUDS), and the Society of Automotive Engineers (SAE) has its J227 driving schedule. These schedules are intended to provide a uniform scenario against which different makes and styles of automobiles can be tested for emissions, mileage, etc. Loosely translated, they say start, accelerate, drive so long at this speed, deaccelerate, stop, etc. If you were to translate these into your own experience and consciously monitor what you do in urban traffic when actually driving your internal combustion engine vehicle, you'd find yourself following a similar cyclical pattern. At some hours of the day in some cities, you spend a lot more of your cycle time stopped than moving.

The better drivers learn how to make the best of this situation by not racing from stoplight to stoplight and avoiding jackrabbit starts in bumper-to-bumper traffic. Everyone learns that when you take your foot off the gas in an internal combustion engine vehicle, it slows down by itself, and this technique can be used to conserve brakes and fuel in traffic.

You drive electric vehicles differently because they have almost zero drag when the electric motor is switched off. If you remove power to a conventional EV, it coasts—in some models it seems nearly forever—until you hit the brakes. So the EV urban driving strategy for better drivers is to give just enough juice to carry them to the next stoplight, or to the car ahead of them in stop and go traffic, etc.

This inherent EV operating mode of pulse and coast also means that the better drivers rarely find the services of regenerative braking action—slowing down by returning energy to the batteries—useful in urban driving unless they live in hilly areas. While regenerative braking can be accomplished electronically down to nearly zero speed, there is little practical energy recovery when driving speed rarely goes above 50 mph, and the vehicle spends a large percentage of its time stopped.

It turns out that electric vehicles are custom-tailored for the urban cycle. You just give them a little juice to get started and most of the time you're coasting until you brake to a stop. And when stopped, your electric motor is not running at all (nor is it polluting!) and you're consuming zero energy. If this sounds like you need a switch rather than a controller, you're exactly right. In fact, this is the premise of the simple battery switch and resistor setup described next.

### Simple switch controller

If you are on a tight budget, you don't have to buy a controller at all, you can make one. How? Use the guidance of the previous section as applied at the top of FIG. 8-1. It doesn't get any simpler than this. If you choose your drivetrain gear ratios carefully and mostly do around-town driving, a simple switch is really all you need. While this method will unquestionably start your EV with a lurch—just like you'd get when riding a subway train—it will impress your friends, passengers, and stoplight acquaintances with its peppy performance too. The downside is that you'll probably wind up shifting gears more than you'd like, and if you ever accidentally have the clutch in with the switch on, you've probably toasted your series dc motor.

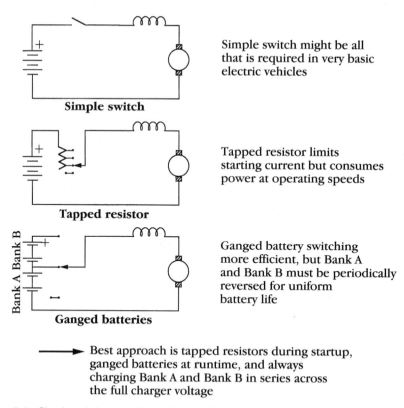

Simple switch

Simple switch might be all that is required in very basic electric vehicles

Tapped resistor

Tapped resistor limits starting current but consumes power at operating speeds

Ganged batteries

Ganged battery switching more efficient, but Bank A and Bank B must be periodically reversed for uniform battery life

→ Best approach is tapped resistors during startup, ganged batteries at runtime, and always charging Bank A and Bank B in series across the full charger voltage

**8-1** Simple switch controller options.

If you go with the tapped resistor plan shown in the middle of FIG. 8-1, you protect your series dc motor on startup, but dissipate power in extra resistive I²R losses. To minimize losses, you step through the tapped resistance from maximum to minimum to none in a few seconds time on startup. This is the startup resistance method that has been used with series dc motors for more than a century, so it works, and your motor is probably happier. Three seconds of dropping half the initial 400-amps current flow is 600 amp-sec of energy, which translates to 0.17 amp-hr to the motor and 0.08 amp hour to the battery. This is not a lot, but if left in the circuit continuously, it seriously decreases your range. The downside is that your startup lurch is gone along with a lot of the snappy performance.

If you go with the ganged batteries plan shown at the bottom of FIG. 8-1, you still produce low starting voltage, and avoid the extra resistive I²R loss. The disadvantage is that Bank A batteries are providing power all the time while Bank B batteries only operate at the high end—your batteries are being discharged at an unequal rate. This setup will lead to early replacement of the Bank A batteries unless you periodically swap Bank A and Bank B via a switch, etc.

The best approach is a combination of tapped resistors and ganged batteries. To keep things simple, you need no more than three resistance levels and two battery settings, because you still have a drivetrain that gives you the extra ratios. All you need is a single switch of the "make-before-break" variety, with five settings—off, resistance 1, 2, 3 (connected to Bank A batteries) and run (connected to all batteries)—that connects to your accelerator pedal. Charging is performed with all batteries in series—the accelerator switch in the run position—and you periodically swap Bank A and Bank B with the simplest electrical switching arrangement you can rig up.

In actual construction, your main five-position switch is connected to a relay or contactor that is physically capable of routinely switching the high currents involved. The switches that less often swap Bank A with Bank B can be of the less exotic knife switch variety or equivalent.

Ford or General Motors or Chrysler cannot put this system into their EVs because of what basically amounts to industry peer pressure. Somebody, somewhere would complain that it doesn't have safety disconnects, motor protection, battery protection, and an onboard computer that monitors status. They would be right. But this is your electric vehicle and, regardless of whether these matter to you or not (you always do want some form of safety disconnect), you can do it your way.

## SOLID-STATE CONTROLLERS

The benefits of the simple switch controller are its low cost and simplicity. But it gives you only discrete operating steps, rather than the smooth range of control that is normally associated with the accelerator pedal of internal combustion engine vehicles.

The true benefit of solid-state controllers is in the electronics. As you read in the motors chapter, electronic controllers can blur the distinction between motor types—a superior controller can make up for any motor deficiency. It's far easier and less costly to implement technology changes with component changes on a circuit board than it is to redesign, stamp, machine, and rewind a motor. The trend in all vehicles is to incorporate more electronics because of their superior monitoring and control benefits. In this regard, the EV is the epitome of this automotive industry-wide cultural trend (the General Motor's Electric Vehicle presentation actually shows an EV as a laptop computer on wheels). An electronics controller that will increasingly be associated with, if not driven by, a computer on a silicon microchip in the future is logically at the center of the action.

Low maintenance and hassle-free usage at minimum cost are the key trade-offs that factor into your EV controller choice. The evolution and availability of today's solid-state electronics components over their earlier counterparts have brought earlier controller design possibilities into practical, everyday products that anyone can call and order. In solid-state controllers, both the "trees" and the "forest" are important, so this section starts with a look at the components used in solid-state controllers, then examines dc and ac controllers in turn.

## Manual switches versus solid-state components

When you draw an electrical circuit with a switch in it, you normally don't associate the switch with a lossy element (i.e., an element whose resistance produces an $I^2R$ loss). But any solid-state component that produces a result equivalent to a switch has an associated voltage drop across it that translates to power consumed. Also, when you close or open a switch, you assume it is instantly on or off, respectively. Not so with solid-state components, which take a small but finite amount of time to turn on or off. And since we are talking about switching hundreds of amps at frequencies anywhere from 2 kHz to 20 kHz or more, this also translates to power consumed. As a result, all solid-state controllers are less efficient than the simple switch arrangement just described—but more than make up for this in convenience, flexibility, and future adaptability to computer control and monitoring.

All solid-state devices make use of the semiconductor effect. Certain borderline conductor, metal-like materials (such as germanium, silicon, gallium arsenide) can exhibit electrical properties: they conduct more negative charges when mixed with a material such as phosphorus, creating an N-type material; they conduct more positive charges (or holes) when mixed with a material such as boron, creating a P-type material. These two materials, when bonded together, form the PN-junction—the fundamental building-block of solid-state electronics. There are an enormous number of solid-state components; we are primarily interested in but a handful. These few devices are the "power" versions of their logic circuit cousins—they are designed to conduct tens or hundreds of amps rather than milliamps. The following devices are the most important to EV converters—their electrical symbol and characteristic curve appear in FIG. 8-2.

**Diode or power diode** This might be viewed as the fundamental PN-junction building block. In theory, application of a positive voltage to the anode or P end of the junction biases the diode in the forward direction and it conducts—it is now switched on. Applying a negative voltage to the anode—or a positive voltage to the cathode or N end—back-biases the diode and it stops conduction. The curve at the top right of FIG. 8-2 shows what happens in the real world. There is a voltage drop associated with the diode conducting in the forward direction that increases with increasing current. In the reverse direction, there is a *peak reverse voltage* at which reverse breakdown occurs and the device ceases to act as a diode but starts to conduct current in the negative direction. Before this point is reached there is a small reverse leakage current associated with the operation of every diode. A certain class of diodes—Zener diodes—are designed to make specific use of the reverse voltage area by having a nearly right-angle knee that limits the reverse voltage to a very specific value.

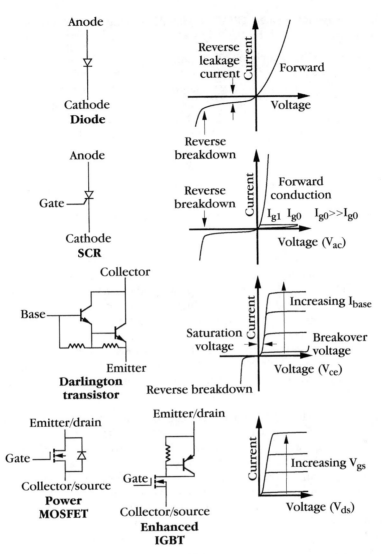

**8-2** Summary solid-state component types—symbols and characteristics.

**SCR or power SCR** The Silicon Controlled Rectifier is a particular type of device from the thyristor family that is characterized by its ability to *latch* into either the on (forward conduction) state or the off (reverse biased) state via an applied gate voltage. Usually modeled as a four layer PNPN diode, once the device is forward-biased (anode positive with respect to cathode) a very small amount of gate current $I_g$ will cause it to break over into forward-conduction and, once conducting, the SCR has an even lower resistance to current flow than a conventional PN diode. It continues to conduct unless the forward current drops below its holding value. Notice its curve in FIG. 8-2 is much steeper in the forward direction and much flatter in the reverse direction—approaching the way a real switch would work. A thyristor-family cousin to the SCR is the five-layer gated TRIAC (Triode Alternating Current switch), whose bidirectional design lends itself particularly well to control of ac circuits.

**Transistor or power transistor** The transistor or bipolar transistor is a three-layer PNP or NPN device. It has the characteristic that a very small amount of current injected into the base can control a very large amount of current flow in the collector-to-emitter path, and increasing amounts of base current $I_b$ generate increasing collector current flow—as shown in FIG. 8-2. In the Darlington configuration, frequently seen in power applications, transistors are cascaded for increased gain or reduction in the base current drive required; these transistors can be on the same chip and packaged inside one case for convenience.

**MOSFET or power MOSFET** The FET (Field Effect Transistor) or unipolar transistor is a cousin to the bipolar transistor whose collector current flow is primarily controlled by the voltage applied to its gate, which generates an internal electric field that controls either its electron or its hole flow—hence unipolar. MOS (Metal Oxide Semiconductor) stands for its typical layer sequence—how it is made. Analogous to the bipolar transistor curve, the collector current flow in a MOSFET is characterized by increasing current conduction levels, this time controlled by increased gate voltage $V_{gs}$ as shown in FIG. 8-2. Unlike the power transistor, the gate is electrically isolated from the source; its input impedance is very high (meaning it can be driven directly from logic circuits); device power gain is very large; and turn-on and turn-off times can be very fast. The unijunction transistor or programmable unijunction transistor is a four-layer distant relation to the conventional FET (its symbol is identical to the SCR's, only the gate is on the anode side of the diode) whose gate voltage, when forward-biased, creates an exceptionally low forward resistance that makes it useful as a trigger device in SCR circuits.

**Enhanced IGBT** The IGBT (Insulated Gate Bipolar Transistor) uses a cell design similar to a MOSFET, but has a P+ instead of an N+ substrate as part of the fabrication process. This results in a device that can be easily driven, like a MOSFET, yet has a high power per square inch of silicon efficiency at high voltages, like a bipolar transistor. The characteristic collector/drain current versus drain-to-source voltage in an IGBT is identical in shape to the MOSFET curve shown in FIG. 8-2—but the specific values are different. Additional positive IGBT benefits over MOSFETs include no increases in losses with temperature (more stable), lower input capacitance (less complicated driver circuits) and greater short circuit capability (more fault tolerant). A new Enhanced IGBT from Motorola, specifically designed for the EV market, extends its capability to include handling high in-rush current during acceleration and lower steady state losses at cruise.

The cost of an EV controller is largely determined by its power devices. Historically, these types of power transistors—high current and high voltage—have not been designed for high-volume manufacturing. As a result, current industrial-rated power devices sell for $100 to $200 each. Plus, the EV application requires high voltage, high current, and high frequency (typically 15 to 20 kHz) and few devices have been manufactured that meet this requirement at all. But times are changing; paralleling older devices or using newer ones makes high performance and high reliability a reality today, and increased volumes will bring these to you at a lower cost in the future. As TABLE 8-1 shows, the newest IGBT devices compare very well with more traditional solid-state power devices, offering low voltage drop (minimum power consumption), fast switching times (high frequency operation), and lower gate voltage drive (least

**Table 8-1  Power device comparison**

| Device Type | Equivalent Quantity | On voltage @ 100 amps | On voltage @ 400 amps | On voltage @ 600 amps | Switch rise Time μsecs | Switch off Time μsecs | Switch fall Time μsecs | Driver Voltage |
|---|---|---|---|---|---|---|---|---|
| 600V Darlington | 1 | 0.95 | 1.5 | 2.0 | 3.0 | 8.0 | 0.6 | 2.7 |
| 600V IGBT | 1 | 1.85 | 2.5 | 3.0 | 0.3 | 1.0 | 0.2 | 15.0 |
| 500V MOSFET | 10 | 1.0 | 4.0 | 6.0 | 0.3 | 0.4 | 0.2 | 20.0 |
| 150V MOSFET | 4 | 0.52 | 2.1 | .15 | 0.2 | 0.2 | 0.1 | 20.0 |

complicated drive circuitry) using the minimum number of components. All these characteristics are desirable for the EV application.[4]

Armed with this component information, you are now ready to look at the motor controller circuits that use them. Let's start with the simplest first.

### dc motor controller—The lesson of the Jones switch

The easiest way to vary the dc voltage delivered to a dc motor is to divide its steady, nonvarying dc component into smaller pieces (FIG. 8-3). The average of the number or size of these pieces will be a resultant dc voltage that the motor "thinks" it is receiving. At the top left of FIG. 8-3, the value of the voltage delivered to the motor is,

$$(1) \quad V_{motor} = ((t_{on})/(t_{on} + t_{off})) \times V_{battery} = (t_{on}/T) \times V_{battery} = K_{dc}V_b$$

This equation shows that the voltage delivered to the motor is proportional to the amount of time the pulse is on versus the total length of the period. As $t_{off} \gg t_{on}$ in the left-hand graph, the average voltage $V_m$ the motor receives is only equal to a small value. But in the right half graph, $t_{on} \gg t_{off}$ and $V_m$ is a much larger value—nearly equal to $V_b$—the battery voltage. If you vary $t_{on}$ while holding T (the total period) constant, you have the principle of *pulse width modulation* or PWM. (If you vary T while holding $t_{on}$ constant, you have *frequency modulation*.) For a fixed ratio, a duty-cycle constant $K_{dc}$ might be defined (equation 1, far right). PWM or *chopper control*, as it is customarily called, allows you easily to control a dc motor by electronically chopping the voltage to it.

The earliest solid-state dc motor chopper controllers used SCR devices. These were technically primitive compared to today's solid-state controllers, but typically offered greater efficiency, faster response, lower maintenance, smaller size, and lower cost than the motor-generator sets or gas tube approaches they replaced. The Jones chopper circuit shown in the middle of FIG. 8-3, adapted from an early GE SCR manual,[5] is one of the simplest variations. Earlier, similar PWM circuits were also built in transformerless style using transistors.[6] The Jones chopper is easy to look at and analyze, and its basic lessons apply to all PWM controllers. It might be viewed as an SCR (SCR$_1$) controlled by a commutating circuit (SCR$_2$ and trigger circuit 2) that controls the $t_{on}/t_{off}$ time and a trigger circuit (trigger circuit 1) that controls the period T or frequency.

The diode D$_2$ connected across the series dc motor's terminals—usually called a *freewheeling diode*—performs two important functions in this and other circuits where the pulsed output is inductive: it "smooths" motor current $I_m$ and enables it to continue to flow when SCR$_1$ is not conducting (top of FIG. 8-3); and prevents the inductive-current-generated high voltage spikes, which could potentially damage the transistors, from appearing across the motor when SCR$_1$ is turned off. Diode D$_1$ prevents the L$_1$–C$_1$

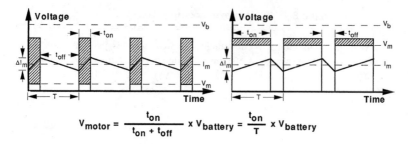

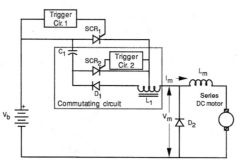

$$V_{motor} = \frac{t_{on}}{t_{on} + t_{off}} \times V_{battery} = \frac{t_{on}}{T} \times V_{battery}$$

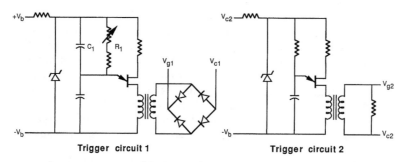

**Simple PWM Jones chopper**

**Trigger circuit 1**

**Trigger circuit 2**

**8-3** Summary of Jones switch PWM chopper controller characteristics.

combination from oscillating. Other than the SCRs and diodes, the key circuit elements are the center tapped autotransformer (both windings have the same number of turns, inductance, etc.), whose inductance is simply labeled $L_1$ in FIG. 8-3, the capacitor labeled $C_1$, and the combined series motor and external inductance labeled $L_m$.

The frequency of the chopper is controlled by the $R_1$-$C_1$ combination (where $R_1$ is the sum of the fixed and variable resistors in that branch of the circuit) in trigger circuit 1—shown at the bottom left of FIG. 8-3. In this case, a unijunction transistor forms the heart of a relaxation oscillator whose period is controlled by the RC time constant such that,

$$(2) \ f = 1/(R_1 C_1)$$

where R is resistance in ohms and C is capacitance in farads.

For a 2-kHz chopping frequency, you might choose $C_1 = 0.1 \ \mu F$,

$$R_1 = 1/(2 \times 10^3 \times 0.1 \times 10^{-6}) = 5 \times 10^3 = 5 \text{ kilohms.}$$

From a design standpoint, the important elements in the circuit are:

- The battery voltage $V_b$.
- The maximum motor current $I_m$ to be commutated.
- The turn off time of $SCR_1$.
- The voltage rating of $SCR_1$.
- The values for $C_1$, $L_1$.
- The value for $L_m$ to minimize the ripple current in the motor armature $\Delta I_m$.

The turn off time of $SCR_1$ is given by,

$$(3)\ t_{off} = 1/((L_1/2)/C_1)^{1/2}$$

In most cases, you know the value of $t_{off}$ you need, and will use the relationship in equation (3) to find the $L_1$ and $C_1$ values. For example, the frequency of 2 kHz used in the previous example translates to a period of $0.5 \times 10^{-3}$ seconds or 0.5 milliseconds. If you wanted a 50 percent duty cycle, this would set $K_{dc} = 0.5$ and establish $t_{off}$ at 0.25 milliseconds.

The voltage rating of $SCR_1$ is controlled by the voltage across capacitor $C_1$ by

$$(4)\ V_c = K_{scr} \times V_b$$

where $K_{scr}$ is a constant defining desired $SCR_1$ voltage.

Let's say in this instance you pick $K_{scr}$ to be 2. With this information and the values for $V_b$ and $I_m$, you now have everything you need to calculate the values for $C_1$ and $L_1$. If $V_b$ was 100 volts and $I_m$ was 400 amps,

$$(5)\ C_1 = (I_m t_{off})/(K_{scr}V_b) = (400 \times 0.25 \times 10^{-3})/(2 \times 100) = 500\ \mu F$$

$$(6)\ L_1 = (2K_{scr}V_b t_{off})/I_m = (2 \times 2 \times 100 \times 0.25 \times 10^{-3})/400 = 250\ \mu H$$

The value of $L_m$ is the key determinant in minimizing the ripple current—$\Delta I_m$ at the top of FIG. 8-3—in the dc motor's armature. The worst case value of $\Delta I_m$ occurs at a duty cycle of 50 percent. The value of $L_m$ to limit $\Delta I_m$ to its worst case value or less is given by,

$$(7)\ L_m = (K_{dc}(1 - K_{dc})V_b T)/\Delta I_m$$

If you wanted to limit ripple current in this case to 0.5 amp, the $L_m$ would be

$$L_m = (0.5 \times 0.5 \times 100 \times 5 \times 10^{-3})/(0.5) = 250\ mH$$

The Jones chopper circuit was a popular early electric vehicle controller. You can still use it today. Like equivalent circuits of its vintage, it is inexpensive and easy to make (with relatively few components). Its disadvantages are that it's relatively unsophisticated, and potentially dangerous to your series dc motor because it has no overcurrent sensing or limiting features; i.e., no recourse or shutdown mechanism if one or both of your SCRs has commutation problems, or other discrete components in the circuit fail outright or drift out of tolerance with temperature and age. Now let's move on to more modern solutions, and address the implied question of what can be done to improve the performance of the PWM circuit.

## PWM controller on an IC chip—The LM 3524 & others

How do you improve the PWM circuit performance? Put all the important components on a single integrated circuit chip. PWM ICs are available today from National Semiconductor, Motorola, Texas Instruments, and several other vendors. At the top of FIG. 8-4 is the National Semiconductor LM 3524 PWM IC. Notice that, compared with FIG.

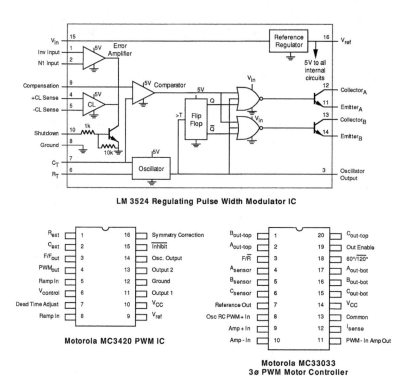

**LM 3524 Regulating Pulse Width Modulator IC**

**Motorola MC3420 PWM IC**

**Motorola MC33033
3ø PWM Motor Controller**

**8-4** Comparison of LM3524 PWM IC with MC 3420 and MC33033.

8-3, commutation, triggering, and relaxation oscillator timing generation functions are now incorporated directly on the chip, and a whole lot more. The savings benefits of parts cost, building time, troubleshooting time, and elimination of mismatched components are obvious. Here are the additional benefits of the LM 3524 PWM IC:

- Front-end amplifier input only requires addition of simple potentiometer to control motor speed directly.
- Front-end comparator has provision to accommodate feedback signal from motor shaft speed detector.
- Front end has circuit to invoke shutdown from any external fault condition: overcurrent, temperature, RPM, etc.
- On-chip oscillator only requires connection of external resistor and capacitor—accommodates frequency adjustment to more than 100 kHz.
- On-chip regulator generates 5 volt, 50 mA reference voltage for internal and external circuit use.
- On-chip current-limit amplifier for external component protection.
- On-chip thermal protection against excessive output current and junction temperature rise.
- Dual-output driver stages permit use of push-pull or single ended driver circuits.
- Duty cycle limit of 45 percent eliminates possibility of destructive simultaneous conduction when used in push-pull or bridge inverter applications.

Quite a collection of benefits. What's the downside? You have to learn to take advantage of all its features, and be sure to buy the chip with the features you want. Not all PWM ICs are created equal—more about that in a moment. Meanwhile, you no longer have to worry about pulse width modulation—that function is now performed inside the chip when it applies its internal oscillator sawtooth wave and error amplifier dc output level (determined by its potentiometer-driven speed control input N1) to the input of the comparator, which then produces duration-modulated pulses. The comparator's output can optionally be driven by the actual motor speed signal fed back into its compensation input. Comparator output pulses fed to multiple-input NOR gates, clocked by an oscillator-driven flip flop, provide a "clean" signal to drive a pair of output transistors that offer the user multiple driving options.

Some PWM ICs offer considerably more capability than others. Contrast the Motorola MC3420 PWM IC pinouts at the bottom left of FIG. 8-4 with those of the National Semiconductor LM 3524 above it. While identical features are offered under different names on different pins (a potential problem—read the PWM IC spec sheets carefully to be sure you get exactly what you want), it offers additional features. Surveying the field, additional PWM IC vendor features can include:

- Output waveform symmetry control by injected dc voltage level.
- External synchronization of oscillator via injected clocking signal.
- Dead time adjust (to prevent simultaneous conduction of push-pull or bridge transistors in inverter applications).
- Inhibit or direct electronic shutdown input pin.

In addition, newer three-phase PWM ICs offer even more capability. Contrast the Motorola MC33033 three-phase PWM IC 20-pin package pinouts (FIG. 8-4 bottom right) with those of both the other 16-pin earlier-vintage PWM ICs. You have still more and better everything. To summarize, PWM ICs are far better than discrete circuits, offering you additional capabilities that you should be incorporating in your controller designs anyway. The newest three-phase PWM ICs offer you the most for your money. Now let's move on to how you actually use the LM 3524 PWM IC and the more modern MC33033 in real motor controller circuits.

## Building a real dc motor controller with the LM3524

As FIG. 8-5 shows, all you add to the LM 3524 PWM IC is:

- External resistor R1 and capacitor C1 to determine oscillator frequency.
- External potentiometer network to control motor speed directly.
- External driver stages to generate required power-stage drive levels.
- External power-stage transistors.

These are the basics. You have a highly modular, very superior solution to the Jones chopper, and you aren't even breathing hard. As newer components are introduced you can take advantage of them just by plugging them into the drive or power stages. You can easily change the frequency or even the motor speed control circuitry if it suits your purposes better.

As for the options, FIG. 8-5 shows two additional circuit boxes:

- Current limit control.
- RPM feedback control.

Current limit control monitors motor armature current $I_m$ and feeds a signal back to the shutdown command input on the PWM chip if it goes beyond a preset level.

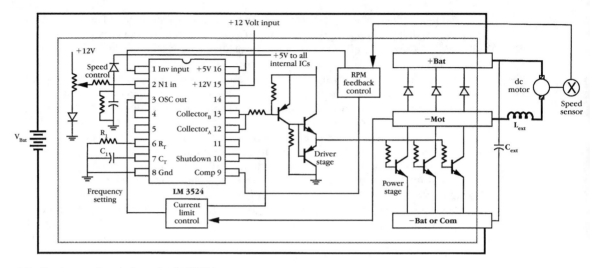

**8-5** Real controller design using LM3524.

The oscillator output is used to sample the signal fed back from the motor current sensor at preset intervals, and generate the shutdown command. This circuit provides the overcurrent sensing and limiting features so vital to your series dc motor's health; when current levels get out of hand it shuts things down, automatically.

RPM feedback control monitors the motor speed sensor and feeds a signal back to the comparator and error amplifier inputs on the PWM chip to make it a *closed-loop* system. The closed-loop system, whose objective is to drive the motor at a speed that produces zero difference signal in the PWM IC's comparator, gives much tighter control over a series dc motor's output. You can adjust it for faster response, smoother pickup, or more stability in the borderline motor control regions—whatever makes the most sense in your case.

The *+Bat, −Mot,* and *−Bat* or *Com* bars (the latter two terms may be used interchangeably) are the heavy-duty terminals that appear on every controller; FIG. 8-5 shows what they connect to inside the controller box as well as outside.

The free-wheeling diode from the Jones chopper example is now multiple diodes connected across the motor between the +Bat and −Mot terminals. In real life, you might use multiple heavy-duty diodes (e.g., 250 amps at 200 volts rating and up). If you're running a high chopping frequency—above the audio band at 15 kHz or more—these diodes also need to have fast recovery times so they can switch quickly back and forth between the conduction and nonconduction states.

The SCRs from the Jones chopper example are now replaced by power transistors connected in series with the motor between the −Mot and −Bat or Com terminals. In real life, you might use multiple heavy-duty power transistors (like the 200-amp Motorola MJ10200 or equivalent). You'll also have to mount these on a heatsink (use the nondrying silicone thermal grease) to keep within each power transistor's maximum junction temperature. This will be covered in detail in the next section. Three critical elements that should appear in all your designs:

*Inductor $L_{ext}$ connected in series with the motor.* Although minimizing inductance inside the controller is a serious issue as you go up in frequency (calling for proper component layout so as to minimize interconnection lead lengths), external inductance has nothing to do with controller inductance, but is used to control motor armature ripple current as in the Jones chopper example. In real life, this means a heavy-duty inductor (e.g., 20 turns of ¹⁄₁₆-inch by ¾-inch copper strap over a 2-inch by 2-inch core) capable of carrying the heavy motor armature currents while meeting the inductance value.

*Capacitor $C_{ext}$ connected across the battery.* The inductor in series with the motor holds the current relatively constant. This capacitor in parallel with the battery holds the voltage relatively constant—minimizing the potentially destructive inductive voltage spikes appearing at the motor-diode-battery +Bat terminal. In real life, this means very heavy-duty capacitors (e.g., 3300 µF at 250 volts rating and up) connected as close as possible to the controller's terminals.

*+12-volt external source.* A separate source of +12 volts is required to power the controller circuitry while a stable +5-volts reference voltage is available from the PWM chip. In real life, you would provide this +12 volts from a stable and isolated external source such as a dc-to-dc converter, and keep it well away from the high-power components. This is covered in detail in the next section.

As you have just learned, using the LM 3524 PWM IC is a great step upward in dc motor controller design. Would it be a great place to start any build-it-yourself motor controller? Yes. Can you do better still? Yes. Let's move on to a build-it-yourself motor controller circuit using the more-modern MC33033 PWM IC.

## Building a modern dc motor controller with the MC33033

The Motorola MC33033 is the three-phase PWM IC of FIG. 8-4. The difference in cost between it and the LM 3524 PWM IC is negligible in real world controller designs, and it is this book's recommended starting point for all modern build-it-yourself motor controller projects.

A sidebar about the good people at Motorola. Motorola has made a substantial commitment to EV control, electronics, and components research, in numerous current research projects. While they are involved with proprietary projects and designs looking five years ahead, the best news for EV converters is that they have made a certain amount of their research available in public documents. Since these people have forgotten more than most of us will ever know about electronic circuit and component design, it is at least prudent to listen to their counsel, and many basic Motorola controller and component recommendations have been incorporated in this section. But real world details change quickly, and if you are seriously interested in a build-it-yourself controller project, contact the Motorola Technical Publications 800 number listed in chapter 5 and ask for the latest revision of their DC Motor Control For Electric Vehicles Application Note.[7]

A modern dc motor controller built using the Motorola MC33033 3ɸ PWM IC is shown in FIG. 8-6. Compare it with the LM 3524 PWM IC design shown in FIG. 8-5. You will notice similarities and differences. At the basic connect-these-to-make-it-work level, two areas remain similar:

- An external resistor R1 and capacitor C1 still determine oscillator frequency.
- An external potentiometer network still controls motor speed directly.

The two other basic areas have changed:

- External driver stages now have opto-coupler and driver ICs, and use IGBTs.
- External power stage now uses IGBTs.

In these two areas, because the MC33033 3ɸ PWM IC-based controller allows you to use integrated circuits in place of discrete components, making your design life

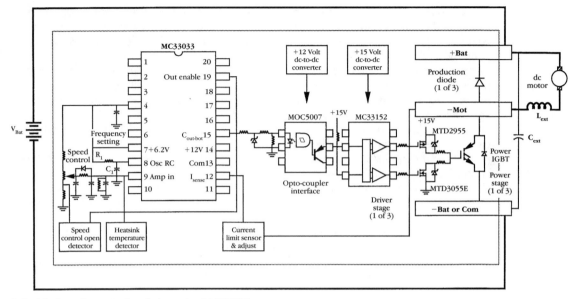

**8-6** Modern dc controller design using MC33033.

even easier, and the newer ICs and discrete components deliver even better performance. You have a highly modular solution, more efficient than the LM 3524 PWM IC-based controller in the basic areas, and you can now concentrate on the other things that really make a difference, such as emergency shutdown circuits in the event of over-current, over-temperature, or open-potentiometer conditions. You can still take advantage of newer component introductions by plugging them into the drive or power stages, and the frequency and motor speed control circuitry can still be changed. But the additional sophistication of the MC33033 3φ PWM IC allows you to monitor and control more motor, controller, and component health and status variables. Let's look at the details.

*Opto-coupler interface.* The purpose of the optical isolation area is to provide galvanic isolation and signal noise immunity. Galvanic isolation means nothing flows— no electrical connection whatsoever—between input and output. This is to isolate the user interface from the power circuitry, and is very important; you don't want to be pulling down 200 amps at 120 volts dc when you step on the pedal connected to the speed control potentiometer. Signal-noise immunity means the motor control circuit environment (high voltage, current, and high frequency) generates tremendous noise, and switching transient levels (1 volt up to 100 volts on the power supply rails) would be typical during operation at 100- to 400-amp current levels. This means the earth ground of the control electronics must be completely isolated from the –Bat or Com terminal of the power electronics so that power circuit noise levels don't swamp the signal levels. In this area, the MC33033 PWM IC logic levels are transformed to gate drive signals. The PWM logic signal goes to the input of the MOC5007 opto-coupler IC via a zener diode-controlled current source that maintains a constant opto-coupler output signal even under varying power supply conditions. The opto-coupler itself is a Schmitt trigger design, which generates a clean output signal from a noisy input signal at the expense of a slight delay time.

*Driver stage.* The purpose of the driver stage is to provide level shifting and signal conditioning. This stage provides the fully floating, optically isolated, 15-volt positive logic drive to the power IGBT. The opto-coupler output signal drives an MC33152 MOSFET driver IC to drive a complementary MOSFET stage; the upper MTD2955 P MOSFET's resistor controls the turn-on time, and the lower MTD3055E N MOSFET's resistor controls the turn off time. These provide a 5- to 7-amp gate current that is capable of switching the IGBT (and its associated gate capacitance) in 500 nanoseconds or less—in other words, fast. In the real world, the driver stage is physically mounted on a printed circuit board directly on top of each IGBT power module to minimize inductance.

*Power stage.* These are the IGBT power devices that actually drive the motor. To specify them you need to specify maximum current, operating current, maximum power dissipation, and maximum forward voltage drop.

The maximum stalled, shorted, or locked armature motor current would be,

$$(8) \ I_{max} = V_{battery}/R_t$$

where $R_t$ is total motor and external wiring resistance.

The resistance of the Advance FB1-4001 model series dc motor selected in chapter 7 is 0.04 ohms, and the combined wiring resistance is 0.01 ohms, so $R_t$ is 0.05 ohms. As $V_{battery}$ is 120 volts,

$$I_{max} = 120 \times 0.05 = 2400 \text{ amps}$$

As this is well beyond the motor's safe operating range, the Motorola controller design limits the maximum operating current to the 300- to 900-amp range.

The maximum IGBT power dissipation in watts is determined from

$$(9) \ Q_{max} = (T_{jmax} - T_{amax})/(TR_{jc} + TR_{cs} + TR_{sa})$$

where $T_{jmax}$ is the maximum allowable IGBT junction temperature in degrees C, $T_{amax}$ is the maximum ambient temperature in degrees C, $TR_{jc}$ is the junction-to-case thermal resistance, $TR_{cs}$ is the case-to-heatsink thermal resistance, and $TR_{sa}$ is the heatsink-to-ambient thermal resistance. Plugging in a maximum allowable junction temperature of 150 degrees C, a maximum ambient temperature of 40 degrees C, and Motorola's IGBT TR values,

$$Q_{max} = (150 - 40)/(0.08 + 0.025 + 0.03) = 815 \text{ watts}$$

The maximum IGBT maximum forward voltage drop is determined from,

$$(10) \ Vce_{max} = Q_{max}/I_{max}$$

Plugging in the previously determined value of $Q_{max}$ and the lowest value of the operating current range,

$$Vce_{max} = (815)/(300) = 2.72 \text{ volts}$$

Three parallel 400-amp IGBT modules (FIG. 8-6 shows only one of the three modules for clarity) can easily meet the 900-amp maximum operating current, and a 2.72-volt forward voltage drop at a 300-amp operating current can be met by several commercially available 400-amp IGBT modules, as you saw in TABLE 8-1.

As of early 1993, the three Motorola-suggested IGBT alternatives were: Toshiba P/N MG400J1US1, PowerX P/N CM400HA-12E, and Fuji P/N 1MB1400L-060. Order up your spec sheets and make your choice. If possible, it is a good idea to match IGBT device characteristics to equalize current sharing among all three units.

*Free-wheeling diode.* Because things happen very fast when switching speeds are specified in the 15-kHz and up range, the free-wheeling diode(s) not only must have the right current and voltage rating but also must have a reverse recovery time similar to the chosen IGBT's turn on time. Because of the motor winding and series inductance time constant with respect to the high switching frequency, the free-wheeling diode is conducting during the off time of the IGBT. If the diode is still conducting during the time the IGBT is switching on, the IGBT and the free-wheeling diode conduct a nasty high current "shoot through" current spike—not a good idea. Motorola's design uses 100-nanosecond "soft" fast recovery diodes (two per IGBT) between the +Bat terminal and the –Mot terminal in close proximity to the respective IGBT collectors.

*Series inductance $L_{ext}$.* The voltage spike generated by the motor's internal and external inductance values will exceed the IGBT breakdown voltage unless checked by the external shunt capacitance. Because $L_{ext}$ is still essential in limiting ripple current through the motor, this value should be determined first, and then the values of IGBT breakdown voltage and $C_{ext}$ chosen accordingly.

*Shunt capacitance $C_{ext}$.* A small amount of inductance can generate a large voltage spike at 400 amps of motor current. At high switching frequencies (15 kHz and up) a capacitor filter is mandatory near the motor's +Bat connection. Since the still higher-frequency "edges" of the switching waveform (times in the area of 200 to 1,000 nanoseconds) also create problems, they too must be capacitor-filtered. In Motorola's design, the high-frequency capacitors are 2-$\mu$F, 400-volt metallized film types (two per IGBT), and the large capacitors are 3,300-$\mu$F, 250-volt electrolytic low-equivalent series resistance types (two per IGBT).

*External +12-volt & +15-volt dc-to-dc converters.* In the real world, protection against reverse battery hookups, excessive charging voltages, and intermittent battery voltages must be accounted for when designing these power sources. At a very minimum they must deliver stable output from a battery input source that could potentially vary widely. As an alternative, Motorola recommends that two +15-volt dc-to-dc converters be used—one for the analog functions and one for the IGBT drivers—to assure motor controller operation as long as the main battery is operational.

*PWM speed control.* As with the other PWM designs, the throttle-control potentiometer is connected to deliver a voltage proportional to its position; in this design from about 0.9 volts at zero speed to 4.1 volts at wide open throttle. These correspond to the 0 percent and 100 percent PWM output signal points, respectively. A low-pass filter on the potentiometer line reduces any noise on the input signal. In this case, the addition of a simple resistor-capacitor network limits the acceleration rate to 0.3 seconds. The resistor $R_1$ and capacitor $C_1$ connected to the Osc RC pin control the oscillator frequency, which is set for 16 kHz in this case. $R_1C_1$ should be chosen for minimum temperature drift over the operating temperature range.

*Speed control fault detector.* Unlike other designs, the MC33033 PWM IC allows you to design around the potentially fatal runaway speed condition caused by a broken or loose wire or other open-potentiometer condition. In this box, you have a comparator that monitors the voltage across a throttle potentiometer resistor. If this voltage drops to zero, the comparator's output toggles low, and a signal is sent to the Out Enable pin on the PWM chip that disables the PWM's output.

*IGBT current limit sensor & adjustment.* The MC33033 PWM IC also allows you to design current-limiting adjustment, protection, and thermal compensation into the controller. In this design, the IGBT on voltage plus diode drop varies from 1.5 volts to 3.5 volts—corresponding to motor currents of 300 amps and 900 amps. In this box, corresponding 1.6-volt and 3.6-volt input levels are delivered to a comparator whose trip point—again disabling the output if current levels are exceeded—is set by its own reference voltage input level. An additional NPN bipolar transistor mounted near the IGBT is used for temperature compensation—the signal level fed back to the comparator reduces the motor current as the IGBT temperature rises, thus reducing the chances of IGBT heat-induced failure.

*Heatsink temperature detector & fan control.* To further control the satisfactory operation of this design over a range of temperatures, this box contains circuitry that monitors heatsink temperature by sensing the base-emitter voltage of an NPN bipolar transistor mounted on the heatsink, using two comparators. One comparator is set to trip at a condition corresponding to a temperature of greater than 20 degrees C and turn on the heatsink fan. This circuit has a 5-degree hysteresis factor in it; the temperature has to drop to less than 15 degrees C before the heatsink fan is shut off again. The other comparator is set to trip if the heatsink temperature rises to greater than 75 degrees C. When this comparator trips it reduces the PWM potentiometer throttle control input signal by 50 percent, thus limiting the PWM's output signal, which in turn limits output current and eventually results in heatsink temperature reduction.

The Motorola controller design described here was actually used in an Arizona Public Service/Southern California Edison-sponsored 1992 Saturn developed by Driesbach Electromotive and, more recently, in Motorola's own third-generation design controller and power module test bed—a 1993 Dodge Dakota pickup—with great results. So if you are contemplating building your own dc motor controller, take advantage of Motorola's experience and contact them for the latest in controller design details and component values.

## Building a modern ac motor controller with the MC33033

ac motor control of three-phase induction motors is more complicated than dc motor control. You have to control motor speed, and you have to generate the three time-varying ac voltage phases spaced 120 degrees apart in time all from a nominally varying dc voltage battery source. There are also more failure modes to design around. With a dc motor, a hiccup in the drive signal causes a slight roughness in the motor signal. With an ac motor, this same hiccup could permanently destroy the system by turning on both the top and bottom control devices at the same time. In general, ac motor controllers require more protection devices to isolate against noisy or extraneous control signals or accidental motor faults. This translates to far more required components, which means greater cost and longer design times, etc. While numerous ac controllers exist in labs and industry today, all these factors tend to restrict building your own ac controller to only the most ambitious and qualified do-it-yourselfers from the EV converter group. But if you have the desire, Motorola can also help you here. Their previously referenced dc Motor article also covers ac, and they have other ac motor control application notes—call the Motorola Technical Publications 800 number and ask for the latest.[8] Let's take a brief look at an actual Motorola ac motor controller design based on the MC33033.

The motor chapter showed that the speed-torque relationship of a three-phase ac induction motor is governed by the amplitude and frequency of the voltage applied to its stator windings (the upper left part of FIG. 8-7 depicts this relationship). The best way to change the speed of an ac induction motor is to change the frequency of its stator voltage. As you can see in FIG. 8-7, a change in frequency results in a direct change in speed, and if you change the frequency in proportion to the voltage (both at ¼, ½, ¾, etc.), you get the speed-torque curves shown.

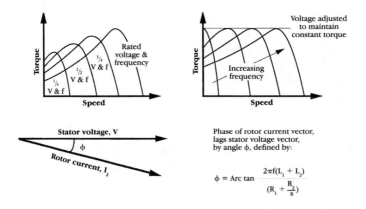

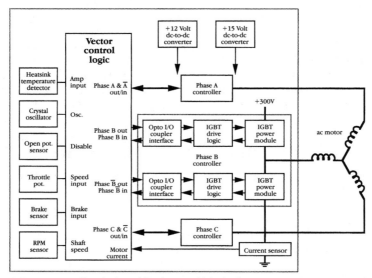

**8-7** ac controller design objectives and block diagram.

The motor chapter's equation (29) illustrated that

$$T_{max} = (3 \times \text{\# of poles} \times V^2)/(4(2\pi f)^2(L_1 + L_2))$$

This equation means that if you maintain the voltage-to-frequency ratio (V/f) constant, maximum torque can be delivered throughout the speed range, as is shown by the upper right graph of FIG. 8-7.

As the motor chapter's equation (27) showed, an ac induction motor's input impedance (for the model used) was given by

$$Z_{in} = (R_1 + (R_2/s)) + j2\pi f(L_1 + L_2)$$

where $R_1$, $R_2$ and $L_1$, $L_2$ are the stator and rotor resistance and inductance, respectively, and s is the slip. This motor's input current would then be described by

$$(11) \quad I_{in} = V/((R_1 + (R_2/s))^2 + (2\pi f)^2(L_1 + L_2)^2)^{1/2}$$

where V is the input voltage. At any instant in time, the varying voltage and current values can be expressed as vector values with an amplitude and phase value. If V is taken as the reference vector as shown in FIG. 8-7, then the phase angle between it and the current vector is given by,

$$(12) \quad \phi = \text{Arc tan } 2\pi f(L_1 + L_2)/(R_1 + (R_2/s))$$

Knowing the voltage and frequency ratio you want to maintain allows you to calculate the voltage, current, and output torque relationship for any values of input voltage and frequency using vector math and lookup tables. In simpler terms, if you feed the speed and torque values you want to some sort of "smart box," it can provide the voltage and frequency necessary to generate the proper motor control signals.

The bottom of FIG. 8-7 shows how this controller would look in the real world using the MC33033 PWM IC. The ac controller application utilizes the full three-phase capability of the MC33033 PWM IC—now located inside the *vector control logic* box. As you can see, complementary output signals are required to control each phase—multiplying the amount of electronics required by a factor of six in the opto, drive, and power stages. In addition, the vector control logic (the "smart box") adds more electronics. The current sensor, throttle control, open potentiometer sensor, heatsink sensor, and oscillator boxes are similar to those for the dc motor case, but the more stringent ac motor requirements necessitate additional RPM and brake sensor electronics. Motorola's recommended voltage to operate this ac motor-controller design is now up to 300 volts—more than your average trunkful of batteries for the dc motor case.

Fortunately, the power of 32-bit microcomputer chips such as the Motorola MC 68332 assists this application area. You can bank on this being an explosive area of future activity, because once you add this extremely fast and smart microcomputer chip, you can do so many other things with it; e.g., monitor battery voltage and current levels and calculate the remaining range of the vehicle (just like today's gasoline vehicle trip computers); and detect and automatically correct out-of-tolerance conditions; and monitor/control other onboard systems.

Unfortunately, unless your last name is Tesla or Steinmetz, all this additional complexity means a lot more sleepless nights in the garage for even the most enthusiastic build-it-yourselfers, and is the real reason why three-phase ac induction motors—despite their enormous advantages—have not reached the do-it-yourself EV converter ranks yet. But stay tuned, that day is rapidly coming. Meanwhile, let's look at the best solution for your EV conversion project of today.

## TODAY'S BEST CONTROLLER SOLUTION

If you've previously read chapter 7 on motors, you already know this book recommends a series dc motor as the best motor solution for today's EV converters. This greatly simplifies our choice of motor controllers—we only have to choose from those in the dc motor universe. But there are still numerous dc controller vendors and models to choose from, and you have to figure out which one to use. The same words of advice apply in the controller area as in the motor area: If someone tells you there is only one controller solution for a given application, ask another person. Like motors, you can probably find three or four good controllers for any application because there are only shades-of-gray controller solutions. And, as with the recommended motor, the controller recommended here is not *the* solution, it is just *a* solution that happened to work best in this case.

## An off-the-shelf PWM dc motor controller wins the race

You can do something today that EV converters of a decade ago could only dream about—pick up the telephone and order yourself a brand-new dc motor controller from any one of a number of sources. You can have it in your hands a few days later, mount it, hook up your EV's electrical wiring and throttle control to it, and be up and running with virtually a 99 percent chance of everything working the first time.

Like series dc motors, today's dc controllers are readily available from many sources, they work great, most of them are of the PWM variety, they are easily installed in different vehicles, and the price is right. A modern, off-the-shelf PWM dc motor controller is not the ultimate, but it's pretty close to the best current solution. More important, it's one that most EV converters will have no trouble in implementing today. After you do your first conversion, are the acknowledged genius in your neighborhood, and know what you really like and don't like, you can get fancy and exotic.

The dc PWM controller recommended here is from Curtis PMC, a division of Curtis Instruments, Inc. of Mt. Kisco, New York. As with the motors, don't read anything into its appearance here. Curtis is only one of a large number of controller manufacturers from the list in chapter 5, and the recommended controller model is only one out of a number they manufacture.

The Curtis PMC model 1221B-7401 dc motor controller, shown in FIG. 8-8, is already very familiar to you. It's characteristics are:

- PWM type controller.
- Based on MOSFET technology.
- Runs at a constant switching frequency of 15 kHz.
- Requires use of an external 5-kilohm throttle potentiometer.
- Has automatic motor current limiting.
- Has thermal cutback at 75 through 95 degrees C.
- Has high pedal lockout (prevents accidental startup at full throttle).
- Has intermittent-duty plug braking.
- Has overvoltage and undervoltage protection.
- Has user accessible adjustments for motor current limit, plug braking current limit, and acceleration.
- Comes in a waterproof heatsink case.

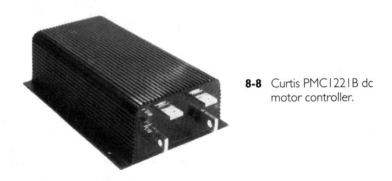

**8-8** Curtis PMC1221B dc motor controller.

The controller is also well-matched in characteristics to the Advance model FB1-4001 series dc motor, particularly in the impedance area (you read about the importance of this for peak power transfer). If the Curtis PMC dc controller characteristics sound familiar, it's because they use the PWM IC technology you've been reading about throughout this chapter, and bring all these benefits to you with additional features in a rugged, preassembled, guaranteed-to-work package at a fine price. The same reputable vendor comments from the motor chapter also apply here.

Installation and hookup is a breeze. If you look closely at the controller terminals, you'll notice the markings M–, B–, B+ and A2 appear (listed clockwise from the lower left when facing the terminals). You already know that the first three correspond to the –Mot, –Bat, and +Bat markings on the terminal bars in FIG. 8-5 and FIG. 8-6. The A2 marking means *Armature 2*, the opposite end of the armature from the *Armature 1* winding that is normally connected to the +Bat, or in this case the B+ terminal. Anything else you might want to know is covered in the Curtis PMC manual that accompanies the controller.

Can you build your own controller? Absolutely. But this off-the-shelf controller gets you up and running quickly, and puts your EV conversion on the road with the least fuss. That's why you will also find a number of professional conversion shops using it as well. Now, let's look at the future.

## TOMORROW'S BEST CONTROLLER SOLUTION

While the series dc motor and PWM controller are unquestionably best for today's first-time EV converters, the bias of this book toward ac induction motors has a definite reason—ac motors are inherently more efficient, more rugged, and less expensive than their dc counterparts. Only the ac controller technology has been lacking. That's why nearly every newly-designed commercial EV today utilizes one or more ac induction motors or its closely-related cousin, the brushless dc motor. What's in the labs today will be available to you in the not-too-distant future and, beyond that, continued improvements in solid-state ac controller technology could put ac motors in every EV conversion of the future. Let's look at developments in two areas—systems and components—that virtually guarantee this outcome.

### AC Propulsion Inc. to the rescue . . . today

While others have only dreamed or talked about it, AC Propulsion Inc. has done it—designed an integrated ac induction motor and controller that has been installed into numerous prototype EVs. In fact, the October 1992 *Road and Track* with the picture of AC Propulsion's Honda CRX smoking its tires has become a collector's item among EV aficionados.[9] AC Propulsion's Cal Tech alumnus co-founders Alan Cocconi (of GM's prototype Impact ac propulsion system fame) and Wally Rippel (of chapter 3's great Great Electric Vehicle Race of 1968 fame—his team won) just had a better idea and did something about it.

The Burbank, California Alternate Transportation Exposition of September, 1992 gave me the privilege of an AC Propulsion Honda CRX closeup. Alan Cocconi stands next to his company's Honda CRX in FIG. 8-9. While AC Propulsion's AC-100 EV controller is complex (and fills the engine compartment), its drivetrain is simple (only 1st gear is installed), and the results are astonishing. Driving it is an absolute breeze and a big surprise. After a small preflight checklist from Alan—"You do have a driver's license, don't you?"—he slipped into the passenger's seat while I took over the wheel. After turning the key, one of the three buttons on the left of the dash  selects forward, neutral, or reverse. Once on the highway, the induction motor behaves like the smoothest automatic transmission imaginable—with one big

**8-9** Alan Cocconi in front of AC Propulsion's Honda CRX EV conversion.

difference. All the motor's torque is instantly available at nearly any speed. After Alan counseled, "Don't be afraid to step on it," I did, and was abruptly pushed back into my seat while silently forming the word "Wow" on my lips. When I turned to look at Alan, he was grinning from ear to ear. Next, Alan directed my attention to the regeneration lever on the right of the dash. All the way in one direction is full regeneration. Take your foot off the accelerator pedal and the vehicle slows down immediately, without even touching the brake. Push the lever all the way in the other direction and you have no regeneration. Take your foot off the accelerator pedal and the vehicle coasts and coasts.

After returning to earth, Alan gave me the walk-around tour, and talked about the batteries (AC Propulsion's Honda CRX uses 28 conventional 12-volt deep-discharge lead-acid batteries that produce 336 volts), controller, motor, charging philosophy, and temporary instrumentation (to monitor performance).

Back at their exhibit-hall booth, AC Propulsion employee Tiffany Mitchell shows off the other benefit of the design (FIG. 8-10)—the incredible small size of the motor! This 4-pole ac induction motor puts out 110 ft-lbs of torque at anywhere from 0 to 5000 RPM, and an astonishing 120-HP maximum between 6500 RPM and 10,000 RPM (Maximum RPM is 12,000), yet it weighs only 100 pounds. Recall that chapter 7's recommended series dc motor put out 70-HP maximum and weighed 143 pounds. Directly behind and above the motor on the booth table is the ac controller. It includes a 100-kVA PWM-style inverter; battery-charging circuitry; control logic; interfaces for control pedals, dash instruments, and lighting; an auxiliary 12-volt dc power supply; and numerous interlocks for operator safety. If you pop the lid on the controller, you can see it uses off-the-shelf components (MOSFET power devices) and standard fabrication techniques.

AC Propulsion's own literature says it best: "The AC-100, combined with presently available lead-acid batteries, allows exceptional acceleration (0 to 60 in

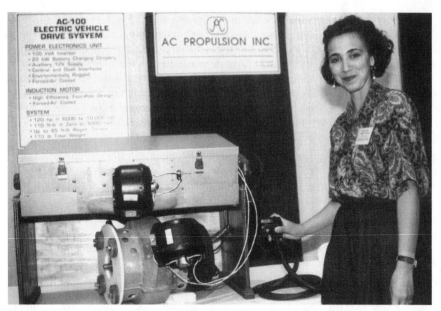

**8-10** AC Propulsion's Tiffany Mitchell shows off ac induction motor in front of ac controller/inverter.

7.8 seconds), gradeability (45 percent grade limit), and range (131 miles at 55 mph) values to be obtained simultaneously for both conversion and 'ground up' type vehicles." The AC Propulsion design also contains its own state-of-the-art battery charger (eliminating the need for an additional external charger), whose 20-kW capability at unity power factor allows you to fully recharge in only two hours from a 240-volt 40-amp ac outlet, and overnight from a conventional 120-volt ac source.

You cannot run out and buy one of these systems for your own EV conversion project today—it costs several times chapter 4's conversion budget by itself, and AC Propulsion deals mostly with OEMs and prime contractors with volume manufacturing potential. But, given the propensity of solid-state devices to double in price performance every few years, the trail AC Propulsion has blazed has an obvious destination. AC Propulsion's Burbank showing was a consciousness-expanding experience of what the future will be like for all EV converters.

## Fuzzy logic, DSP, & intelligent modules

While innovative systems firms such as AC Propulsion have made the best of today's available components, developments are coming down the road that will assist them (and others like them) even more. Three areas stand out: fuzzy logic, DSP, and intelligent modules. Let's briefly look at each in turn.

**Fuzzy logic**  Fuzzy logic was coined from the *fuzzy set* theory developed by UC Berkeley professor Lofti A. Zadeh in 1965.[10] The Japanese jumped on it immediately, and present-day cameras, trains, air conditioners, and vacuum cleaners bear the fruits of their labors. Automakers have been much slower to pick up on this field, but discover it they have, and it's an industry buzzword today.

In simple terms, computers think in binary logic: 1s and 0s, on and off, yes and no. Fuzzy logic gives you a mathematical way of saying, "Yeah, that seems about right" and implementing it with electronics. Rather than complex feedback systems that home in on the results you want, you can implement a simple logic approach that moves you there directly. As a result, you save components and dollars. Camera-makers, with whom miniaturization has been raised to an art form, were ecstatic with the technology and have been using it for several generations now. What fuzzy logic means to future EV converters is smaller and less expensive electronics. So there will definitely be a fuzzy in your future.

**DSP (digital signal processing)** Optical data processing can whip digital data processing hands down because it handles everything in parallel as opposed to the serial, bit by bit approach of today's microcomputer chips. While the mathematical description of the process is awesome (Fourier integrals, etc.), the simple description is, you transform matters from the time domain into the frequency domain, do your analysis, then transform the results back.

While even the smallest of optical data processors occupies a sizable part of a laboratory laser bench, a subset of the optical data processing technique—digital signal processing—can be put in a box and used to augment today's microcomputers. When the design is perfected, the digital signal processor itself can be put on a chip. Golfers ought to know that the heart of today's highly realistic golf simulator (that you can rig up in your basement) is nothing other than a DSP board.[11]

Recall the ac vector controller of the previous section? Its complicated conversions require a significant number of components to make it happen in the real world. Augment it with a few DSP chips and you will some day be able to hold an entire space vector PWM controller in the palm of your hand. What DSP means to future EV converters is smaller and less expensive electronics. So there will definitely be DSP in your future.

**Intelligent modules** If decentralized management is the epitome of management efficiency, then decentralized intelligence should do the same for a solid-state controller. You've already seen the principle at work in today's computer world; networks of decentralized microcomputers have replaced the centralized mainframe.

Applied to the controller world, it means that when something happens at one of its extremities—say it's one of the IGBT power modules—you want to deal with the matter at that extremity (shutdown, voltage reduction, etc.) rather than having to go all the way to the controller's central "brain" for a decision. Motorola's trademarked SMART-MOS and SMALLBLOCK intelligent module technologies have already made significant strides in power device gate drive and short circuit protection. What intelligent module technology means to future EV converters is more reliable, powerful, and less expensive electronics. So there will definitely be intelligent modules in your future.

# 9
# Batteries

*More than 85 percent of an average car's gasoline energy
is thrown away as heat.*

Dr. Paul MacCready, *Discover*, March 1992

Today's batteries, motors, and controllers are all superior to their counterparts a decade ago. Contrary to those who say you'll need a different type of battery before EVs are suitable at all, today's conventional lead-acid batteries of the *deep-discharge* variety are perfectly adequate for your EV conversion.

Improved lead-acid batteries are routinely available from numerous suppliers at a good price. Assuming a proper system design, if you install batteries correctly—and maintain them conscientiously—you don't have to worry about replacing them for tens of thousands of miles. Future batteries will be lighter and more powerful, but can hardly be more convenient than they are today.

In this chapter you'll learn about how batteries work and the language used to discuss them. You'll be introduced to the different battery types, and their advantages and disadvantages. Then we'll look at the best type of battery for your EV conversion today, the lead-acid type used in chapter 11's conversion, and look at probable future battery developments.

## BATTERY OVERVIEW

Your EV's chassis involved mechanical aspects, and its motors and controllers dealt with electrical ones. Its batteries will now take you into the chemical area. While there are all sorts of battery developments going on in the labs, the objective here is to give you a brief battery background, and introduce the lead-acid batteries you'll be working with on your EV conversion. Many good books about batteries are available both at introductory[1] and the more advanced levels[2] for those who want more data.

Because your EV battery *pack*—a collection of 16 to 24 6-volt (or 12-volt equivalent) individual lead-acid batteries—represents the single largest replacement cost item, and quite possibly was also your largest initial expense item, it's worth spending some time learning about batteries so you can choose and use them wisely.

Batteries are the life's breath of your EV, and every EV converter should be familiar with them on three levels. To graduate from battery class you need to:

- Understand what goes on inside a battery.
- Become familiar with a battery's external characteristics.
- Learn the pros and cons of working with real world batteries.

Knowledge in these three areas is a sound business investment that can save you time and money. Using the tried-and-true, 100-year-old, lead-acid battery as an example, let's look at each of the three areas in turn, starting with what goes on inside a battery.

## INSIDE YOUR BATTERY

A battery is a chemical factory that transforms chemical energy into electrical energy. You don't need to know all the detailed inner workings of your battery, but you should have a basic understanding of its elements and processes. We'll start with a little history and overview, then get into the pieces and parts.

Your first acquaintance with chemical batteries probably occurred in high school biology class when your instructor reproduced Luigi Galvani's 1786 experiment by placing a voltage across a frog's leg and making it twitch. Alessandro Volta "went to school" on the Galvani phenomenon. He reasoned that if a voltage across two dissimilar metals produced a reaction in the frog's leg, two dissimilar metals in a conductive solution would produce a voltage, and the first battery—his *Voltaic pile*—was born in 1798.

Battery improvements have steadily occurred ever since, but the basic principles have remained unchanged. Battery action takes place in the *cell*—the basic battery building block—that transforms chemical energy into electrical energy. A cell contains the two active materials or *electrodes* and the solution or *electrolyte* that provides the conductive environment between them. There are two kinds of batteries: in a *primary* battery, the chemical action eats away one of the electrodes (usually the negative), and the cell must be discarded or the electrode replaced; in a *secondary* battery, the chemical process is reversible, and the active materials can be restored to their original condition by *recharging* the cell. A battery can consist of only one cell, as in the primary battery that powers your flashlight, or several cells in a common container, like the secondary battery that powers your automobile starter.

### Active materials

In chemical jargon, the active materials are defined as *electrochemical couples*. This means that one of the active materials, the positive pole or *anode*, is electron deficient; the other active material, the negative pole or *cathode*, is electron rich. The active materials are usually solid (lead-acid) but can be liquid (sodium-sulphur) or gaseous (zinc-air, aluminum-air). TABLE 9-1 gives a snapshot comparison of a few of these elements.

When a load is connected across the battery, the voltage of the battery produces an external current flow from positive to negative corresponding to its internal electron flow from negative to positive. The observed voltage in a galvanic cell is the sum of what is happening at the anode and cathode. To make an ideal battery, you'd choose the active material that gave the greatest *oxidation potential* at the anode coupled with the material that gave the greatest *reduction potential* at the cathode that were both supportable by a suitable electrolyte material. This means you'd like to pair the best reducing material—Lithium (+3.045 volts with respect to Hydrogen as the ref-

**Table 9-1  Element oxidation voltages comparison**

| Periodic Table Group | 1A Light Metals | 2A Light Metals | 8 Heavy Metals | 8 Heavy Metals | 8 Heavy Metals | 1B Low-Melting Metals | 2B Low-Melting Metals | 4A Low-Melting Metals | 6A Nonmetals | 7A Nonmetals |
|---|---|---|---|---|---|---|---|---|---|---|
| Element-Symbol | Lithium  Li | Beryllium  Be | | | | | | | | Fluorine  F |
| Number-Voltage | #3 +3.045 | #4 +1.85 | | | | | | | | #9 −2.87 |
| Element-Symbol | Sodium  Na | Magnesium  Mg | | | | | | | Sulfur  S | Chlorine  Cl |
| Number-Voltage | #11 +2.714 | #12 +2.37 | | | | | | | #16 +0.51 | #17 −1.36 |
| Element-Symbol | Potassium  K | Calcium  Ca | Iron  Fe | Cobalt  Co | Nickel  Ni | Copper  Cu | Zinc  Zn | | Selenium  Se | Bromine  Br |
| Number-Voltage | #19 +2.925 | #20 +2.87 | #26 +0.44 | #27 +0.277 | #28 +0.246 | #29 −0.337 | #30 +0.763 | | #34 +0.78 | #35 −1.065 |
| Element-Symbol | Rubidium  Rb | Strontium  Sr | | Rhodium  Rh | Palladium  Pd | Silver  Ag | Cadmium  Cd | Tin  Sn | Tellurium  Te | Iodine  I |
| Number-Voltage | #37 +2.925 | #38 +2.89 | | #45 −0.6 | #46 −0.987 | #47 −0.7995 | #48 +0.403 | #50 +0.136 | #52 +0.92 | #53 −0.536 |
| Element-Symbol | Cesium  Cs | Barium  Ba | | | Platinum  Pt | Gold  Au | Mercury  Hg | Lead  Pb | | |
| Number-Voltage | #55 +2.923 | #56 +2.90 | | | #78 −1.2 | #79 −1.68 | #80 −0.854 | #82 ≈0.126 | | |

erence electrode)—with something that just can't wait to receive its electrons, or the best oxidizing material—Fluorine (–2.87 volts with respect to Hydrogen)—with something that just can't wait to give electrons to it.

In practice, many other factors enter the picture, such as availability of material; ease in making them work together; ability to manufacture the final product in volume; and cost. As a result of the trade-offs, only a few electrochemical couple possibilities make it into the realm of commercially-produced batteries that you will meet later in the chapter.

## Electrolyte

The electrolyte provides a path for electron migration between electrodes and, in some cells, also participates in the chemical reaction. The electrolyte is usually a liquid (an acid, salt, or alkali added to water), but can be in jelly or paste form. In terms of chemistry, a battery is electrodes and electrolyte operating in a cell or container in accordance with certain chemical reactions. Figure 9-1 shows the chemistry of a very simple lead-acid battery cell that will be examined in the next few sections as it undergoes the four stages: fully charged, discharging, fully discharged, and charging. It consists of an electrode made of sponge lead Pb, another electrode made of lead peroxide $PbO_2$, and an electrolyte made of a mixture of sulfuric acid $H_2SO_4$ diluted with water $H_2O$.

## Overall chemical reaction

Combining active material elements into compounds that further combine with the action of the electrolyte significantly alters their native properties. The true operation of any battery is best described by the chemical equation that defines its operation. In the case of the lead-acid battery, this equation is given by

$$(1)\ Pb + PbO_2 + 2H_2SO_4 \leftrightarrow 2PbSO_4 + 2H_2O$$

The left side of the equation represents the cell in the charged condition, the right side represents the discharged cell. In a charged lead-acid battery, its positive anode plate is nearly all lead peroxide $PbO_2$, its negative cathode plate is nearly all sponge lead Pb, and its electrolyte is mostly sulfuric acid $H_2SO_4$—the top of FIG 9-1. In a discharged condition, both plates are mostly lead sulfate $PbSO_4$, and the acid electrolyte solution used in forming the lead sulfate, becomes mostly water $H_2O$—the bottom of FIG. 9-1.

## Discharging chemical reaction

The general equation gives a more accurate view when separately analyzed at each electrode. The discharging process is described at the anode by

$$(2)\ PbO_2 + 4H + SO_4^{--} + 2e^- \rightarrow PbSO_4 + 2H_2O$$

The discharging process is described at the cathode by

$$(3)\ Pb + SO_4^{--} - 2e^- \rightarrow PbSO_4$$

When discharging, the cathode acquires the sulfate $SO_4$ radical from the electrolyte solution and releases two electrons in the process. These electrons are acquired

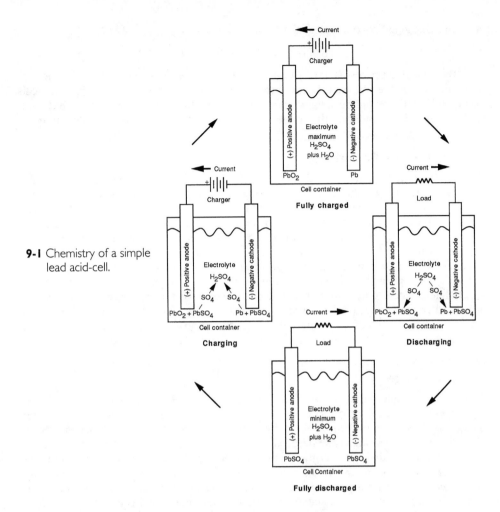

**9-1** Chemistry of a simple lead acid-cell.

by the electron-deficient anode. The electron flow from negative cathode to positive anode inside the battery is the source of the battery's power and external current flow from positive anode to negative cathode through the load. In the process of discharging (right of FIG. 9-1), both electrodes become coated with lead sulfate $PbSO_4$—a good insulator that does not conduct current—and the sulfate $SO_4$ radicals are consumed from the electrolyte. At the same time the physical area of the sponge-like plates available for further reaction decreases as it becomes coated with lead sulfate; this increases the internal resistance of the cell, and results in a decrease of its output voltage. At some point—before all the sulfate $SO_4$ radicals are consumed from the electrolyte—there is no more area available for chemical reaction and the battery is said to be fully discharged.

## Charging chemical reaction

The charging process is described at the anode by

$$(4)\ PbSO_4 + 2H_2O - 2e^- \rightarrow PbO_2 + 4H^+ + SO_4^{--}$$

The charging process is described at the cathode by,

$$(5)\ PbSO_4 + 2e^- \rightarrow Pb + SO_4^{--}$$

The charging process (left of FIG. 9-1) reverses the electronic flow through the battery and causes the chemical bond between the lead Pb and the $SO_4$ sulfate radicals to be broken, releasing the sulfate radicals back into solution. When all the sulfate radicals are again in solution with the electrolyte, the battery is said to be fully charged.

## Electrolyte specific gravity

The specific gravity of any liquid is the ratio of the weight of a certain volume of that liquid divided by the weight of an equal volume of water. Or the specific gravity of a material can be expressed as its density divided by the density of water, because the density of any material is its mass-to-volume ratio. Water has a specific gravity of 1.000. Concentrated sulfuric acid has a specific gravity of 1.830—1.83 times as dense as water. In a fully-charged battery at 80 degrees F, water and sulfuric acid mix in roughly a four-to-one volume ratio (25 percent sulfuric acid) to produce a 1.275 specific gravity, and the sulfuric acid represents about 36 percent of the electrolyte by weight.

While specific gravity is not significant in other battery types, it is important in lead-acid batteries, because the amount of sulfuric acid combining with the plates at any one time is directly proportional to the discharge rate (current × time—usually measured in ampere hours), and is therefore a direct indicator of the state-of-charge.

## State-of-charge

Battery voltage, internal resistance, and amount of sulfuric acid combined with the plates at any one time are all indicators of how much energy is in the battery at any given time. Frequently this is given as a percentage of its fully charged value; e.g., 75 percent means that 75 percent of the battery's energy is still available and 25 percent has been used.

Traditionally, the specific gravity of the electrolyte was used as a measurement. Today, because voltage can be used to determine a battery's state-of-charge and a hydrometer—the device used to measure specific gravity—can introduce inaccuracy and contaminate a battery's cells, state-of-charge is determined electronically.

## Gassing

As charging nears completion, another phenomenon takes place: hydrogen gas $H_2$ is given off at the negative cathode plate and oxygen gas $O_2$ is given off at the positive anode plate. This is because any charging current beyond that required to liberate the small amount of sulfate radicals from the plates ionizes the water in the electrolyte and begins the process of electrolysis (separating the water into hydrogen and oxygen gas). While most of the hydrogen and oxygen gas recombines to form water vapor (the main reason that periodic replenishing of the water is needed in this battery-type), the presence of flammable and potentially explosive hydrogen gas strongly indicates that charging be conducted in a well-ventilated area, and that you avoid lighting a cigarette.

## Equalizing

Over time, the cells of a lead-acid battery begin to show differences in their state-of-charge. Differences can be caused by temperature, materials, construction, electrolyte, even by electrolyte *stratification* (the tendency of the heavier sulfuric acid to sink to the lower part of the cell) causing premature aging of the plates in that part. The only cure for these differences is to use a controlled overcharge, equalizing the characteristics of the cells by raising the charging voltage even higher after the battery is fully charged, and maintaining it at this level for several hours until the different cells again test identical. Obviously, this produces substantial gassing, so the precautions of well-ventilated area and no smoking definitely apply.

### Electrolyte replacement

Adding sulfuric acid to a discharged lead-acid cell or battery does not recharge it—it only increases the specific gravity without converting the lead sulfate of the plates back into active material (another reason why measuring voltage is preferred). Only passing a charging current through the cell restores it to its fully charged condition. If you add sulfuric acid to a battery, always pour the acid into the water and stir the solution (with a nonmetal stirring device) so the acid does not settle to the bottom; then pour the solution into the battery.

### Sulfation

The finite life of the battery is caused by the fact that all of the $SO_4$ sulfate radicals cannot be removed from the plates upon recharge. The longer the sulfate radicals stay bonded to the plates, the harder it is to dislodge them. To postpone the inevitable as long as possible, the battery should be kept in a charged state, and equalizing charging should be done regularly.

## OUTSIDE YOUR BATTERY

When you hook up to the closed container of a battery, it exhibits certain external physical and electrical properties. You should be relatively familiar with these because they are useful to you. We'll review the terms you've already become familiar with from chapters 7 and 8, and then move into the battery-specific areas.

### Basic electrical definitions

You recall the example of the gallon plastic drinking water jug from chapter 7. Let's relate it to the key electrical definitions:

*Voltage.* The battery's force (i.e., its *potential* or *voltage*) corresponds to the water's force (its pressure or potential to do work): the height of the water in the water jug example. When you hook up a light bulb to a battery, the bulb lights. When you hook two batteries in series to double the voltage, the bulb lights even brighter.

There is another important aspect to the voltage of a battery. In the water jug analogy, the pressure of the water coming out of the jug goes down as you take more and more water out of the jug. In the same way, battery voltage goes down as you use the battery—as you use up its capacity. This important battery characteristic will be covered in more detail later in the section.

*Current.* The current (the rate of electron flow) corresponds to the rate of flow of the water coming out the bottom of the jug. When you double the voltage, you sent twice as much current through the wire and the light bulb became brighter.

*Resistance.* The resistance corresponds to the size of the hole controlling the rate of flow of the water coming out the bottom of the jug. A battery's voltage is directly related to current flow by resistance via the Ohm's law equation (1) you met in chapter 7:

$$V = IR$$

where V is voltage in volts, I is current in amps, and R is resistance in ohms. Actually, there are two resistances: the external resistance of the load (the light bulb in this case) and the internal resistance of the battery. The battery's internal resistance is important in battery efficiency (heating losses), power transfer, and state-of-charge determinations.

*Power.* Electrical power is defined as the product of voltage and current:

$$(6)\ P = VI$$

where V is voltage in volts, I is current in amps, and P is the power in watts. To use a 100-watt light bulb instead of a 50-watt light bulb requires twice the amount of power from the battery—twice the current at the same battery voltage. If the Ohm's law equation is substituted into equation (6),

$$(7)\ P = I^2R$$

This equation defines the power losses in the resistances in the circuit—either external load or internal battery.

*Efficiency.* Battery efficiency is

$$(8)\ Efficiency = Power\ Out/Power\ In$$

The principal battery losses are due to heat. These come from resistance and chemical sources: internal resistance of the battery determines its heating or $I^2R$ losses when charging and discharging; chemical reaction between the lead and the sulfuric acid produces heat (called an exothermic reaction) during charging; and chemical reaction absorbs heat (called an endothermic reaction) during recharging.

While $I^2R$ losses are present whether charging or discharging—because they are proportional to the square of current flow—battery heat rise is higher during charging (because $I^2R$ heating losses add to the internal heat-generating chemical reaction) and lower during discharging (because $I^2R$ heating losses are balanced by the internal heat-absorbing chemical reaction). Given the $I^2R$ relationship, charging or discharging at a lower current rate obviously contributes to keeping battery losses lower.

## Battery capacity & rating

Capacity and rating are the two principal battery-specifying factors. Capacity is the measurement of how much energy the battery can contain, analogous to the amount of water in the jug. Capacity depends on many factors, the most important of which are:

- Area or physical size of plates in contact with the electrolyte.
- Weight and amount of material in plates.
- Number of plates, and type of separators between plates.
- Quantity and specific gravity of electrolyte.
- Age of battery.
- Cell condition—sulfation, sediment in bottom, etc.
- Temperature.
- Low voltage limit.
- Discharge rate.

Notice the first four items have to do with the battery's plates and electrolyte—its construction; the next two items concern its history; and the last three depend on how you are using it at the moment. We'll get into all the details, but keep in mind that the most truthful thing you can say about battery capacity is—it depends.

Battery capacity is specified in ampere-hours. A battery with a capacity of 100 ampere-hours could in theory deliver either 1 amp for 100 hours or 100 amps for 1 hour. This doesn't help you any more than would drawing a straight line on a map if someone asked you for a destination. You need the second coordinate, the second factor—rating.

A battery's rating is the second specifying factor. It refers to the rate at which it can be charged or discharged. It is analogous to how fast the sink will fill up with the water from the jug. In equation form,

(9) Battery Rating = Capacity/Cycle Time

In this equation, the rating is given in amperes for a capacity in ampere-hours and a cycle time in hours. In practical terms, a battery with a capacity of 100 ampere hours that can deliver 1 amp for 100 hours (known as a C/100 rate) would not necessarily be able to deliver the much higher 100 amps for 1 hour (known as a C/1 rate). You can only get the water out of the jug so fast.

Requesting 10 amps from a fully charged 100 AH capacity battery reflects a C/10 rate; this same request reflects a much lower C/40 rate from a 400 AH battery. In other words, smaller batteries have to deliver energy faster in relation to their size, or larger batteries have lower discharge rates in relation to their capacity.

Capacity of commercial batteries is standardized by the Battery Council International (BCI) into several usable figures. Two figures, a *20-hour capacity* and a *reserve capacity* are usually given for every battery depending on its application.

*20-hour capacity*. This is a battery's rated 20-hour discharge rate—its C/20 rate. Every battery is rated to deliver 100 percent of its rated capacity at the C/20 rate—if discharged in 20 hours or more. If a battery is discharged at a faster rate, it will have a lower ampere-hour capacity.

*Minutes at 25 amps reserve capacity*. This is the number of minutes a fully charged battery can produce a 25-amp current. This is the automotive starting battery rating that tells you how long your starter battery will power your automotive accessories if your fan belt breaks and disconnects the alternator; i.e., how many minutes you have to get to the nearest gas station.

*Minutes at 75 amps reserve capacity*. This is the number of minutes a fully charged battery can produce a 75-amp current. This is the golf cart battery rating—because 72 minutes translates to about the amount of time it takes to play two rounds of golf. So this figure tells you how long your batteries will power your golf cart: two rounds, three rounds, etc.

*Three-hour reserve capacity*. This is the BCI standard currently coming into vogue covering EV users. It is defined as 74 percent of the 20-hour rate. Because three hours translates to the average amount of time an EV might be in daily use, commuting, shopping, etc.:

(10) 3-Hr Reserve Capacity = 0.74 × 20-Hr Reserve Capacity

## The gentle art of battery recharging

The objective with batteries is to maintain a balance. How fast batteries are filled and emptied are critical factors determining both their immediate efficiency and ultimate longevity. Where batteries are filled and emptied—relative to their state of charge—are equally critical factors.

Because urban driving patterns for EVs are highly intermittent, battery discharge rates will vary all over the map. While energy is drawn out of your battery pack a lot harder than C/20 on startup and acceleration, you're only doing this momentarily, and the urban driving cycle usually implies that an EV's battery pack is given a certain amount of "rest" between discharge requests. The bottom line is:

- Avoid placing continuous, heavy, C/1-type loads on your batteries anywhere in their state-of-charge cycle. A battery pack that can deliver 100 percent of its capacity when discharged in $X$ time might only deliver 50 percent of its capacity when discharged in $X/3$ time. Remember the example of the water flowing out of the jug—the faster you take it out, the less pressure there is to push out the remaining amount.
- Avoid over-discharging your batteries when they're below 20 percent state-of-charge. High-rate discharging below the 20 percent state-of-charge can greatly reduce battery life or even destroy them.

Unlike discharging, you can control the destiny of your batteries during the charging process. In fact, it's vital that you do, because both overcharging and undercharging shorten battery life. Continually overcharged or too rapidly charged batteries can be destroyed; constantly undercharged batteries become sulfated and inefficient. Chapter 10 covers modern battery rechargers that can help you. The top of FIG. 9-2 shows the ideal battery charging curve.

Confine heavy charging within the 20 percent to 90 percent of the state-of-charge range, because a lead-acid battery's ability to store energy is reduced when almost full or nearly empty. Below 20 percent and above 90 percent, C/20 is the most efficient rate (divide the capacity of your battery in ampere-hours by 20) to charge your batteries. In the 20-to-90 percent range, C/10 delivers the fastest rate at which it's efficient to charge a lead-acid battery; it wastes more heat than at the C/20 rate, but saves time. Below 90 percent, control charging by limiting the current so as not to charge nearly empty batteries too rapidly. Above 90 percent, limit voltage so as not to overcharge the batteries (or possibly damage other attached electronic devices).

## Temperature determines performance for lead-acid batteries

Because the energy stored in a lead-acid battery is dependent on a chemical reaction, everything about it is affected by temperature: capacity, voltage, current, etc. Because the lead-acid chemical reaction is most efficient at 78 degrees F, most battery manufacturers rate their batteries at this "standard" temperature. The middle left diagram in FIG. 9-2 shows that the output of a battery is strongly affected by the temperature. Notice that only about 70 percent of a battery's capacity is available at 32 degrees F, while about 110 percent of battery capacity is available at 110 degrees F. The obvious conclusion here is that EV converters in colder climates need to opt for the next larger model battery model in the line, while their sunbelt counterparts can either enjoy their extra power, or downsize a notch if they live in the desert.

Notice also that batteries don't freeze at 32 degrees F because of the concentration of sulfuric acid $H_2SO_4$ in the electrolyte. This concentration increases (the specific gravity is higher) with increasing temperature and vice versa, so it's important to keep low-temperature operation batteries near fully charged at all times; letting the electrolyte freeze in a lead-acid battery can result in permanent battery cell damage.

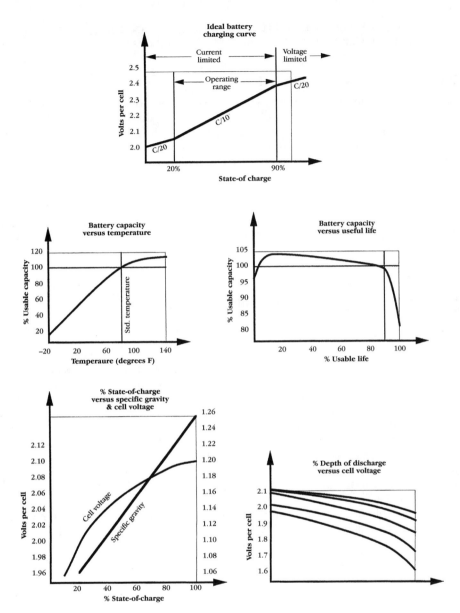

**9-2** Lead-acid battery charging and discharging characteristics.

## Age determines performance for lead-acid batteries

Battery capacity is also highly dependent on age. The middle right diagram in FIG. 9-2 shows that the battery's capacity starts at about 95 percent when brand-new, rises to about 105 percent after it has been used for about 20 percent of its lifetime, stays fairly level, then drops off rapidly after 90 percent of its lifetime. One observation is that a brand new EV battery pack will not give you as good a result as one that's been used awhile. Another observation is that once you begin to see battery performance go down significantly, it's time to think about buying another set.

## Battery charge—Use it or lose it

Because every battery has an internal resistance, it will discharge by itself if it sits around doing nothing. Temperature and battery age are the two main determinants of how rapidly this takes place—increasing temperature and age hasten the process. A 5 percent capacity loss per week is the average for lead-acid batteries (or 50 percent in 10 weeks) whether you use them or not, so periodically recharge the batteries in your stored EV. This is also the reason why buying the batteries is the last step in your EV conversion process.

An examination of FIG. 9-2's temperature curve suggests its corollary—that batteries kept at cold temperatures don't discharge themselves at all. True, but you also can't get the energy out of the battery for your own use at cold temperatures! If you are planning to store your batteries for a few months, 40 degrees F is the best temperature. Fully charge your batteries before storage and warm them up before using them.

## State-of-charge measurement—Volts or specific gravity

The lower left graph of FIG. 9-2 shows why specific gravity has been used as a battery state-of-charge indicator for so long—it's an easy-to-use-and-understand straight line. Unfortunately, it doesn't show you that temperature directly affects specific gravity (specific gravity measures higher at lower temperatures). In addition, the device used to measure it—the bulb hydrometer—is prone to calibration, compensation, and read-out errors because you're typically measuring in the range from 1.100 to 1.300 to three decimal places. And if you use a hydrometer on a regular basis, it's virtually guaranteed that you'll contaminate one or more battery cells.

A digital voltmeter readout accurate to 3½ digits (at least to 0.1 volt) is today's preferred method for measuring state-of-charge. Thanks to modern electronics, you can observe voltage levels, current levels, and/or have the voltmeter readout drive the battery charger electronics directly. You can even monitor single-cell voltages if your battery type has external cell straps, all without the trouble of opening your batteries, dealing with sulfuric acid and hydrometers, etc. Since voltage also varies with temperature (lower temperatures produce lower voltages), you can either make a little chart to help you with your individual readings or rig a circuit to do it for you or your charger automatically.

Whether you use specific gravity or the voltage method, to get the most accurate state-of-charge reading let the batteries "rest" for several hours (two hours minimum, six hours is better, 24 hours is probably optimum, if you can afford it) before taking measurements. Monitor one or a few batteries rather than the whole pack; check ambient temperature and odometer reading at the same time; and keep a logbook. Do it at a convenient time, do it consistently, and make a simple graph. Your diligence will reward you with a beautiful record of your EV's battery health.

## Discharge not in haste

The maxim "Haste makes waste" is absolutely true when discharging batteries. The lower right graph of FIG. 9-2 shows that the faster you discharge your batteries, the lower the voltage (the less capacity) you have. If you take more and more from less and less, eventually you wind up with nothing at all—a polite way of saying over-discharging kills battery life.

A corollary of this action (*depth-of-discharge*) affects the number of charge/discharge cycles your batteries can deliver. The number of cycles you can expect from your batteries is approximately given by the equation,

$$(11)\ \text{Battery Life Cycles} = K_d/\text{Depth-of-discharge in \%}$$

This equation says that the number of battery life cycles is inversely proportional to the depth-of-discharge ratio. If you consistently discharge your batteries to 90 percent, you're going to get less cycles out of them than running them down to the 50 percent depth-of-discharge area. In numbers, $K_d$ might be about 12,000 for starting batteries, 24,000 for deep-cycle batteries, and 30,000 or more for industrial batteries. These values reflect the fact that heavier-duty batteries deliver more cycles or support heavier depth-of-discharge rates better than starting batteries.

What this equation doesn't say is $K_d$ will vary with each and every individual user—because every user's application is different. If a manufacturer's literature mentions they obtained 750 cycles out of one of their batteries, it's no guarantee that you will. On the other hand, you might even do better.

Because of gassing and loss of plate material as you go above the 90 percent charging point (10 percent depth-of-discharge), liability for battery damage as you go below the 20 percent charging point (80 percent depth-of-discharge), and the fact that every lead-acid battery has a finite lifetime, the best operating guidance translates to operating your deep cycle batteries at the middle of this range—roughly the 40 to 60 percent depth-of-discharge range—for optimum balance between cycle life, depth of discharge, and the actual physical (calendar) battery life. Heavier-duty industrial batteries can target the 60 to 80 percent maximum depth-of-discharge range for most efficient operation.

## LEAD-ACID BATTERIES

The practical aspects of lead-acid batteries affect all EV converters. You need to be intimately familiar with:

- Characteristics you should be aware of when buying.
- Steps you should take during installation.
- Maintenance you should perform during ownership.

The intent of this section is not to make you a battery professional, but to provide you with practical knowledge so that you're prepared to buy, install, and maintain your batteries.

### Battery types

As you learned at the outset, there are two major classes of batteries: primary or nonrechargeable, and secondary or rechargeable. Unless your EV's task is to operate on the Moon (like the Lunar Rover you read about in chapter 3) or some other specific mission, you are unlikely to require the services of a nonrechargeable battery.

Among rechargeables, there are lead-acid batteries and there are all the rest. In a nutshell, there are no alternatives to the lead-acid battery for the casual EV converter today, because the disadvantages of the other two choices far outweigh the benefits.

*Nickel-cadmium batteries.* These are the type you'd use in your portable computer, shaver, or appliance, and are unquestionably better than lead-acid batteries in their ability to deliver twice as much energy pound-for-pound; they also have about 50 percent longer cycles. But the Nickel-Cadmium electrochemical couple delivers a far lower voltage per cell (1.25 volts)—meaning you need more cells to get the same voltage. It is far more expensive (four times as much and up). There are fewer sources for the heavy-duty EV-application batteries (cadmium itself is harder to obtain and has generated environmental concerns). Finally, most of the Nickel-Cadmium technology development is taking place overseas (England, France, Germany, Japan).

*Nickel-iron batteries*. The "Edison-battery" used in early 1900s EVs are even a poorer choice. They offer a higher cycle life (about twice as many), deliver slightly more energy pound-for-pound (about a third more), and are very rugged mechanically. But the nickel-iron electrochemical couple delivers only slightly more voltage per cell than a NiCad (about 1.3 volts) and has a high internal resistance and self-discharge rate (10 percent per week). Its performance degrades significantly with temperature (both above and below 78 degrees F). It's far more expensive (four times as much and up), there are few sources for them (they're only made in Europe and Japan), and there is little technology development taking place.

All the battery development going on in the labs (which we'll look at briefly later in the chapter) is great, but you can't buy one. Your choice boils down to the good old lead-acid battery. But all lead-acid batteries are not created equal. Confining our discussion to the larger sizes suitable for the heavy-duty EV application, you have three types to choose from:

*Starting batteries*. These are the kind used to start the engine, found in every internal combustion engine vehicle in the world today. The average starting battery spends only a few seconds of time turning over your vehicle's electric starter motor and the rest of its time being recharged by the alternator under the lightest of loads (unless you are driving at night in the rain with all the electrical accessories on). While they are great for this "high-power output for a short period of time" application, they are not suitable for use in your EV (other than for powering its accessories) because this battery type has thin plates that are only lightly loaded with active material. Used in an EV, it would give you only the shortest deep-cycle discharge life— you'd be lucky to get 100 cycles out of it. Even on a brief trip, if you tromped down too hard (or for too long) on the accelerator pedal, you'd be lucky to make it back to your own driveway.

*Deep-cycle batteries*. These are what you need. The low end of the capacity range might go into a golf cart-type vehicle. The upper end of the capacity range goes into your EV. They can also be found in manufacturer's catalogs under the marine heading. Any of these are a step up from starting batteries; they have much thicker plates and are specifically designed for a deep-discharge cycle life in the 400 to 800 cycles (and up) ballpark.

*Industrial batteries*. These monsters go into forklift pallet and stationary wind- or solar-generation applications. While they give great depth-of-discharge results on paper, have 1000 cycles and up cycle-life, and make great counterweights for forklifts, their weight and size generally make them unsuitable for EV applications.

Your mission is to go after the deep-cycle batteries that might be found under the golf cart, marine, or electric vehicle catalog headings.


## Battery construction

From a manufacturing viewpoint, a lead-acid battery is one of the most efficient things going. While lead is definitely something you don't want in anything you drink or consume (you don't even want it in the paint on the wall inside your home), the EPA loves lead-acid batteries because more than 97 percent of all batteries are recycled and 100 percent of every battery is recyclable.

Battery construction makes this possible. Used lead batteries are gathered at collection points, then sent to smelting specialists where they are disassembled. The lead is melted, refined, and delivered to battery manufacturers and other users; the plastic is ground up and sent to reprocessors who make it into new plastic products; and the acid is collected and either reused or treated.

How a battery is constructed affects which battery you buy. Figure 9-3 shows the details:

*Plates.* Battery plates are formed on a wirelike grid of lead alloy (antimony is sometimes used to stiffen the lead); a mud-like lead oxide sulfuric acid and water paste is applied to them and allowed to harden. An expander is added to the negative cathode plate that prevents it from contracting in use. The plates are then "cooked" in a dilute sulfuric acid solution by sending a forming charge through them that changes the positive anode plate to a highly porous, chocolate-brown lead dioxide material, and changes the negative cathode plate to a gray sponge lead. The positive and negative plates are assembled into a "sandwich" with separators—thin sheets of electrically insulating material that is still porous to the electrolyte—and held in place inside the battery by the plate straps. How thick and heavy these plates are, combined with the efficiency of their design (how much area is exposed), and how efficient the separator is all collectively determine the capacity of your battery. While the initial way to tell the capacity of a battery is to pick it up, the only real way to tell is by using it.

*Case or container.* This is a plastic or hard rubber one-piece rectangular container with three or six cells molded into it. Each cell has molded-in ribs running across the width of its bottom or down the long dimension of the battery (FIG. 9-3). The plates are mounted at right angles to the ribs, whose multifold purpose is to stiffen the case, support the plates in a non-electrically-conductive manner, and act as collection chan-

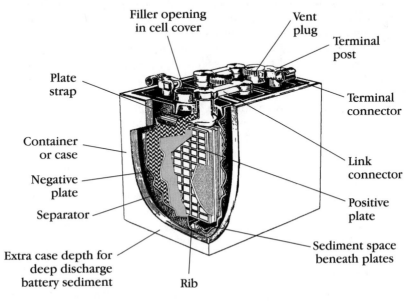

**9-3** Lead-acid battery construction.

nels for the active material shed from the plates. A battery is usable until the active material it sheds makes a pile that eventually reaches the plates and shorts them out. All other things being equal, a deeper-case battery will outlive a shallow-case battery because it allows a higher pile of active material to accumulate. Larger industrial batteries are cost-effectively rebuilt just by opening them, dumping the used active material, cleaning out any cell residue and replacing the plates, separators, and electrolyte.

*Cell connectors or links.* These connectors can be inside the battery (through or over the cell partitions) or outside the battery via the link connector (FIG. 9-3). External connectors found in older and larger battery types allow you access to individual cell voltage measurements; improved battery reliability has made individual cell measurements less necessary with modern batteries.

*Filler opening and vent plugs.* These openings allow you to refill your battery with distilled water or electrolyte solution. Vent plugs are baffled to allow gas to escape but not an accidental electrolyte splash. Individual vent plugs are threaded to screw into the vent well but might also be ganged into multiple opening press-fit caps.

*Terminal post.* These might be taper top, side terminal, or L-type on starting batteries, but they are usually of the stud type on deep-discharge batteries (a heavy duty post with a bolt and a washer). This is because high currents soften the lead terminal post, and taper top terminals might shrink away from their connectors and cause increased resistance and intermittent problems.

## Battery distribution & cost

Cost has been scrupulously avoided until this point, because a battery's "suggested retail price" is like your automobile sticker price—it bears little relationship to the real world price.

Battery manufacturers have multitiered international distribution networks. This means they have wholesale and retail distributors across the USA and around the world. They sell to these distributors in wholesale bulk—750 batteries at a time and up, etc. Wholesale distributors or battery specialists might in turn sell to retailers like golf cart dealers, etc. There are shipping costs involved at each step, along with whatever prevailing local rules or conditions exist.

The point is, price is negotiable, and you should expect a discount of 25 to 30 percent from suggested retail list price—especially for your 16- to 24-battery order. As with any other large ticket item, it pays to shop around. While you can open your telephone directory and find battery retailers in virtually any medium size and up city in the United States, it might pay to drive the extra 100 miles or so to deal with the large-volume battery dealer in the biggest city near you. On the other hand, if your local dealer's largest deals are six batteries at a time and you come in with your bid for 20 batteries, you might have some real negotiating leverage.

Get at least three bids on the same battery model from the same manufacturer, and compare them. If you want to get more creative at the expense of spending more time, compare several manufacturers this way. Be sure to add in all shipping, setup, deposits (we'll cover this next), and tax charges so you have an apples-to-apples comparison in your bids. For example, shipping charges from a nonlocal source might cancel out their advantage over a higher-priced local source.

## Battery core deposit

Certain states and localities do more than just suggest you recycle your batteries—they make it mandatory. Dealers are required to charge you a "core deposit" on your batteries that gets refunded to you when you bring them in for recycling—a larger version of the soda bottle deposit. Obviously, this adds to the price of your battery purchase. Another reason why one price cannot be quoted.

The flip side of this is your no-longer-usable batteries have an intrinsic scrap value. Like your used car when you buy a new one, your used batteries have a small but well-defined trade-in value. The lead commodity price quote from your local newspaper times the weight of your batteries gives you an approximate figure to deal with.

### Battery installation guidance

If you think of what you will be doing with your batteries, it will help you during the installation process. In general, it comes down to three areas:

*Safety.* This always has to be number one. You do not want a loose battery in the back seat becoming a missile in the event of a sudden stop, or Uncle Fred's cigar torching your EV as he casually examines it while it's recharging, or periodic battery maintenance overfills burning acid holes in your interior upholstery. You do want to mount your batteries in a location (or locations) where they can be securely tied down and well ventilated; where they are not harmful to anyone or anything and vice versa; and where than can be reached easily for servicing. This is the advantage of choosing a pickup chassis as an EV conversion platform—it already takes care of all these aspects.

*Tightness.* All electrical and mechanical battery connections should be tight when you finish installation, and you should periodically recheck them. Loose electrical connections mean reduced efficiency, invite corrosion, and create other needless intermittent electrical problems. Clean all connections before tightening them down. Loose mechanical connections mean premature aging of your battery by vibrating it to death. (Tight, by the way, means snug tight, or snug tight with a lock washer—something that doesn't break off later when you try to remove it in corroded form.)

*Measurement and watering.* Ideally you want to install a system to minimize your service labor at the time you install your batteries. Even if you don't do this at first—plan your installation to accommodate it as a future retrofit. Electrical measurement is relatively easy—you just need convenient voltage and current pickoff points and an analog meter, digital meter or automatic charging circuit to help you do the rest. Automatic watering is a little trickier but at a minimum involves being able to conveniently get at your battery tops—aka, your batteries are not strewn all over the inside of your EV conversion in random fashion.

### TLC & maintenance for your batteries

The tender loving care and maintenance of your batteries will reward you many times over. All it really takes is a plan, a schedule, and the discipline to do it. The plan starts with a notebook or logbook. The data you need are voltage from a battery (or two or three—not the whole pack), the odometer reading, the date, and a comments section listing the water you added and anything unusual you noticed. The schedule is weekly or bi-weekly. Other than the log sheet, four areas are involved:

*Safety.* This appears again because it is important. Always recharge your batteries in a well-ventilated area; don't smoke, light a match, or make an electrical spark around them; and strongly encourage all your friends and visitors to do the same. Working around batteries means you need to invest in protective eyeware (safety glasses), protective rubber gloves, and old clothes whose increasingly holey (yup, acid makes holes in them) appearance doesn't bother you.

*Distilled water.* Steam-distilled water is the only kind of water you ever want to put in your batteries. Just cover the plates; don't overfill the cells till the electrolyte overflows the battery top and causes a mess. Ideally, you want a tube from your water jug, or a small cup or glass with a pouring spout, or a clean funnel—something that makes it easy for you to pour without spilling. If any electrolyte is spilled, clean it up immediately and neutralize the spill area and its surroundings with baking soda per the next section's instructions. Remember that the battery electrolyte is a strong acid that eats metal, upholstery, clothing, shoes, and people without discrimination.

*Corrosion and tightness.* Make a visual inspection of your batteries, the battery connections, and the battery compartment. Look at, touch, and pull on things. The battery tops (and anything else in the battery compartment) should be kept clean of dust, dirt, corrosion, and splashed battery acid. Nip any one of these in the bud immediately. An old toothbrush and a box of baking soda works wonders. Use a solution comprising two tablespoons of baking soda added to a small glass of water (one pound per gallon is the ratio), applied to the battery tops and terminals. Never use it in the battery cells—be sure to keep them tightly capped during cleaning. Diligence with baking soda and toothbrush will neutralize any acid and keep the batteries clean. Touch and pull to check that none of your electrical connections have worked loose. Tighten any loose connections immediately.

*Measurement.* Use a digital voltmeter or hydrometer to give you a readout of the battery state-of-charge. Remember to monitor your batteries in a "rested" condition, and try to shoot for the same rest period in all your measurements, or make a note of any discrepancies.

## TODAY'S BEST BATTERY SOLUTION

You already know this book recommends lead-acid batteries as the best solution for today's EV converters. You also know what type of lead-acid battery to buy and a lot about its characteristics. Your choice is made even more easy because there are only a certain number of battery vendors in your immediate geographic area to choose between. Unlike buying motors, controllers, and other parts, you're not likely to be ordering your batteries by mail. Your choice basically comes down to who offers the best price on the batteries you want, and what capacity, rating, voltage, size, and weight you need.

In a slight departure from the previous chapters, we're going to recommend one manufacturer, then look at several alternative offerings from their line to give you the flavor of the real choices you will encounter. The batteries recommended are from the Trojan Battery Company of Santa Fe Springs, California. As with the motors and controllers, don't read anything important into their appearance here. They are only one of a large number of battery manufacturers. A list of these appeared in chapter 5, but in this case, which battery distributors are operating in your geographic area is the more important factor.

Before getting into the actual batteries, let's add a few more definitions to your already expanded battery vocabulary:

*Power density (or gravimetric power density).* Also known as *specific power,* this is the amount of power available from a battery at any time (under optimal conditions), measured in watts per pound of battery weight. It translates directly to the acceleration and top speed performance your EV can get out of its batteries.

*Energy density (or gravimetric energy density).* Also known as *specific energy,* this is the amount of power available from a battery for a certain length of time (under optimal conditions), measured in watt-hours per pound of battery weight. It translates directly to the range performance your EV can get out of its batteries.

*Volumetric power density.* This is a factor more of interest to the technical battery community working across different battery chemistry types. It is power density measured in watts per gallon or watts per cubic foot—volume rather than weight.

*Volumetric energy density.* Ditto here. This is energy density measured in watt-hours per gallon or watt-hours per cubic foot—again volume rather than weight.

You will find these useful both for this section's comparisons as well as those made in the future batteries section. Now let's look at the winning batteries.

## Five Trojan Battery solutions

Trojan Battery Company has been innovating golf cart battery solutions since the 1950s; their appearance here should not surprise you. Electrical vehicle batteries today are substantially superior to those of only a decade ago. You can pick from 6-volt or 12-volt solutions, and the distribution network has evolved to give you more service at better prices.

We're going to look at three 6-volt and two 12-volt alternatives from Trojan. The Trojan T-125 model—one of the 6-volt alternatives—is shown in FIG. 9-4. Notice its rugged construction, and the stud-type terminal posts with bolts and nuts. This case and construction is common to all family members in this 6-volt line. Figure 9-5 shows you a 12-volt unit, the 5SHP model case mockup previewed by Trojan at the September, 1992 Burbank Alternate Transportation Expo. You might (or might not) have the EV label on the batteries you buy from your distributor.

**9-4** Trojan T-125 6-volt deep-cycle battery.    **9-5** Trojan 5SHP 12-volt deep-cycle battery.

TABLE 9-2 gives you the details of this lineup of five recommended EV battery choices from Trojan. Other than suggested list price—an area that we'll save for special discussion—this is all from published data that you can get from your local dealer.

It lays out everything you need, but doesn't quite give it to you in the form you need it—yet. Figure 9-6, also drawn from published data, shows the actual capacity versus time performance charts; notice the similar performance of the 6-volt and 12-volt data groups. You can use this data to determine the results of applying actual loads to any of the batteries you choose.

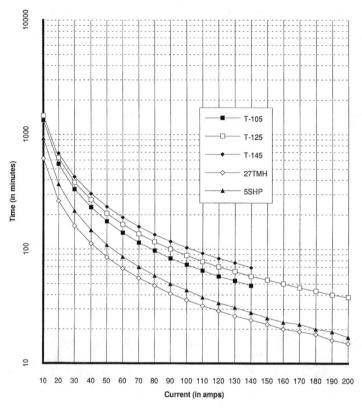

**9-6** Trojan 6-volt and 12-volt deep-cycle battery family time vs. current curves.

Figure 9-7 is from actual Trojan data on the T-105 model battery calculated 6/29/92. It is the real-life example of equation (11) shown earlier:

(11) Battery Life Cycles = $K_d$/Depth of Discharge in %

In this figure $K_d$ is around 28,000, so it shows that Trojan technology is pushing its deep-cycle batteries into the industrial battery area. In other words, the T-105 model and its other family members are heavy-duty deep discharge batteries.

To figure out how many batteries you need, first determine the voltage at which you are going to operate your EV conversion. This voltage is established from your chassis, motor, and controller trade-offs, and heavily influenced by ultimate use— longest range or fastest acceleration.

Our objective here was to pick the best battery for chapter 11's actual pickup conversion, so the operating voltage of 120 volts was selected. Assum-

### Table 9-2 Comparison of recommended Trojan electric vehicle batteries

| Trojan Battery Model | Nominal Voltage | 20 AH Capacity | Minutes @ 25 Amps | Minutes @ 75 Amps | 3 AH Capacity | Weight in Pounds | Energy Density wtt-hours/lb | Length | Width | Height | Suggested List Price |
|---|---|---|---|---|---|---|---|---|---|---|---|
| T-105 | 6 volts | 217 | 419 | 107 | 161 | 61 | 15.5 | 10.375 | 7.125 | 11.1875 | 107.76 |
| T-125 | 6 volts | 235 | 477 | 125 | 174 | 66 | 15.6 | 10.375 | 7.125 | 11.1875 | 115.67 |
| T-145 | 6 volts | 244 | 530 | 145 | 181 | 71 | 15.0 | 10.375 | 7.125 | 11.5 | 168.85 |
| 27TMH | 12 volts | 117 | 200 | 50 | 87 | 60 | 15.2 | 12.75 | 6.75 | 9.75 | 106.86 |
| 5SHP | 12 volts | 165 | 272 | 78 | 122 | 86 | 14.2 | 13.5625 | 6.75 | 11.5 | 220.50 |

### Table 9-3 Comparison of Trojan electric vehicle battery trade-offs

| Trojan Battery Model | Nominal Voltage | Quantity in Vehicle | Vehicle Voltage | Battery Total AH Capacity | Battery Total Watt-Hours | Battery Total Weight | Battery Total Cubic Feet | Battery Total Cost @ 70% |
|---|---|---|---|---|---|---|---|---|
| T-105 | 6 volts | 20 | 120 | 217 | 26040 | 1220 | 9.57 | 1508.64 |
| T-125 | 6 volts | 20 | 120 | 235 | 28200 | 1320 | 9.57 | 1619.38 |
| T-145 | 6 volts | 20 | 120 | 244 | 29280 | 1420 | 9.84 | 2369.90 |
| 27TMH | 12 volts | 10 | 120 | 117 | 14040 | ~600 | 4.86 | 748.02 |
| 5SHP | 12 volts | 10 | 120 | 165 | 19800 | 860 | 6.09 | 1543.50 |

ing you want all your batteries of the same type, and also that you're not going to use any tricky series-parallel wiring combinations, this means you'll either require 20 of the 6-volt batteries or 10 of the 12-volt batteries, all wired in series to obtain the 120 volts. When you wire your batteries in series, the total capacity available—the total ampere-hours—is the same as that available from any one battery. The total watt-hours is simply the total voltage times the total ampere-hours. The total weight, cubic feet, and cost is simply the sum of the individual 10- or 20-member battery set. TABLE 9-3's summary data is laid out in much more usable format.

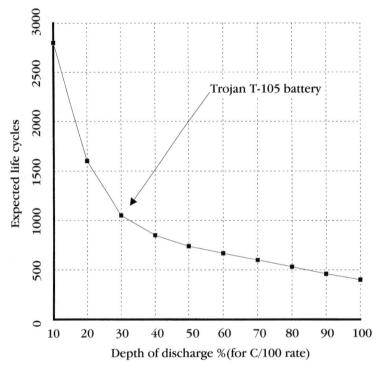

**9-7** Depth of discharge versus expected life cycles for Trojan T-105 battery.

Your choices are actually quite simple and logical at this point. Notice the 6-volt options give you from 26.0 to 29.3 kWh of energy—total on-board battery kilowatt hours. The two 12-volt choices give you either 14.0 or 19.8 kWh. Looking at the weight column, the 6-volt choices weigh from 1220 to 1420 pounds, while the 12-volt choices weigh either 600 or 860 pounds. In total cubic feet, although this figure would obviously translate to a larger actual mounting space required, the 6-volt batteries require from 9.6 to 9.8 cubic feet, while the 12-volt batteries need either 4.9 or 6.1 cubic feet.

Cost is another matter. Some figure had to be used here, so the manufacturer's suggested list price as of March 1993 was plugged in, discounted 30 percent (70 percent of list) and multiplied by either the 10 or 20 number appropriate for the battery string. As mentioned earlier, your costs will vary from these figures, so look at the cost here only for comparative purposes.

With the cost disclaimer out of the way, we can proceed. If you want the best range, choose the highest onboard energy. If you want the most acceleration choose the lowest weight. If cost is a factor, choose the lowest cost. As all of these usually are factors, dividing energy and cost by weight is usually a good way to see what gives you the most for the least. TABLE 9-4 does this and gives you two clear winners—the T-125 for the 6-volt group, and the 27TMH for the 12-volt side. Notice the T-125 gives you about 1 kWh less energy, saves you 100 pounds, and costs you $750 less than its bigger T-145 brother—it's the best deal in the 6-volt group and the best choice for chapter 11's pickup conversion. On the 12-volt side, the 27TMH gives you 40 percent less energy at 14.0 kWh, but saves you 260 pounds and costs less than half that of its bigger 5SHP brother. It would be the ideal solution for a lighter-weight vehicle. Or if you just wanted to tool around your neighborhood in chapter 11's pickup truck and win all the local drag races, ten 27TMH batteries weighing in at 600 pounds save you a cool 720 pounds over the 20 6-volt T-125 battery solution.

### Table 9-4  Trojan battery final trade-offs

| Trojan Battery Model | Total kwh/lbs | Energy Decision Criteria | Total $/lbs | Cost Decision Criteria |
|---|---|---|---|---|
| T-105 | 21.34 | | 1.237 | |
| T-125 | 21.36 | Highest 6-volt energy/lb | 1.227 | Lowest 6-volt cost/lb |
| T-145 | 20.62 | | 1.669 | |
| 27TMH | 23.40 | Highest 12-volt energy/lb | 1.247 | Lowest 12-volt cost/lb |
| 5SHP | 23.02 | | 1.795 | |

## TOMORROW'S BEST BATTERY SOLUTION

If you pick up a battery book (or read a battery chapter in any book) from any decade of the 1900s, it's interesting to note that every one states the best battery of the future is "just around the corner." The reality is a little different. The reality is things move very slowly in the battery world—events are usually measured in decades, not years. So while this section will talk about the latest and greatest in batteries, don't expect any of these at your battery dealer soon—if at all.

While lead and sulfuric acid would not be my initial choice for any construction project—one of the heaviest elements mated with one of the nastiest compounds—reliability, performance, and cost all weigh heavily in the lead-acid battery's favor. In fact, lead-acid's suitability for so many applications has greatly diminished even the need to search for alternatives.

But the 1990s save-the-environment, reduce-oil-dependency, and let's-try-electric-vehicles changed consciousness alters the pattern. Government, industry, and university laboratories all over the planet reflect this change. Pouring money on a problem never guarantees a solution, but it does guarantee that a lot more will be happening, and that some of what happens will be usable and good. Let's look at future battery trends starting with the consortium that's pushing the outside of the battery envelope—the USABC.

### USABC to the rescue

In 1991, Ford, General Motor, and Chrysler (joined by the Department of Energy and the Electric Power Research Institute) had a better idea—the United States Advanced Battery Consortium (USABC). In short, the Electric and Hybrid Vehicle Research, Development and Demonstration Act of 1976 recognized the need for battery development; the Department of Energy defined and funded it; and the USABC focused the efforts.

The near term result was that a plethora of projects was honed down to just three high-energy battery research areas that could deliver significant vehicle range and power advantages: lithium polymer, lithium metal sulfide, and nickel metal hydride. Contracts for these were let in late October 1992.

## Future batteries—The big picture

TABLE 9-5, adapted from an SAE paper,[3] shows the entire story at a glance. Notice that eleven different battery technologies are on the list, and they are not equal. In very general terms, higher energy density and power density are desirable and look easy to do—on paper. Getting both at the same time, along with high cycle-life and low cost in a battery that operates efficiently over a range of temperatures, can be manufactured and supported by infrastructure, and causes no harm to people or the environment, has proven to be a bit more elusive. Notice that none of the batteries developed thus far—even the tried and proved lead-acid battery—even approaches its theoretical specific energy value. We still have a long way to go.

*Lead-acid.* Don't expect this battery to go away soon, if at all. The big money involved in the lead-acid battery business will make the the lead-acid batteries of the early 2000s as superior to today's as today's are to their 1980s counterparts. Along the way to higher specific energy and specific power, lead-acid batteries will also evolve variants like: *sealed* (higher cost, lower efficiency versus convenience of not watering); *flow-through* (the conventional type you're accustomed to, but improved by greater plate thickness, improved separators, and higher specific gravity electrolyte solution); *tubular* (electrode improvement) and *gelled* (electrolyte improvement).

*Nickel-cadmium.* Close on the heels of lead-acid today, this battery type promises to be even better in the future. Its advantages over lead-acid today (less pronounced decrease in capacity under high-discharge currents, higher cycle life, slower self-discharge rate, better long-term storability, and improved low-temperature performance) will continue to provide markets to fund development aimed at improving its disadvantages over lead-acid: higher cost and environment-related cadmium issues.

*Nickel metal hydride.* The environmentally benign alter ego to the nickel-cadmium battery is also flat-out superior to it in specific energy and specific power comparisons, and should become the preferred alkaline battery of the future. The USABC certainly thinks so and is investing its research and development dollars on it.

*Lithium-polymer.* This battery's specific energy and specific power numbers evoke nothing but envy from its competitors and mouth-watering anticipation from its advocates. The whole lithium group has shown great promise in the labs, and smaller lithium polymer batteries have greatly impressed users in the computer industry. But lithium still has to deliver on its promise when packaged in the giant, economy suitable-for-powering-EVs size. Stay tuned for future developments here—the USABC certainly will.

*Sodium-sulfur.* This technology has strong proponents in England, Germany, Japan, Canada, and the United States. This is good because its opponents have a continuous field day merely by labeling it and using scare tactics. Yes, sodium is combustible in air. Yes, sulfur is also on the head of matches. Yes, maintaining a "thermos bottle" at 350 degrees C to contain its molten sodium and sulfur electrodes and beta-alumina electrolyte is inefficient. Yes, it can explode and/or do nasty things if punctured in an accident. So can internal combustion engine vehicles. On the other hand,

**Table 9-5  Comparison of future electric vehicle battery trade-offs**

| Battery Type | Nominal Cell Voltage (volts) | Operating Temp Range (degrees C) | Life Cycles | Theoretical Spec Energy (wh/lb) | 3 Hr Rate Spec Energy (wh/lb) | Energy Density (kwh/cu ft) | Specific Power 30 sec pulse (watts/lb) | Power Density 30 sec pulse (kw/cu ft) |
|---|---|---|---|---|---|---|---|---|
| Lead Acid | 2.1 | 35 - 70 | 600 | 79.5 | 15.9 | 2.55 | 72.7 | 8.50 |
| Nickel Cadmium | 1.25 | 30 - 50 | 2000 | 99.1 | 25.0 | 3.40 | 86.4 | 9.35 |
| Nickel Metal Hydride | 1.4 | 20 - 60 | 600 | 84.1 | 29.5 | 4.96 | 68.2 | 11.33 |
| Nickel Zinc | 1.6 | 40 - 65 | 250 | 155.0 | 27.3 | 2.83 | 59.1 | 2.83 |
| Nickel Iron | 1.25 | 40 - 80 | 800 | 121.4 | 22.7 | 3.40 | 52.3 | 6.51 |
| Sodium Sulfur | 2.08 | 300 - 400 | 350 | 345.5 | 38.6 | 3.26 | 54.5 | 5.10 |
| Sodium Nickel Chloride | 2.59 | 250 - 350 | 1000 | 360.0 | 59.1 | 4.81 | 76.3 | 6.37 |
| Zinc Bromine | 1.8 | 0 - 45 | 500 | 194.5 | 31.8 | 1.98 | 38.6 | 3.26 |
| Zinc Air | 1.62 | 25 - 65 | 70 | 595.5 | 59.1 | 1.84 | 22.7 | 1.84 |
| Lithium Iron Disulfide | 1.66 | 400 - 450 | 500 | 295.5 | 75.0 | 6.80 | 170.5 | 1.56 |
| Lithium Polymer | 3.5 | 0 - 100 | 300 | 248.2 | 72.7 | 7.36 | 90.9 | 5.95 |

it continues to be one of the most promising advanced battery systems for EV propulsion; its specific energy and specific power numbers are greatly superior to those for lead-acid batteries; and pilot plant battery production has already started. While many problems remain to be solved, perhaps low-cost production the largest among them, sodium sulphur is a hot technology in more ways than one.

*Sodium metal chloride.* If sodium sulfur is great, then all its advantages at a lower operating temperature with still better specific energy and specific power numbers has to be greater yet. And sodium metal chloride battery technology is. Throw in cells that can be assembled in the discharged state, have higher open circuit voltage and better freeze/thaw and failure-mode characteristics than sodium sulfur's, and you have a real winner. This one's a comer.

*Lithium iron disulfide.* The promise of lithium-something batteries on the high temperature side is equally mouth-watering. This battery's specific energy numbers are the best of all, and its specific power numbers simply leave all others in the dust. Double check TABLE 9-5 to become a believer. USABC is already a believer, although it is funding the lithium metal sulfide variant. Stay closely tuned here also.

## Batteries . . . back to the future

While the lithium solution is the holy grail for battery makers, gasoline is still the nemesis because

$$SP = (125,000 \text{ BTU/gal} \times 7.48 \text{ gal/cu ft})/(3413 \text{ BTU/kWh}) = 274.2 \text{ kWh/cu ft}$$

This equation means that the energy density for gasoline is more than 107.5 times that of lead-acid batteries and is even 37.3 times that of lithium polymer batteries. Because batteries provide only one percent as much power per weight as gasoline, most of your attention should be paid to minimizing weight and maximizing efficiency elsewhere in your EV.

On the other hand, maybe the best solution of all is the one from the car in the movie *Back to the Future*—if you're going to have lead in your EV, it might as well be in the form of shielding for an ultra efficient, tiny nuclear reactor that generates all the electricity you need on demand.

# 10

# The charger & electrical system

*An efficient charger is an indispensable part of any electric vehicle.*

The charger is an attached and inseparable part of every electric vehicle battery system. Discharging and recharging your batteries are opposite sides of the same coin; you cannot have one without the other. As you learned in the battery chapter, how you recharge your batteries determines both their immediate efficiency and ultimate longevity. As with motors, controllers, and batteries, technology has also made today's chargers superior to their counterparts a decade ago.

Because your motor, controller, batteries, and charger are also inseparable from the electrical system that interconnects them, it too is covered in this chapter, along with the key components needed for its high-voltage, high-current power side and its low-voltage, low-current instrumentation side.

In this chapter you'll learn about how chargers work and the different types, meet the best type of charger to choose for your EV conversion today (used in chapter 11's conversion), and look at likely future charging developments. You'll also look at your EV's electrical system in detail and learn about its components so when you meet them again during chapter 11's conversion process, they will be familiar to you.

## CHARGER OVERVIEW

The battery chapter dealt with discharging and recharging; now we'll take a closer look at the recharging side. It's a wise business decision to invest a few hundred dollars in a battery charger that gets the most out of a battery pack that can cost a thousand dollars or more and might be replenished several times during your EV ownership period. An efficient charger is an indispensable part of any EV.

The objective here is to give you a brief background and get you into the recommended battery charger for your EV conversion with minimum fuss. You have three battery charger choices today: build your own, buy an *offboard* charger, or buy an *onboard* charger. We'll look at each area in turn and give our recommendations. Let's start with a look at what goes on during the lead-acid battery discharging and charging cycle to understand what has to be done by the battery charger.

## BATTERY DISCHARGING & CHARGING CYCLE

As you already know from the battery chapter, batteries behave differently during discharging and charging—two entirely different chemical processes are taking place. Batteries also behave differently at different stages of the charging cycle. Let's start with a look at an actual battery, then look at the discharging and charging cycle specifics.

### What you can learn from a battery cycle life test

Figure 10-1 shows Cycle Life test results for the Trojan 27TMH deep cycle lead-acid battery we looked at in chapter 9. Two parameters are being monitored versus number of cycles: the minutes at 25 amps capacity, and the *end of charge current* (EOCC).

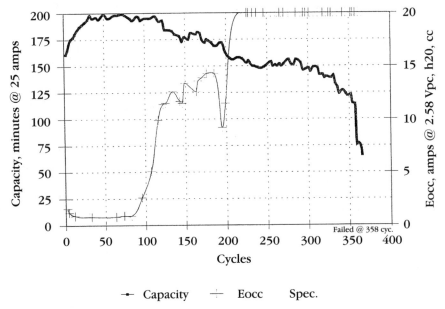

**10-1** Cycle life test results for Trojan 27TMH deep-cycle lead-acid battery.

The capacity parameter measurement is an actual version of the graph you saw in FIG. 9-2 (the right middle one). There are more wiggles in the real battery's data curve, but the resemblance between the two graphs is striking, and there should be no revelations for you here. This battery didn't actually "fail" at the end of 358 cycles; that's just a name assigned (by a battery test engineer) to the point at which this battery dropped below 50 percent of its rated capacity.

The end of charge current (EOCC) might be new to you. Notice it's quite low early in the battery's cycle life (around one amp) but rises steadily until—at some point around mid-life—it shoots up to its limit value (around 20 amps in this graph). What does this mean to you? It means that a battery's charging current fluctuates widely over

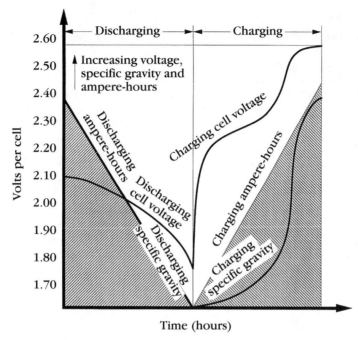

**10-2** Graphic summary of battery discharging and charging cycles.

its lifetime. It means that you can quickly kill a new battery that only requires a small amount of current to kick in its charging cycle by placing an unregulated voltage source without any current control across it. It means you have to crank up the voltage and current when charging a more mature battery. Both of these mean that you cannot plug a charger into your battery, set it, and forget it—because a battery's charging needs also change from cycle to cycle.

Sealed lead-acid batteries, with a small amount of calcium added to eliminate the need for rewatering, don't exhibit this characteristic; their EOCC is relatively flat so you can be a little more tolerant with them (they are also more expensive—you pay your maintenance costs up front).

## Battery discharging cycle

Let's observe the discharge cycle first, to contrast what is happening to the parameters with what goes on during charging. Capacity, cell voltage, and specific gravity all decrease with time as you discharge a battery. Figure 10-2 shows how these key parameters change (a standard temperature of 78 degrees F is presumed):

*Ampere hours.* The measure of the battery's capacity and percent state-of-charge (the area under the line in this case) is shown decreasing linearly versus time from its full charge to its full discharge value.

*Cell voltage.* Cell voltage predictably declines from its nominal 2.1-volt fully charged value to its fully discharged value of 1.75 volts.

*Specific gravity.* Specific gravity decreases linearly (directly with the battery's discharging ampere-hour rate) from its full charge to its full discharge value.

## Battery charging cycle

Battery charging is the reverse of discharging. Figure 10-2 again shows you how the key parameters change:

*Ampere hours.* This is the opposite of the discharging case, except that you have to put back slightly more than you took out (typically 105 to 115 percent more) because of losses, heating, etc. The area under the line increases linearly versus time from its fully discharged to its fully charged value.

*Specific gravity.* Specific gravity increases wildly over time as a battery is charging, so making specific gravity measurements during the charging cycle is not a good idea. At the early part of the charging cycle, specific gravity increases slowly because the charging chemical reaction process is just starting. Specific gravity increases rapidly as the sulfuric acid concentration builds, and gassing near the end of the cycle contributes to its rise.

*Cell voltage.* Voltage also increases wildly over time as a battery is charging, so making voltage measurements during the charging cycle is not a good idea either. Notice cell voltage jumps up immediately to its natural 2.1-volt value; slowly increases until 80 percent state-of-charge (approximately 2.35 volts); increases rapidly until 90 percent state-of-charge (approximately 2.5 volts); and then builds slowly to its full charging value of 2.58 volts.

## The ideal battery charger

Battery charging is the reverse of discharging, but the rate at which you do it is critical in determining its lifetime. The basic rule is: Charge it as soon as it's empty, and fill it all the way up. The charging rate rule is: Charge it slower at the beginning and end of the charging cycle (below 20 percent and above 90 percent).

When a lead-acid battery is either almost empty or almost full, its ability to store energy is reduced due to changes in the cell's internal resistance. Attempting to charge it too rapidly during these periods causes gassing and increased heating within the battery—greatly reducing its life. Ideally, you limit battery current during the first 90 percent of the charging cycle and limit battery voltage during the last 10 percent of the charging cycle. Either method by itself doesn't do the job (FIG. 10-3 shows why).

The graph at the upper left in FIG. 10-3 shows constant current charging in the ideal case. If you have a 200 ampere-hour capacity battery, you know that feeding it 20 amps for 10 hours will about get the charging job done. Unfortunately, with no restrictions on voltage, it can rise far beyond its natural cell voltage of 2.1 volts, and the resultant overcharging can damage the batteries and any attached circuit loads.

The graph at the upper right in FIG. 10-3 shows constant voltage charging in the ideal case. The constant voltage—usually set at the level where gassing begins—causes a decrease in current flow through the battery with time as the battery charges. Unfortunately, with no restrictions on current, this method allows far too much current to flow into an empty battery—feeding 100 amps or more of charging current to a fully discharged battery can damage it or severely reduce its life.

Let's look at the ideal approach during all four state-of-charge phases—0 to 20 percent, 20 to 90 percent, 90 to 100 percent and above 100 percent. Figure 10-3 shows the results.

**Charging between 0 & 20 percent** The first 20 percent of a fully-discharged battery's charging cycle is a critical phase and you want to treat it gently. You learned in chapter 9 that all batteries have a BCI-standardized 20-hour capacity rating—every bat-

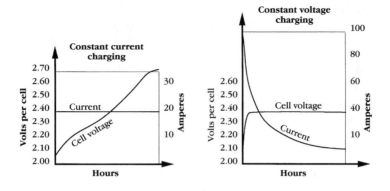

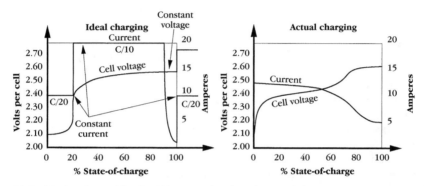

**10-3** Ideal and actual lead-acid battery charging characteristics.

tery is rated to deliver 100 percent of its rated capacity at the C/20 rate. During the first 20 percent of the charging cycle, you ideally want to charge a battery at no more than this constant-current C/20 rate. To determine the first 20 percent charging current,

$$(1) \text{ Charging current} = \text{Battery capacity}/\text{Time} = C/20$$

For a 200 ampere-hour capacity battery, charging current would be

$$\text{Charging current} = 200/20 = 10 \text{ amps}$$

In other words, you would limit this battery's initial charging current to 10 amps. You can blast your battery with 200 amps and charge it in 1 hour, but you will prematurely age and kill it—it will not deliver its full, useful life to you. The graph at the lower left in FIG. 10-3 shows the result of current-limited C/20 rate charging during the first 20 percent part of the charging cycle. The voltage rises gently and your battery is very happy.

**Charging between 20 to 90 percent** In the middle of the charging cycle, you can charge at up to the C/10 rate. This is the fastest rate that efficiently charges a lead-acid battery. This rate is not as efficient as the C/20 rate—more energy is wasted in heat if you recall the I²R losses—but it gets the charging job done faster. At even less efficiency (and more risk to your batteries), you can bump it up to higher C/5 or C/3 rates during this period of recharging if time is essential to you, and if you closely monitor the battery's temperature so its operating limits are not exceeded and you don't wind up "cooking" it. Charging current would be 20 amps at the C/10 rate for the 200 ampere-hour battery. Figure 10-3 shows that voltage, after a step increase when current settings were changed, rises slowly to its 90 percent state-of-charge value of approximately 2.50 volts.

**Charging between 90 & 100 percent**  At this point, you want to drop back to the C/20 rate or, ideally, switch to a constant-voltage method. If you switch to constant voltage set at the deep cycle battery's full charging value of 2.58 volts, FIG. 10-3 shows the result—current provided to the battery drops rapidly during this last 10 percent of charging and your battery is very happy while receiving its full charge.

**Charging above 100 percent (equalizing charging)**  You learned about equalizing charging in chapter 9. It is needed to restore all cells to an equal state of charge (to "equalize" the characteristics of the cells); to keep the battery operating at peak efficiency; to restore some capacity of aging batteries; to restore float-charged or shallow-charged batteries to regular service; and to eliminate effects of sulfation in idle or discharged batteries. Equalizing charging is controlled overcharging at a constant current C/20 rate with the charging voltage limit raised to 2.75 volts. It is done after the battery is fully charged, and maintained at this level for 6 to 10 hours. Equalizing charging should *not* be done at rates greater than C/20. Equalizing charging should be done every 5 to 10 cycles or monthly (whichever comes first); only in well-ventilated areas (with no sparks or smoking) as it produces substantial gassing; and only while close attention is being paid to electrolyte level, as water consumption is substantial during rapid gassing periods. Figure 10-3 shows the step increase in the voltage to 2.75 volts and the increase in current back to the C/20 level.

Now let's look at the time involved in using the ideal approach to charge our hypothetical 200 ampere-hour capacity battery:

$$
\begin{array}{ll}
10 \text{ amps (C/20) for 5 hours} & = 50 \text{ AH} \\
20 \text{ amps (C/10) for 7 hours} & = 140 \text{ AH} \\
10 \text{ amps (C/20) for 1 hour} & = 10 \text{ AH} \\
\text{Totals: 13 hours} & = 200 \text{ AH}
\end{array}
$$

This approach requires 13 hours to charge a 200 ampere-hour capacity battery. Provided you don't exceed battery temperature, you could charge at the C/5 rate during the middle of the cycle (40 amps for 3.5 hours) and reduce total time to 9.5 hours.

### The real world battery charger

The starting battery in your internal combustion engine vehicle is recharged by its engine-driven alternator, whose output is controlled by a voltage regulator. The starting battery is discharged less than 1 percent in its typical automotive role. The entire rated output power of the alternator is placed across it, and the voltage regulator makes sure voltage doesn't climb above 13.8 volts—simple. Why not use the same setup to recharge your deep cycle EV batteries?

The answer is voltage, current, and "boom." This approach is not used because the voltage of your EV's deep-cycle battery pack string is probably 96 volts or more, but the alternator is typically set up for driving a 12-volt starting battery. Assuming you could adjust your alternator or buy the correct voltage model, and had a suitable electric motor around to turn it, the full alternator output applied to a completely discharged deep-cycle battery pack would deliver too much current and charge it far too quickly. You'd damage and/or destroy your batteries in short order—that's the "boom" part. The voltage regulator would not only be useless in stopping this, but would prevent raising the voltage to 2.58 volts per cell for the final 10 percent of the cycle, and would not allow the voltage to be further raised to 2.75 volts for the equalizing charge process.

The solution is ac power, a transformer, a rectifier, a regulator of either the "electronically smart" or "manually adjustable" variety, and a timer: an accurate description of today's EV battery charger.

The × pattern in the graph at the lower right in FIG. 10-3 shows what most actual battery chargers deliver. Using a variation or combination of constant-current, constant-voltage, tapering, and end-of-charge voltage versus time methods, all battery chargers arrive at a method of current reduction during the charging cycle as the cell voltage rises. Fortunately, you can buy something off-the-shelf to take care of your needs. But you have to investigate before buying to make sure a given battery charger does what you want it to do.

Battery chargers are "sized" using the formula,

(2) Charging current = (Battery capacity × 115 percent)/(Time) + dc load

In this equation—very similar to equation (1)—the charging current determines the size charger you need, the 115 percent is an efficiency factor to take losses into account, the dc load is whatever else is attached to the battery (this is zero, assuming you disconnect your batteries from your EV's electrical system while recharging). You're already familiar with battery capacity and time. You can plug chargers up to 20 amps into your standard household 120-volt ac outlet. Higher current capacity chargers require a dedicated three-phase 240-volt ac circuit—the kind that drives your household electric range or clothes dryer. Let's take a look at your actual options: build-your-own, and several different types of off-the-shelf chargers you can buy.

## Build your own battery charger

You don't have to be a rocket scientist to build your own battery charger. In fact, some EV converters prefer their own simple designs because it gives them better control over the charging process at a relatively low cost. Figure 10-4 shows you the basic components you need: timer, variac, and the battery charger itself.

Inside the simplest battery charger you have a *transformer* (for a 120-volt battery pack, its turns ratio should give you about 130 volts out for 120 ac volts in), and a *diode bridge* (something like a Motorola MDA 3504 works fine—it's rated at 35 amps and 400 volts). A *digital voltmeter* (Radio Shack Micronta 22-185A or equivalent DVM—3½-digit or greater accuracy) measurement across the battery gives you its voltage; a DVM measurement across a shunt in series with the battery gives you the current.

If you want to get fancier, add circuitry inside the current-regulation and voltage-regulation boxes to give you more control. At its simplest, solid-state, discrete-component, potentiometer-adjustable voltage- and/or current-regulation circuitry can provide you the output levels you need. At the next step up in complexity, you can add some sensing circuitry to invoke automatic shutoff when a certain level is reached or exceeded. Beyond that, with the addition of a few IC chips, you can program the entire charging profile you want—voltage, current, and timing.

Back at the simple level, a *variac* or *autotransformer* (a model with a 0 to 150-volt ac output for a 120-volt ac input rated at 30 amps works great) is useful. By vary-

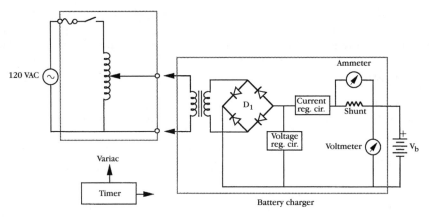

**10-4** Block diagram of a basic battery charger.

ing its adjustment knob, even the simplest of chargers becomes a precision tool, and you can dial up any charging current you need.

The timer—it could be a giant economy-sized model that everything plugs into or a very small one in the heart of the battery charger's electronics—puts you into the big time. At the simplest level, you dial up the C/20 charging level for the capacity rating of your battery pack, set the timer, and go off and do something else. The timer trips the charger off at the preset time, and the battery pack sits there waiting for the next step in the cycle. At the exotic level, the IC chips of your custom-designed timer are an integral part of the charger's control electronics and you can program how many amps for how long at how many volts, etc., for each step in the entire charging profile.

Battery state-of-charge is determined by a cell voltage or specific gravity measurement. Battery rate-of-charge is determined by how much current you use. Monitoring battery current and voltage are vital and essential to the charging process. A modern DVM gives you a simple yet accurate digital readout. You can also rig a simple switching arrangement, so you can use a single DVM to tell you either the voltage or the current reading. Or splurge, buy two DVMs, and have a Star Trek-style charger with knobs, switches, potentiometer dials, and two digital readouts.

## Lester battery charger

The name Lester was a household word in battery chargers in the 1970s, and still is for EV converters of the 1980s and 1990s. Lester manufactures more than 350 charger models, and their ferroresonant chargers of today are far superior in performance to their SCR and rectifier-based ancestors.

Lester's Model 9387 charger is a 12/96-volt ferroresonant model that can be driven from a 120-, 208-, or 240-volt ac source. (It is one of several models in the line—check for availability of a 120-volt model.) Lester identified two desirable consumer preferences: power from multiple ac sources, and dual-mode dc outputs. Multiple ac input options lets you plug it into whatever outlet's convenient; dual dc outputs let you charge either your main battery pack or the auxiliary battery that powers your lights, accessories, etc. Lester recommends an auxiliary battery instead of tapping one of the batteries in your main pack string. As you learned in the controller chapter, pulling one (or two) of the batteries to a much lower state of charge level prematurely ages them.

Lester's Model 9387 charger is a two-step constant current charger. Its output current pattern over time—shown in FIG. 10-5—makes it a 35/8 charger, giving 35 amps at the high leg that drop to 8 amps at a predetermined time. While everyone agrees that you should drop the current as the battery reaches a higher state-of-charge, not everyone agrees with starting out charging at a C/20 or less rate. Clearly, Lester's engineers believe starting at slightly less than the C/5 rate (35 amps) is a worthwhile trade-off versus minimum charging time.

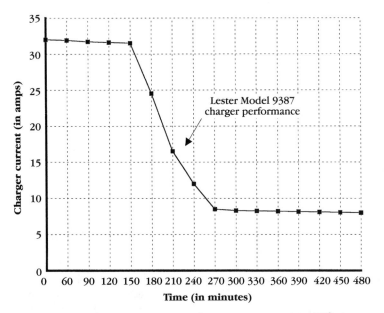

**10-5** Output current 35/8 pattern for a Lester model 9387 charger (12/108 volt).

Lester, the most well-known name, is a good example of one of many different approaches used by commercial battery charger manufacturers, and illustrates why EV converters need to investigate before they invest in a commercial charger. On the other hand, Lester would probably be happy to build a custom EV market charger (programmable IC chips and all) if enough users asked them—it's their business.

## K & W battery charger

At the other end of the battery charging spectrum is K & W. This company started with the premise that most EV users would prefer to carry their battery charger with them for "opportunity charging" and designed a charger to meet those specifications. The result is its Model BC-20. Its transformerless design makes it light in weight (9.9 pounds). But its ground-fault interrupter protects you from line shocks; it is powered from universally available 120-volt ac; and it features dc outputs from 48 to 108 volts (programmable via an internal resistor change).

The K & W Model BC-20 charger is a triac-controlled type that provides constant current to the gassing point and constant voltage to the finish. An optional Model LB-

20 line booster is available to enable the BC-20 to charge a 114 or 120-volt battery pack from a 120-volt ac outlet at reduced output current.

Whether to have an on-board or an off-board charger is another consideration. Onboard gives you driving convenience (charge whenever you like); is light in weight; and is low in power consumption. Some commercial EV controllers, like the AC Propulsion model discussed in chapter 8, even share components between the on-board inverter-controller and battery charger for minimum parts count and maximum efficiency (current flow decides the circuit function automatically—nifty!). Offboard gives you high power capability that translates to minimum charging time in a permanent charging station, which can incorporate many additional features.

## TOMORROW'S BEST BATTERY CHARGING SOLUTIONS

Against the numerous trade-offs available to you with today's commercially available battery chargers, along with the additional options of the build-your-own approach (using today's advanced electronics components), are the still newer developments coming down the road for tomorrow and beyond. These include rapid charging techniques, induction charging, replacement battery packs, and infrastructure development. Let's take a brief look at each.

### Rapid charging

A number of modern papers have discussed the alternatives of rapid charging. The Japanese are perhaps the most forward-thinking in this area.[1] In short, if charging your EV's battery pack in eight hours is good, then accomplishing the same result in four hours is better. Fortunately, you can use pulsed dc current, alternating charge and discharge pulses, or just plain high-level dc to accomplish the results. Unfortunately, you need to start with at least 240-volt ac (480-volt ac is better), and you probably overheat and shorten the life of any of today's lead-acid batteries in the process. However, if you design your lead-acid batteries to accommodate this feature (a larger number of thinner lead electrodes, special separators, and electrolyte), it becomes easy. Even nickel-cadmium batteries can be adapted to the process if your wallet is bigger. You'll hear more about this idea as time goes by.

### Induction charging

Many alternatives have been proposed for induction charging. General Motor's Hughes division is probably the most forward-thinking in this area.[2] Figure 10-6—taken at the September 1992 Burbank Alternate Transportation Expo—shows the Hughes concept. You drop the rear license plate to expose the EV's inductive charging slot, drop the Hughes "paddle" into the slot, and turn on the juice. With other inductive floor-mounted or front-bumper-mounted approaches, you had to position your vehicle accurately. The Hughes approach makes it even easier than putting gasoline into a vehicle. There are no distances to worry about, no spills, and no risk of electrical shock. The process of putting electrical energy into your EV can be computerized with a credit card. EV charging kiosks—or whatever shape they assume—will someday be as familiar as telephone booths, and Hughes-style inductive charging will probably lead the way.

### Replacement battery packs

Many attractive alternatives have also been proposed in this area.[3] The EV racers at the Phoenix Solar & Electric 500 have used it for years. When they wheel into the

**10-6** Hughes inductive charging system paddle.

pits, saddle-bag-style battery packs are dropped from their outboard mounting positions and fresh battery packs attached in their place. This same approach, with neighborhood "energy stations" replacing gasoline stations also has a role in the future. Future EV designs could be standardized with underbody pallet-mounted battery packs. You wheel into the energy station, drop the old battery pack, raise the new one into place and latch it down, pay by credit card (probably a deposit for the pack plus the energy cost for the charge), and you're on your way within a few minutes time.

### Beyond tomorrow

Nothing stops you from using rapid charging, induction charging, and/or replacement battery pack techniques today. The first takes a special heavy-duty lead-acid battery. The other two can be done by an individual, but obviously require infrastructure development for widespread adoption. Articles have even been written about how you can charge your EV from a solar source today.[4]

But serious infrastructure development is necessary for the real prize: roadway-powered electric vehicles. Numerous papers and articles have been written—all because of the mouth-watering appeal of the idea.[5] A simple lead-acid-battery-powered EV has more than enough range to carry you to the nearest interstate highway. Once there, you punch a button on the dashboard, an inductive pickup on your EV draws energy from the roadway thru a "metering-box" in your vehicle (so the utility company knows who and how much to charge), and you get recharged on your way to your destination without all the pollution, noise, soot, and odors.

If the roadway and your EV are both of the "smart" design, you can even get traveller information (weather, directions, etc.) while the roadway guides your vehicle in a hands-off mode, and you read the morning newspaper or scan the evening TV news. All it takes is infrastructure development—read that as M-O-N-E-Y.

## YOUR EV'S ELECTRICAL SYSTEM

To say that the electrical system of an EV is its most important part is not an oxymoron—far from it. The idea is to leave as much of the internal combustion vehicle instrumentation wiring as you need intact, and to carefully add the high-voltage, high-current wiring required by your EV conversion.

Do a great job on your EV conversion's high-voltage, high-current wiring and it will reward you with years of trouble-free service. Do it in a sloppy manner and, aside from the obvious boom or poof (accompanied by smoke), you open yourself to a world of strange malfunctions. The hidden beauty of any home-built EV conversion is that there are only two places to blame: you can look in the mirror and you can look under the hood.

This section will cover the electrical system that interconnects the motor, controller, batteries, and charger along with its key high-voltage, high-current power, and low-voltage, low-current instrumentation components. Figure 10-7 shows you the system at a glance. We'll look at the components that go into high-current and low-current side separately, then discuss wiring it all together.

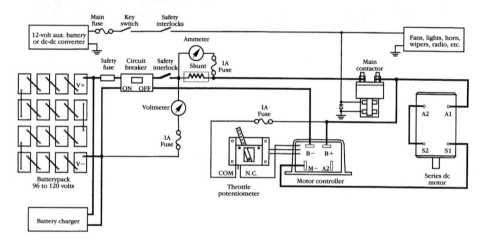

**10-7** Electric vehicle basic wiring diagram.

### High-voltage, high-current power system

The heavier lines of FIG. 10-7 denote the high-current connections. When you put the motor, controller, battery, and charger together in your vehicle, you need contactor(s), circuit breakers, and fuses to switch the heavy currents involved. Let's take a closer look at these high-current components.

**Main contactor** A contactor works just like a relay. Its heavy-duty contacts (typically rated at 150 to 250 amps continuous) allow you to control heavy currents with a low-level voltage. A single-pole, normally open main contactor—shown in FIG. 10-8—is placed in the high-current circuit between the battery and the controller and motor. When you energize it—typically by turning the ignition key switch on—high-current power is made available to the controller and motor.

**10-8** Main contactor—single pole.

**Reversing contactor**  This contactor is used in EVs when electrical rather than mechanical transmission control of forward-reverse direction is desired. The changeover contacts of this double-pole contactor—shown in FIG. 10-9—are used to reverse the direction of current flow in the field winding of a series dc motor. When this contactor is used, a forward-reverse-center off switch is added to the low-voltage wiring system after the ignition key switch.

**10-9** Reversing contactor—double pole, cross-connected.

**Main circuit breaker**  A circuit breaker is like a switch and a resetting fuse. The purpose of this heavy-duty circuit breaker (typically rated at 300 to 500 amps) is to instantly interrupt main battery power in the event of a drive system malfunction, and to routinely interrupt battery power when servicing and recharging. For convenience, this circuit breaker is normally located near the battery pack. The switchplate and mounting hardware are useful—the big letters immediately inform casual users of your EV of the circuit breaker's function.

**Safety fuse** The purpose of the safety fuse is to interrupt current flow in the event of an inadvertent short-circuit across the battery pack. In other words, you blow out one of these before you arc-weld your crescent wrench to the frame and lay waste to your battery pack in the process.

**Safety interlock** There is an additional switch that some EV converters incorporate into their high-current system (usually in the form of a big red knob or button on the dashboard)—an emergency safety interlock or "kill switch." When everything else fails, punching this will pull the plug on your battery power.

### Low-voltage, low-current instrumentation system

The remaining lines in FIG. 10-10 comprise the low-voltage instrumentation wiring. The instrumentation system includes key switch, throttle control, and monitoring wiring. Key switch wiring—controlled by ignition key—routes power from the accessory battery or dc-to-dc converter circuit to everything you need to control when your EV is operating: headlights, interior lights, horn, wipers, fans, radio, etc. Throttle control wiring is everything connected with the all-important throttle potentiometer function. Monitoring wiring is involved in remote sensing of current, voltage, temperature and energy consumed, and routing to dashboard-mounted meters and gauges. Let's take a closer look at these low-voltage components.

**Throttle potentiometer** This is normally a 5-kilohm potentiometer, but has a special purpose and important safety function. The Curtis model, designed to accompany and complement its controllers, and to use the existing accelerator foot pedal linkage of your vehicle, is shown in FIG. 10-10. The equivalent model, for replacement use or for ground-up vehicle designs not already having an accelerator pedal, is shown in FIG. 10-11. With either of these, the Curtis model provides a *high pedal disable option* that inhibits the controller output if the pedal is depressed; i.e., you cannot start your EV with your foot on the throttle—a very desirable safety feature. As the Curtis controller also contains a fault input mode that turns the controller off in the event an open potentiometer input is detected (e.g., a broken wire)—a con-

**10-10** Curtis throttle potentiometer with high pedal disable switch (note 3 switch contacts at left).

**10-11** Curtis replacement accelerator pedal throttle potentiometer.

dition that would result in a runaway—you are covered in both instances by using the Curtis controller and throttle with high pedal disable option. Figure 10-7 shows that the throttle potentiometer wiring goes directly to the controller inputs, with the interchangeable potentiometer leads and the common and normally closed contacts wired as shown.

**Auxiliary relays** Figure 10-12 shows these highly useful auxiliary control double-pole, double-throw relays in both 20-amp-rated 12-volt dc (on right) and 120-volt ac coil varieties. A typical use for the ac coil type would be as a charger *interlock*. Wired in series with the onboard charger, ac voltage sensed on the charger's input terminals would immediately disable the battery pack output by interrupting the auxiliary battery key switch line, which, in turn, opens the main contactor. dc coil uses are limited only by your imagination: additional safety interlocks; voltage, current, or temperature interlocks; and controlling lights, fans, and instrumentation.

**10-12** Auxiliary relays—120-volt ac (left) and 12-volt dc.

**Terminal strip** Whatever you're doing in the electrical department, a simple terminal strip like the one shown in FIG. 10-13 makes your wiring easier and neater. Using one or more of these as convenient tie-off points not only reduces error possibilities in first-time conversion wiring, but also makes it simpler to trace down your connections later if needed. Of course, it's only as valuable as your hand-drawn sketch of what function is on which terminal . . .

**10-13** Terminal strip.

**Shunts** Shunts are precisely calibrated *resistors* that enable current flow in a circuit to be determined by measuring the voltage drop across them. Two varieties are shown in FIG. 10-14: the left measures currents from 0 to 50 amps; the right measures currents from 0 to 500 amps.

**10-14** Ammeter current shunts—50-amp (left) and 500-amp.

**Ammeter** The most useful of all your EV onboard instruments is your ammeter. The dual-scale model shown in FIG. 10-15 (top left) delivers 0 to 50 amp or 0 to 500 amp monitoring ranges at the flip of a switch. The higher range enables you to determine

**10-15** Instrumentation meters—ammeter (top left), voltmeter (top right), battery indicator (center), temperature (bottom left), rotary switch (bottom right).

your motor's instantaneous current draw; it functions much like a vacuum gauge in an internal combustion engine vehicle—the less current the higher the range, etc. The lower range functions the same as the ammeter that might already be on the instrument cluster of your internal combustion conversion vehicle—it tells you the amount of current your 12-volt accessories are consuming.

**Voltmeter**  The second most useful EV onboard instrument has to be your voltmeter. The dual expanded-scale model (FIG. 10-15, top right), delivers 50- to 150-volt or 9- to 14-volt ranges at the flip of a switch. Expanded scale means only the required voltmeter range is used—the entire scale is expanded to fill just the range of voltages you use. The higher range enables you to determine your battery pack's instantaneous voltage (it functions like the fuel gauge in your internal combustion engine vehicle)—the less voltage the less range remaining, etc. The lower range functions like the voltmeter that might already be on the instrument cluster of your internal combustion conversion vehicle—it tells you the status of your 12-volt system.

**Battery indicator**  Users seem either to like the battery indicator a lot or find it redundant (with the voltmeter) so investigate before you invest. The Curtis Model 900 version (FIG. 10-15 center) indicates battery pack state-of-charge as of your last full chargeup using a 10-element LED readout. The battery indicator is wired directly across the battery as if it were a voltmeter. Proprietary circuitry inside the module then integrates the voltage state into a readout of remaining energy that's displayed on one of the 10 LED bars. While this is not as useful for those employing onboard charging, and certainly does not come close to being a battery energy management system, it is useful for those who charge only from a fixed site as a guide to when the next pit stop is required.

**Temperature meter**  While your lead-acid-battery-powered EV won't run at all if it's very, very cold, you can use it again when the temperature rises. At the high-temperature end of the scale, you can cause permanent damage to your batteries, controller, or motor if their temperature limits are exceeded. A temperature gauge certainly falls into the nice-to-have rather than the mandatory category, but if you are so inclined, an easy way to keep tabs on temperature is by using a thermistor and a temperature gauge like the one shown in FIG. 10-15 (bottom left). Or you can use multiple thermistors—one bonded to each object of interest (battery pack, controller, motor, etc.)—and monitor all by switching between them.

**Rotary switch**  Rotary switches such as the four-pole, two-position switch shown in FIG. 10-15 (bottom right) are ideal companions to your instrumentation meters for range, sensor, and function switching. While you can opt for only an ammeter and a voltmeter in your finished EV, you might want to check voltage and current at numerous points during the testing stage. A handful of rotary switches helps you out in either case.

**Fans**  Fans fall into the mandatory category for keeping temperature rise in check for engine compartment components, or for keeping the battery compartment ventilated during charging. Whether dc-powered full time from the key switch circuit, dc-powered intermittently via relay closure, or powered from an ac outlet, the spark-free brushless 12-volt dc motor fans and 120-volt ac motor fans are the type you want to choose.

**Low-voltage protection fuses**  All your instrumentation and critical low-voltage components should be protected by 1-amp fuses (the automotive variety work fine) as shown in FIG. 10-7. Whenever 25 cents can save you up to $200, it's a good investment.

**Low-voltage interlocks** Many EV converters prefer to implement the kill switch referred to earlier in this section on the low-voltage side. Often it's easier because there are a number of interlocks already there—seat, battery, impact, etc. In addition, a low-voltage implementation takes just a simple switch, possibly a relay, and some hookup wire—a few ounces of weight at the most, while a high-current solution takes several pounds of wire plus bending and fitting, etc.

**dc to dc converter** Figure 10-16 shows you two dc-to-dc converter options. Most of today's EV converters will opt for the 128-volt to 12-volt model on the left (it operates from 78 volts to 126 volts input and delivers a nominal 13.5-volt output). The advantages using one of these 25-amp units to power the key switch accessories, throttle potentiometer, and instrumentation boils down to one word: weight. Using a dc-to-dc converter in place of an auxiliary battery saves you 50 pounds. And most dc-to-dc converters give you a nice, stable 12-volt output, even with widely varying battery pack voltage swings. By this time, you already know it's inadvisable to draw power from anything less than all the batteries in the pack's string (or risk reduced battery life, etc.), so choose your dc-to-dc converter accordingly—no 12-volt to 12-volt models, please.

**10-16** Sevcon dc to dc converters— 72/12 model (left) and 128/12 model.

**Auxiliary battery charger** Figure 10-17 shows you the other reason for using a dc-to-dc converter: you won't need this auxiliary battery charger. An auxiliary battery also has to be recharged, so you would need a separate charger supplying 12 volts (unless you opt for a dual-voltage charger like Lester's 12/108 model). If it's a deep-discharge accessory battery, using any automotive charger just won't do. On balance, the battery and charger wind up costing you the same as the dc-to-dc converter, but take continuing attention and deliver a less steady voltage with an added weight penalty.

**10-17** Auxiliary battery charger for onboard 12-volt instrumentation battery.

**Wattmeter** The wattmeter shown in FIG. 10-18 should seem familiar to most homeowners—it's identical. No, you don't put this in your weight-conscious EV. You mount

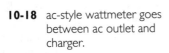

**10-18** ac-style wattmeter goes between ac outlet and charger.

it on the wall of your garage, hooked up between your battery charger and its electrical service. Keep a logbook nearby. You can record the wattmeter's reading, charge your EV's battery pack, do your tooling around, then come back and repeat the process. Over time, the wattmeter tells you your EV's energy use patterns, and can quickly tell you if something is amiss (dragging brake shoe, etc.) by deviations from the pattern. Plus you can use the results to show your wife/husband, friends, and neighbors just how much money you saved compared to an internal combustion engine vehicle.

## Wiring it all together

Five things are important here—wire and connector gauge, connections, routing, grounding, and checking. We'll cover them in sequence.

**Wire & connectors** This might be one of the last things you think about, but it's by no means the least important. While your wire size and connector type choices on the instrumentation side are not as important as the connections you make with them, *all* of these are important on the power side.

Working with AWG 2/0 cable gauge wire is not my favorite pastime—think of it as involuntary aerobic exercise—but its minimal resistance guarantees you a high-efficiency EV as opposed to the world's greatest moving toaster.

Minimal resistance means how the connectors are attached to the wire cable ends is equally important to the overall result. Crimp the connectors onto the cable ends using the proper crimping tool (ask your local electrical supply house or cable provider) or have someone do it for you. A dinky triangle contact crimp—that you can easily get away with when working in AWG 18 hookup wire—is fatal to your round ferrule AWG 2/0 connector. It will cause you a hot spot that sooner or later will melt (or be arc-welded) by the routine 200-amp EV currents. Meanwhile, you will get poor performance. If you are getting 20 miles per charge and your neighbor is getting 60 miles per charge with the identical setup, and you checked for the obvious mechanical-motor-controller-battery reasons, chances are it's in your wiring. Treat each crimp with loving attention and craftsmanship—as if each was your last earthly act—and you will be in heaven when it comes to your EV's performance.

**Connections** On the power side, connections are important. These occur when your AWG 2/0 wire connectors attach to motor, controller, batteries, shunts, fuses, circuit breakers, switches, etc. Check to ensure surfaces are flat, clean, and smooth before attaching. Use two wrenches to avoid bending flat-tabbed controller and fuse lugs. Torque everything down tight, but not gorilla-tight. Check everything and retighten battery connections at least monthly.

**Routing**  Aim for minimum length routing on the power side. Leave a little slack for installation and removal, and a little more slack for heat expansion; then go for the line that's the shortest distance between the two points. On the instrumentation side, it's neatness and traceability that count: you want it neat to show off to friends and neighbors, you want it orderly so that you (or someone else) can figure out what you did.

Figure 10-19 shows the ideal layout of Don Moriarty's custom sports racer—everything neatly laid out on a giant heatsink backing plate (⅛-inch to ¼-inch aluminum—reinforced and cross-braced, etc.). This shelf can be hinged at the back to the firewall and pinned in the front (or vice versa). Gas shocks (of the rear trunk deck variety) can be added to make its 30 to 50 pounds easy to lift up for access—a user-friendly touch. In chapter 11, Jim Harris' "magic box" gives you another page for your ideabook—a different approach that produces the same desirable results: minimum length combined with neatness and traceability.

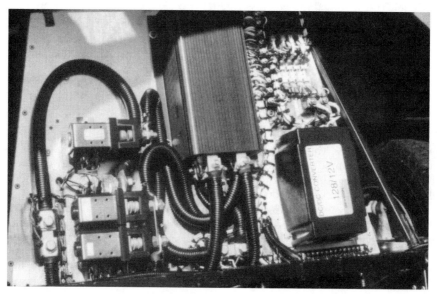

**10-19**  Don Moriarity's nearly ideal layout of electric vehicle control wiring and components.

**Grounding**  The secret of EV success is to be "well-grounded" in all its aspects. Well-grounded in electrical terms means three things:

*Floating propulsion system ground.* No part of the propulsion system (batteries, controller, etc.) should be connected to any part of the vehicle frame. This minimizes the possibility of being shocked when you touch a battery terminal and the body or frame, and of a short circuit occurring if any part of the wiring becomes frayed and touches the frame or body.

*Accessory 12-volt system grounded to frame.* The 12-volt accessory system in most EV conversions is grounded to the frame—just like the electrical system of the internal combustion vehicle chassis it utilizes. The body and frame is not connected to the propulsion system, but it can and should be used as the ground point for the 12-volt accessory system—just as the original vehicle chassis manufacturer did.

*Frame grounded to ac neutral when charging.* The body and frame should be grounded to the ac neutral line (the green wire) when an onboard or offboard ac charger is attached to the vehicle. This prevents electrical shock when the batteries are being charged. To guarantee shock-free performance, transformerless chargers should always have a ground fault interrupter, and transformer-based chargers should be of the isolation type.

**Checking** This is not a paragraph about banking. It's a paragraph about partnership. Whatever system you decide to use—continuity checking, verbal outcry, color coding, matching terminal pairs to a list, etc.—at least have one other human being help you. It'll make it go faster, and chances are you'll find something that you alone might have overlooked. Speaking of chances, be sure your EV's drive wheels are elevated the first time you hook up your batteries to your newly-wired-up creation so it doesn't accidentally "wander" through your garage door.

# 11
# Electric vehicle conversion

*Pickup trucks, the best selling vehicles, are also the best EV conversions.*

Whether you've read through the whole book thus far or just picked up the book and skipped ahead to this chapter title, you've finally made it to the chapter that tells you how to do it.

The conversion process is what ties your electric vehicle together and makes it run. You've learned that choosing the right chassis to meet your driving needs and goals is a key first step. You've also learned about motor, controller, battery, charger and system wiring recommendations. Now, it's time to put them all together. A carefully planned and executed conversion process can save you time and money during conversion, and produce an efficient vehicle that's a pleasure to drive and own after it's completed.

This chapter goes through the conversion process step-by-step with the assistance of a professional, and introduces the type of chassis to choose for your EV conversion today—the compact pickup truck. You'll discover that after the simple act of going through the conversion process once, your subsequent conversion efforts go much more smoothly, and the results look and perform even better—you are now an expert.

## CONVERSION OVERVIEW

What do you do for a fresh point of view when constructing something mechanical— even if it's an electric vehicle? You go to a mechanic. We went to a master mechanic and machinist, Jim Harris, who has had more than 20 years experience, owned his own auto repair shops, and was able to offer a completely fresh point of view on how to build an EV.

What do you do first? Start with a pickup truck. Pickup trucks are the most popular single-person transport vehicle for commuter use, and are useful for carrying many other loads as well. They make an excellent platform for EVs because they isolate the batteries from the passenger compartment very easily; the additional battery weight presents no problems for a vehicle structure that was designed to carry the weight anyway; and the pickup is far roomier in terms of engine compartment and pickup box space to do whatever you want to do with component design and layout. Pickup trucks, the best selling vehicles, are also the best EV conversions.

Keep in mind that, while we are talking about a step-by-step Ford Ranger pickup truck EV conversion here, the principles will apply equally well to any sort of EV conversion you do. And while the Ford Ranger pickup truck used in this chapter's conversion is also used by other conversion professionals, the Chevy S-10 and Dodge Dakota platforms also have their proponents.

The neatest result of this project is the public reaction Jim Harris received on his first vehicle—the 1987 Ford Ranger shown in FIG. 11-1. His enjoyment of it led him to do a second conversion—the 1993 Ford Ranger shown in FIG. 11-2—and to start up a new company offering conversion kits or ready-to-run EV conversions. Perhaps your experience will be the same.

**11-1** Jim Harris' 1987 Ford Ranger EV conversion.

The objective here is to get you into a working EV of your own with minimum fuss, converting from an internal combustion engine vehicle chassis. For those who are into building from the ground up and kit-car projects, there are other books you can read, and the techniques discussed here can be adapted.[1] The actual process for your EV conversion is straightforward:

- Before conversion (planning)—who, where, what, when.
- Conversion (doing)—chassis, mechanical, electrical and battery.
- After conversion (checking)—testing and finishing.

Figure 11-3 shows you the entire process at a glance. Let's get started.

## BEFORE CONVERSION

A little effort expended before conversion, in the who-where-what-when planning stage, can pay large dividends later because you've thought out what you need be-

**11-2** Jim Harris' 1993 Ford Ranger EV conversion.

forehand and don't have to go running around at the last minute. Let's look at the individual areas.

## Arrange for help

This is the "who" part. Help comes in two flavors: inside help and outside help.

*Inside help.* Whether it's lifting out the engine, pulling the high-current wiring, installing the batteries, or just halving the assembly time and having someone to talk with while you work, an inside helper can work wonders. It's strongly suggested you schedule one for all the heavy tasks and for any/all others as it suits you.

*Outside help.* This involves subcontracting out entire tasks to professionals who are more competent and faster at doing their specialty. Excellent candidates for outside help would be in internal combustion engine removal and fabrication of electric motor mounting components.

## Arrange for space

This is the "where" part. Your work area can be owned or rented or borrowed, but it has to be large enough, clean enough, heated or cooled enough, with light and appropriate electrical service available to take care of your needs during the length of time your conversion requires. Let's run down the checklist of what you should look for.

**Ownership or rental** You'll need the space for 80 to 100 hours of work time or more—one to three months of calendar time and up. If you're using your own two-car garage for the project, household rules of companionship dictate you don't tell your wife or husband, "I just need it for a little while honey—you don't mind parking your Porsche in the street do you?" Plan ahead. It's going to take at least a couple of months.

An alternative solution is to rent the space you need. This can be anything from a neighbor's unused garage to an oversized public storage-locker to space at a local garage. At some point your conversion is going to look like a dinosaur with its bones strewn all about; at another it's going to be very messy and greasy; at another you might need to attach a winch to an overhead beam that can support the weight of the

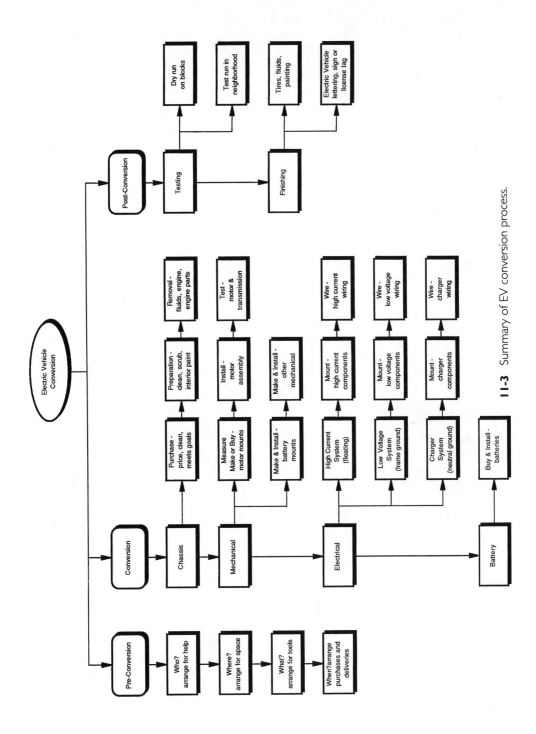

**11-3** Summary of EV conversion process.

engine you are removing; and at most times you are going to have upwards of several thousand dollars worth of components sitting around—consider all these needs.

Another alternative is to do it as a two-step. If you have the internal combustion engine parts removed by a professional, then tow your chassis body to your work area for the conversion, there is less need for a large work area and heavy ceiling joists, and it won't get as dirty.

**Size**  Realistically, an EV conversion will overflow the needs of any single car garage or storage locker. While it can be done, you will be cramped, so why not do it right to begin with. Unless you opt for the two-step approach just mentioned, you need at least a two-car garage or equivalent amount of space.

**Heating, cooling, lighting, & power**  Suit yourself here. You're going to be working in this spot for several months—why not do it in comfort and convenience.

### Arrange for tools

This is the "what" part. What kind of tools are you going to need to do the job? If you have to get into areas and tools that you are unfamiliar with, maybe you are best served by subcontracting those tasks to an outside professional. A little forethought in this area lets you quickly sort out those tasks you want to farm out versus those you want to do yourself—all just by using the tools criteria.

### Arrange for purchases & deliveries

This is the "when" part. Ideally, you have a just-in-time arrangement—the exact part you need comes magically floating through the door exactly when you need it. Reality might fall somewhat short of this, but nothing stops you from setting it up as your goal by thinking about what you are going to need by when in advance.

## CONVERSION

Conversion planning can even pay greater dividends. As you saw from FIG. 11-3, there are four parts to the conversion or doing stage and each is further subdivided:

- Chassis—Purchase, preparation, removal of internal combustion engine parts.
- Mechanical—Motor mount fabrication, motor installation, battery mounts, and other mechanical parts fabrication and installation.
- Electrical—High-current, low-voltage, and charging system components and wiring.
- Battery—Purchase and installation of batteries.

A simple way of looking at the procedure is: buy and clean up the chassis, remove all the internal combustion engine parts, make or buy the parts to mount the motor and batteries, mount and wire the electrical parts, then buy and install the batteries. Let's look at the individual areas.

## CHASSIS

The chassis part involves the purchase, preparation, and removal phases. In other words, you do everything necessary to get the chassis you're going to convert ready for conversion. Let's take a closer look at each step.

## Purchase chassis

The first step in a step-by-step Ford Ranger EV conversion is to purchase the Ford Ranger internal combustion engine pickup truck you're going to convert to electric. As an added incentive, take another look at FIG. 11-1 and FIG. 11-2. These are the finished conversion photographs. Notice, in each case, the vehicle is absolutely stock on the outside (Jim Harris did add a smooth nose in place of the 1993 model's grille). The only giveaway is the lettering. If you're going for acceleration, make sure to make the "E-L-E-C-T-R-I-C" letters on the back of your tailgate large.

The chassis purchase details were covered in chapter 6; the bottom line is to get the most for the least. You want the stripped 4-cylinder (6-cylinder in some models and years), manual transmission, least weight version. Ideally, you don't need the engine so you can do a tradeout with the dealer or selling party on the spot—just make sure everything else is as close to perfect running condition as possible—less work for you later. Most of all, make sure there's little or no rust; clean is nice, too. Mechanical parts you can replace, but a rusted body is nearly impossible to deal with; dirt and crud just exacerbate the condition and possibly hide additional problems.

On the other hand, you can start with a showroom new vehicle like that shown in FIG. 11-4A and own a conversion you'll be real proud of. In addition, removing everything from its crowded engine compartment—as shown in FIG. 11-4B—is easier because all the parts are clean and without accumulated road grime.

**11-4A**  When you start with a new vehicle . . .

**11-4B** . . . you start with a full but clean engine compartment.

## Purchase other components

Once you pick your chassis, you can make other part decisions based on your overall performance goals: high mileage, quick acceleration, or general-purpose commuter. The easiest and quickest way to do your conversion is to order all of the parts you need in one kit from someone like KTA Services Inc. Their kits typically contain the parts shown in FIG. 11-5; you get nearly everything you need to complete your conversion, the parts are matched to work together, and you have someone to turn to in case you need additional assistance. Buying a prepackaged kit will greatly simplify your conversion both in terms of what it takes to get the job done and the time it takes to run around and get all the parts you need from various vendors, make sure they match, etc.

Parts to build this particular pickup truck conversion (and pickup trucks in general) are available from Jim Harris' company—Zero Emission Motorcars, Inc. Jim's company provides the built-up control-box and motor-to-transmission adapters that really save you time and guarantee results, or the entire kit for a midrange or low-end pickup truck of your choice, which includes all heavy duty coil springs, battery brackets, cables, mounting hardware, etc.

The other items you want to order at this time are the detailed shop maintenance and electrical manuals for your chassis. These might also be available in the library or in a technical bookstore in a larger city. Study these before starting your conversion.

## Prepare chassis

The next step is to clean the chassis and make a few measurements. The first chassis-cleaning step is to give its engine compartment a good steam cleaning. Well-used chassis might require extra scrubbing to remove accumulated dirt and grime. This has a twofold purpose: it minimizes the grease and gunk you have to deal with in parts re-

**11-5** Typical KTA Services Inc. electric vehicle kit.

moval, and you can *see* what you need to look at and measure. You might even want to repaint the engine compartment area at this time.

The measuring step involves determining the position of the transmission/drivetrain or transaxle in your internal combustion engine vehicle and reproducing this position as closely as possible in your EV. Your internal combustion engine chassis will have one of three possible engine-drivetrain arrangements: lateral-mount front engine, transmission, and driveshaft; transverse-mount front engine with transaxle; and rear engine with transaxle (VW-style). Pickup truck conversion chassis will nearly always deal with one of the first two variations. Two measurements are important: vertical height from transmission housing to floor (measured on a level surface); and a vertical and horizontal reference point on transmission housing to body or frame.

Vertical clearance from the floor is important in any conversion where your motor mounts also support your transmission. You want to reproduce this as closely as possible so your tranny is not running uphill or downhill from the way it started.

A firewall or frame horizontal and vertical alignment mark is also important—it helps in obtaining proper mechanical alignment of the mating parts later on. Scribe this mark accurately with an awl or nail and highlight it with a dab of spray paint or touch-up paint so you can later find it.

## Removing chassis parts

The next step is to drain the chassis liquids, remove all engine-compartment parts that impede engine removal, remove the engine, and then remove all other internal combustion engine-related parts.

The parts removal process starts with draining all fluids first: oil, transmission fluid, radiator coolant, and gasoline. Remember to dispose of your fluids in an environmentally sound manner or recycle them. Draining gasoline from your tank is particularly dangerous and tricky—put out your cigarette before you attempt it—drain as much as you can before you physically remove the tank. Discharging an air-conditioning system, if your conversion vehicle has one, is a job best left to a professional.

Next, disconnect the battery and remove all wires connected to the engine. Carefully disconnect the throttle linkage—you will need it later—and set it aside out of harm's way. Then remove everything that might interfere with the engine lift-out process. The hood is a good item to start with, followed by radiator, fan shroud, fan, coolant and heater hoses, and all fuel lines. Disconnect the manifolds and remove the exhaust system at this time, too.

If your work area isn't equipped for heavy lifting—the smallest 4-cylinder engine with accessories attached can weigh around 300 pounds—you might be best served by letting a professional handle the heavy work. Engine overhaul shops are usually glad to handle the job and you might be able to cut a deal with them for the engine and parts that results in a net gain on the transaction. Plus there are a number of other aspects—draining fluids, storing parts, not breaking cables and wires—plus the physical act of actually removing the engine without damage to you, your chassis, or the engine. To use an analogy, it's like changing your own motor oil at home versus at the JiffyQuick. It takes you several hours and you have to clean up a mess (yourself and your spot). The JiffyQuick people, who do it for a living, take 10 minutes and neither they nor you get dirty.

If you do the job yourself (schedule an "inside" helper for this step), pick a sturdy joist to attach your lifting winch to—not a two-by-four—or rent an engine-lifting dolly. Cover the vehicle's fenders and sides with moving pads to protect them during lift-out. Attach the chain or cable to the engine at two different points. Remove the bolts from the engine mounting brackets and those attaching the engine to the transmission bell housing. The disconnected engine slides straight back away from the transmission, then up and out (in theory). In practice, since you have to both pull down on the winch cable and pull back on the engine block, another set of hands helps greatly. The objective of the whole process is to remove the 300 pounds of engine without banging, bending, or damaging anything inside the bell housing (transmission spline shaft, etc.) or on the now exposed business-end of the engine (clutch-flywheel assembly, etc). Carefully set the removed engine down in an out-of-the-way yet protected part of your work area.

Finally, remove everything else associated with the internal combustion engine: gas tank, gas lines, muffler, exhaust pipes, ignition, cooling and heating systems.

## MECHANICAL

The mechanical part involves all the steps necessary to mount the motor, and install the battery mounts and any other mechanical parts. In other words, you next do all the mechanical steps necessary for conversion. You follow this sequence because you want to have all the heavy drilling, banging, and welding—along with any associated metal shavings or scraps—well cleaned up and out of the way before tackling the more delicate electrical components and tasks. Let's take a closer look at the steps.

### Mounting your electric motor

Your mission here is to attach the new electrical motor to the remaining mechanical drivetrain. The clutch-to-flywheel interface is your contact point. Figure 11-6 gives you an overview of your task in generalized form. Four elements are involved:

- The critical distance—flywheel to clutch interface.
- Rear support for electric motor.

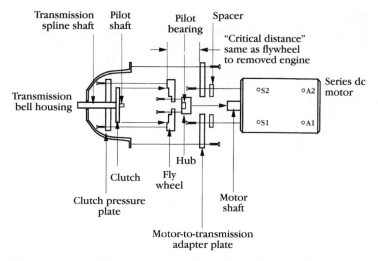

**11-6** Elements of the electric motor to mechanical transmission connection process.

- Front support—motor-to-transmission adapter plate.
- Flywheel to motor shaft connection via the hub or coupling.

We'll cover what's involved in each of these four areas in sequence. Understand that this discussion has to be generalized because there are at least a dozen good solutions for any given vehicle. So we're going to talk in general terms here—you'll have to translate them to your own unique case. And if your skills do not include precision machining of automotive metal parts, this is another good area to enlist the services of a professional such as KTA Services or Jim Harris' company or a local machine shop.

**The critical distance—Flywheel to clutch interface** After you remove the engine and before you take the flywheel off, carefully and accurately measure the distance from the front of the engine to the face of the flywheel (the part the clutch touches). This is the critical distance you want to reproduce in your electric motor mounting setup. In the 1987 Ford Ranger conversion, this critical distance measurement is 1.750 inches—but it will vary for different vehicles. You're going to put an adapter plate on the electric motor, put a hub or coupling on the electric motor's shaft, and attach the flywheel to this hub. When you've done this, the critical distance from the front of the flywheel to the front of the motor adapter plate (FIG. 11-6) should measure exactly the same.

Knowing your goal—the critical distance—makes it easier to navigate towards it. Whether you have a front engine plus transmission or front or rear engine plus transaxle, this goal will be the same, although the specifics will differ.

Your original flywheel might weigh 24 pounds and have an attached ring gear (for the starter motor). Remove the ring gear that fits around the flywheel's outside edge and have the flywheel machined down to 12 pounds or so by removing metal from its outside edge and rear (the part that faces the motor). Don't touch the flywheel's face (unless it has obvious burrs or other defects). These steps don't hurt your flywheel at all, yet save weight—the most important factor in your EV conversion's performance.

This is also a good time to give the clutch (the inside layer of the clutch pressure plate/clutch/flywheel "sandwich") a thorough exam. If the clutch is old (used for more than 30,000 miles) or obviously worn, now is the time to replace it. Also take a look at transmission seals and mounts and replace them if excessively worn. Keep in mind you're going to be using your clutch in an entirely new way (*not* using it is a better description, because there's nothing to start up—you put it in gear first), so this new clutch ought to last you 200,000 miles or more.

**Rear support for electric motor**  This is a fairly straightforward area. You basically have two possibilities: support the motor around its middle or support it from the end opposite its drive connection to transmission or transaxle.

Figure 11-7 shows some of the middle-style motor mounts available from KTA Services—accommodating different electric motor diameters. The bottoms of the mounts have bolt holes that attach to your vehicle's motor mounts. The two halves of the curved steel strap go around the motor and hold it securely in place. This is the preferred and most common method of mounting electric motors—particularly in the larger sizes.

**11-7**  Typical KTA Services Inc. electric motor mounts for 7-inch through 9-inch motor diameters.

End mounts are similar motor faceplate adapter mounts—they bolt on to the motor's end face through mounting holes, and are then tied off to the frame through heavy rubber (old tires) shock isolators. This approach was very popular with early VW conversions that used smaller electric motors already securely attached to the transaxle housing at their front.

**Front support—Motor-to-transmission adapter plate**  Here's where the fun begins. Figure 11-8 shows you two motor mounting kits from KTA Services—from among many possibilities. A simple VW Bug adapter kit is shown in the upper part of FIG. 11-8; a more complex VW Rabbit adapter kit is shown at the bottom. With either kit, the large plate in the background that the other parts are resting on is the motor-to-transmission adapter plate.

Notice there are two bolt hole ring-patterns in the mounting plates. As shown in FIG. 11-6, the inner-ring pattern with its countersunk mounting holes allows flat-head mounting bolts to attach the mounting plate to the motor and to be flush with the plate's surface on the inside when tightened. As also shown in FIG. 11-6, the outer-ring pattern allows hex-head mounting bolts to be used to attach the mounting plate to the transmission bell housing and to be torqued down tight. Notice in FIG. 11-8 that the

**11-8** Typical KTA Services Inc. electric motor mounting kits for VW Bug (left) and VW Rabbit (right).

center hole in each adapter is either sized to accommodate a collar around the motor, or the outside diameter of the coupling. This can vary widely from case to case.

The spacer shown in FIG. 11-6 and the depth to which the hub fits on the motor shaft control the critical distance measurement; adjust these as needed.

If you are building your own, 1-inch thick aluminum is the preferred building stock. Cut your cardboard pattern or template out to completely cover the bell housing front, exactly locate and mark the center of the transmission shaft (it might or might not be in the center area of your template), locate and mark the bolt hole locations on its front, trace the bell housing's outline on its back, and trim your pattern for 1-inch or less overhang all around (to aesthetically suit your taste). Then transfer the electric motor mounting bolt hole patterns to this template using the transmission pilot shaft center as a reference. Take this completed template or pattern to the machine shop and ask them to reproduce it in metal.

If you are in a let's-get-it-done mode, Jim Harris' company makes a complete motor-to-transmission assembly with adapter plate, and precision-cast flywheel and hub/bushing parts that guarantees perfect results for Ford Ranger conversion projects.

Also needed is a torque rod to support/stabilize the motor-transmission combination and prevent excessive rotation under high acceleration loads. This adjustable length rod normally attaches between one of the adapter plate to transmission bolts and the frame.

**Flywheel to motor shaft connection** This is the least standardized area. The VW Bug hub or coupling (FIG. 11-8, left) is a four-bolt affair with a keyway (square notch in central shaft opening) and set screw opening (hole in outside of coupling). The VW Rabbit hub or coupling (FIG. 11-8, right) is a six-hole affair also with keyway and set screw opening. The coupling is press fit onto the electric motor's shaft, and the set screw further secures it. Figure 11-9 shows the physical positioning of the six-bolt coupling in front of the electric motor to give you a better idea.

The pilot bearing shown in FIG. 11-6 fits into the hub and mates with the transmission's pilot shaft (a transaxle does not require a pilot bearing). The clutch attaches to the transmission spline shaft, and the clutch pressure plate fits over it and bolts to the flywheel to complete the "sandwich."

Hubs (with or without pilot bearing, etc.) are going to differ widely in size and shape for different vehicle, transmission, and motor types, yet their function is the

**11-9** Close-up of KTA Services Inc. VW Rabbit hub or coupling shown in front of electric motor.

same. You must determine what you need in your particular case. Call KTA Services, Jim Harris, or your local mechanic to help you if you're not sure. This is not the time and place to be shy.

**Mounting & testing your electric motor** Depending on your accumulated skill and luck, your motor installation can either go smoothly the first time or involve several cut-and-try iterations. The basic approach is to mount the hub on the motor, attach the motor to its adapter plate, attach the flywheel to the hub, measure the critical distance to make sure it's exactly the same as it is for the internal combustion engine (add spacers as needed, etc.), and then attach this completed assembly to the transmission. The winch used to take out the internal combustion engine is probably the easiest way to install your motor assembly into your vehicle (the completed assembly weighs upwards of 150 pounds with a 20-HP series dc motor), but a floor jack (elevate the front of the vehicle on jack stands first) or engine lifting dolly will also work. The installation step is basically the reverse of the removal step: attach the assembly at two points, then slide the motor up and toward the transmission until the shafts are in alignment. When everything fits exactly, tighten the bolts on the transmission and on the motor mounts.

For a quick test, first jack up the vehicle so its front or rear drive wheels are off the ground. Then put a 2/0 cable in place across the motor's A2-S2 terminals. Attach another 2/0 cable from a battery's negative terminal to S1—you can borrow the 12-volt starter battery you just removed for this purpose—and attach another 2/0 cable to the battery's positive terminal, but don't connect it yet. With the transmission in first gear, briefly touch the positive cable from the battery to A1 and do two things:

- Look to see if the rear (or front) wheels move.
- Listen for any strange or grinding noises, etc.

If the wheels move, good. If the wheels move and there are no strange grinding sounds this is doubly good, and you can go on to the next step. If you hear something strange or the wheels don't turn (in this case, first ensure that the battery is charged), you need to unbutton your motor assembly from the tranny and look into the problem.

## Fabricating battery mounts

Jim Harris' 1987 Ford Ranger pickup conversion uses 20 6-volt batteries. Jim elected to mount four of his batteries in the engine compartment area just vacated by the radiator's removal, and the 16 remaining batteries (in a four-by-four array) in the pickup bed area. In Jim's 1987 Ford Ranger pickup conversion, the rear battery bracket is attached directly to the frame for maximum rigidity and strength and lowest center of gravity. In FIG. 11-10A, Jim points to the chassis frame member underneath the cut-away pickup box floor. In FIG. 11-10B, Jim holds up the sturdy pieces of 2-inch by 2-inch by ³⁄₁₆-inch steel angle iron that make up the battery mounting frame. The outside frame dimensions are slightly larger than the dimensions of the four-by-four battery array to allow for battery expansion. The batteries rest on a ¾-inch-thick marine-grade plywood base, and are wedged in place inside the front and rear frames by strips of wood.

**11-10A** 1987 Ford Ranger rear frame rails.

## Additional mechanical components

Jim wound up adding heavier rear leaf springs for better handling to his 1987 Ford Ranger pickup conversion, and to his 1993 Ford Ranger pickup conversion he added a smooth nose in place of the stock grille for better airflow characteristics. In practice, both of these would be added at this point in the assembly process. Any metal fabricated parts inside the engine compartment (air dams, etc.) are better done at this stage because they are easier to get at before the electrical components and wiring are in-

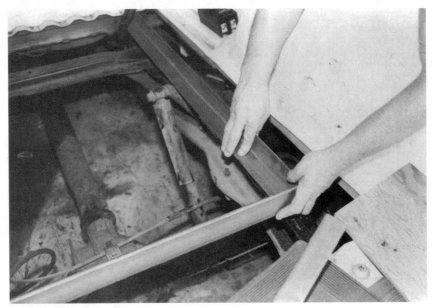

**11-10B**  Battery mounting frame in place.

stalled. Shocks, coil and leaf springs, and external body parts are also best added at this time while you still have your heavy tools out.

## Cleanup from mechanical to electrical stage

It's a good idea to use an air hose to clean up the engine compartment and rear pickup box areas of your conversion after the mechanical phase is completed. Follow this with a broom sweep-up of the work area and/or a damp mop-up. The reason for this cleanup is to minimize the chance of metal strips or shavings finding their way into your electrical components or wiring during the electrical phase.

## ELECTRICAL

The electrical part involves mounting the high-current, low-voltage, and charger components and doing the electrical wiring to interconnect them. To do the electrical wiring requires knowledge of your EV's grounding plan: the high-current system is floating, the low-voltage system is grounded to the frame, and the charging system ac neutral is grounded to the frame when in use. Doing the electrical wiring also involves knowledge of your EV's safety plan: appropriate electrical *interlocks* must be provided in each system to assure system shutdown in the event of a malfunction, and to protect against accidental failure modes. The electrical wiring is also greatly facilitated by using a *junction box* approach, so wiring is neat (no wires running every which way inside the engine compartment) and you can later trace your wiring. Let's take a closer look at the steps.

## High-current system

First you attach the high-current components, then pull the AWG 2/0 cable to connect them (another step where scheduling an inside helper is appropriate). Refer back to

FIG. 10-10. Notice there are seven components in the high-current line (in addition to batteries and charger—these we'll save for later):

- Series dc motor.
- Motor controller.
- Circuit breaker.
- Main contactor.
- Safety fuse.
- Ammeter shunt(s).
- Safety interlock.

You've already mounted the motor. In Jim's 1987 Ranger conversion, he mounted the controller on an aluminum heatsink plate directly above the motor in the engine compartment. Figure 11-11A shows the driver-side view of Jim's series dc motor attached to the transmission with its A2 to S2 jumper installed, and additional power cables attached to its A1 terminal and clearly-marked S1 terminal. Figure 11-11B shows the passenger-side view of Jim's series dc motor attached to the transmission (notice the earlier-mentioned torque arm between motor and frame), with its S1 power cable going up to the controller and motor, and controller power cables (with standoff clip to separate them) going down the chassis frame rail toward the rear battery pack. Figure 11-12 shows a transverse mounted motor from Bill Williams' 1976 Honda Civic wagon for comparison. Notice the commonality of the components—motor, adapter, controller, torque arm, and wiring— but with specific differences in wiring, attachment method, and size.

**11-11A** 1987 Ford Ranger electric motor power cabling from driver side.

Figure 11-13 (top) shows the power cabling to the mounted controller viewed from the front of the engine compartment. The M-destination of the earlier-referenced S1 power cable is clearly visible as is the B+ destination of the A1 power cable.

**11-11B** Ford Ranger electric motor power cabling from the passenger's side.

**11-12** Bill Williams' 1976 Honda Civic transverse-mounted electric motor.

Figure 11-13 (bottom) shows Jim measuring position prior to mounting the circuit breaker in the rear pickup box area near the batteries. This circuit breaker has two inputs and two outputs. Other EV converters prefer to mount this circuit breaker within reach of the driver. The trade-off is you have to pull heavy AWG 2/0 power cables be-

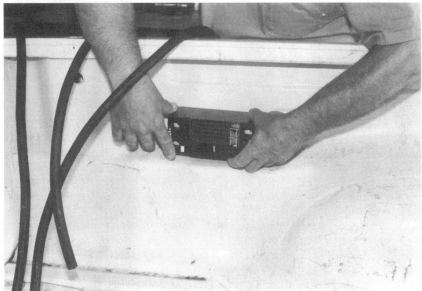

**11-13** 1987 Ford Ranger controller power cabling from front (top) and rear pickup box mounted main circuit breaker (bottom).

hind the dash or some other location inside the driving compartment. The benefit is the peace of mind afforded by knowing that if everything goes haywire (contactor and relay contacts weld shut, etc.) there is one switch you can reach over and touch to shut everything down.

The high-current safety fuse, ammeter shunt(s), and main contactor are all located inside the junction box that will be discussed in a moment. Jim chose to implement a safety interlock on the low-voltage rather than the high-current side. Those who insist on a dashboard-mounted high-current safety or kill switch (usually with a big red knob on it) have to endure the pain of pulling AWG 2/0 cable over, under, around, and through the rear of the dashboard—not my idea of a good time. By the way, the AWG 2/0 being discussed here is stranded (solid wire would be nearly impossible to work with in this size), insulated copper (never aluminum) wire—and it's about ¾ inch in diameter so you have your hands full.

The high-current system has a floating ground. This means that the negative terminal of the battery pack is not connected to the frame or body at any point—it floats instead. This eliminates the possibility of an accidentally dropped tool arc welding itself to the body or chassis or, worse yet, causing your vehicle to bolt forward or backward while simultaneously causing battery pack meltdown. It also eliminates the possibility of your receiving a 120-volt electrical shock while casually leaning over the fender to measure battery voltage.

## Low-voltage system

On the low-voltage side, the idea is to blend the existing ignition, lighting, and accessory wiring with the new instrumentation and power wiring. There are six main components on the low-voltage side:

- Key switch.
- Throttle potentiometer.
- Ammeter, voltmeter, or other instrumentation.
- Safety interlock(s).
- Accessory 12-volt battery or dc-to-dc converter.
- Safety fuse(s).

Every EV conversion should use the already-existing ignition key switch as a starting point. In an EV, the key switch serves as the main on-off switch with the convenience of a key—its starting feature is no longer needed. You should have no problem in locating and wiring to this switch.

In Jim's 1987 Ranger conversion, he mounted the throttle potentiometer on the driver's side fender well (FIG. 11-14, top). Figure 11-14 (bottom) shows it mounted in place and wired up.

Instrumentation wiring is simple; just be sure to observe meter polarity markings—the plus (+) marking on the meter goes to the positive terminal on battery. The ammeter is connected across the shunt(s) already wired into the high-current system. The voltmeter goes across the battery. The best solution is to wire the voltmeter so that it is always on, giving you a continual readout of battery status. You don't have to worry about draining the battery because a modern voltmeter's internal resistance is high enough to cause only a miniscule current drain (an order of magnitude less than the battery's own internal self-discharge rate). The battery indicator or state-of-charge meter also goes across the battery, but it should operate only when the key switch is on, so wire it to the on-off side of the main contactor (away from the battery). The temperature meter wiring is not particularly critical; just observe proper polarities and grounding, and make sure the thermistors are securely attached to whatever you are measuring. Jim utilized only ammeter and voltmeter instrumentation, with a range switch mounted on the 1987 Ranger's dashboard (FIG. 11-15, top).

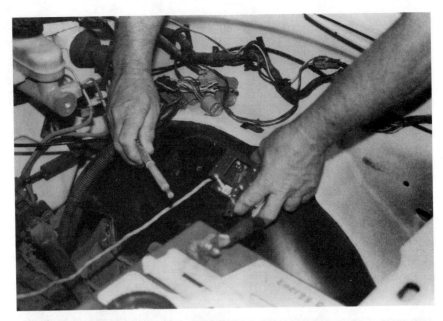

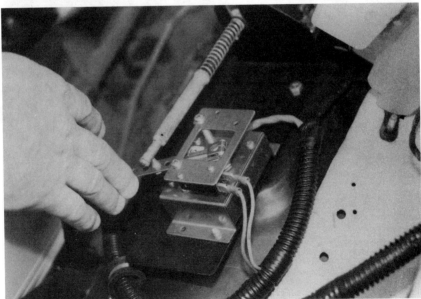

**11-14** 1987 Ford Ranger throttle potentiometer positioning (top) and mounted and wired throttle potentiometer (bottom).

If you wish to utilize a more modern digital voltmeter readout in place of the analog meters, you need to adjust the DVM's sample-and-hold circuit (it memorizes the value at any instant) either to display the average of the last few moment's sample-and-hold values, or to give a steady readout when a read button is pressed. Otherwise the rapidly changing voltage or current will be hard to interpret.

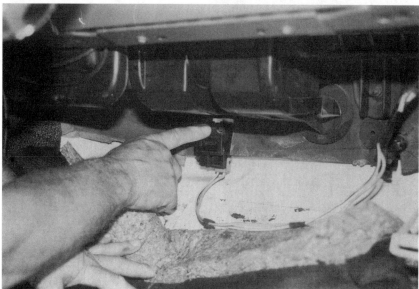

**11-15** 1987 Ford Ranger ammeter and voltmeter instrumentation (top) and impact cutoff switch (bottom).

The subject of safety interlocks is an important one. Jim's design uses three—all wired in series on the low-voltage 12-volt key-switch line: a fuel injection impact switch, a main safety cutoff switch, and a charger cutoff switch (to be covered in the charger section). The fuel injection impact switch's normal role is to shut off the fuel system in the event of a crash impact. Jim points to its location under the passenger's side of the 1987 Ranger's dashboard in FIG. 11-15 (bottom). The main safety cutoff switch is a highly accessible, dashboard-mounted switch wired in series with the key switch. Punching it immediately removes energizing voltage from the main high-current contactor. A few EV converters also use a seat interlock switch that latches closed when the driver's presence in the seat is detected. You might wish to consider this as an option.

Jim opted to use a battery as the source of 12-volt accessory system power. You can do the same or utilize the dc-to-dc converter shown in chapter 10 that's driven from the main battery pack voltage. If you opt for the dc-to-dc converter, now is the time to install it and wire it in place—its input side goes directly across the main battery pack plus and minus terminals. Its output side provides +12 volts at its positive terminal, and its negative terminal is wired to the chassis. If you elect to use a 12-volt deep-cycle accessory battery, do the wiring for it now but wait until the battery phase to purchase, install, and connect it up.

Try to use AWG 12 (20-amp rating) or AWG 14 (15-amp rating) stranded insulated copper wire for the low-voltage system. The instrumentation gauges can be wired with AWG 16 or even AWG 18 wire.

Safety fuses of the 1-amp variety should be wired across the potentiometer and all delicate instrumentation meters. The key-switch circuit can utilize the original fuse panel but don't use the original wiring for any loads greater than 20 amps—the main fuse should be of the 10-, 15-, or 20-amp variety.

Unlike the high-current system, the low-voltage system is grounded to the frame—the negative terminal of the 12-volt battery (or dc-to-dc converter) is wired directly to the frame or body. Most internal combustion engine chassis come this way. You eliminate rewiring, extra wiring, and potential ground loops by using the existing negative-ground-to-the-frame convention.

## Junction box

A good junction box design cleans up the hodgepodge of instrumentation wiring running every which way inside the engine compartment, enables you (on anyone else!) to later retrace your wiring, and provides convenient mounting and tie-off points for various components. However, not all junction boxes are created equal. Jim's "magic box" is more equal than most—it combines simple design and layout with high utility (FIG. 11-16). The high-current safety fuse (in the center, behind the power cable), the ammeter shunts (large one on center of back wall, small one in front of main contactor), and the main contactor (on left side of box) form—along with the terminal strip—the "backbone" from which all interconnections are made. Notice all the safety fuses are located in one convenient area at the right rear of the box. Jim's objective was to simplify, so his design has only two high-current cables (rightmost cable exits to motor A1, left cable exits to battery positive terminal) and three low-voltage instrumentation cable bundles. You'll see the continuous design evolution of Jim's magic box in later photographs as the conversion progressed.

## Charger system

The benefits of the onboard charger are convenience and the ability to take advantage of on-the-road charging opportunities as they are presented. The dual objectives in wiring the onboard charging system are to prevent the charging routine from becoming a "shocking" experience (via proper grounding), and to prevent momentary distractions from causing you to drive away while the charging cable is still attached (via a charger safety interlock). There are four main components to the charger system:

- Compact onboard charger.
- Lightweight line booster (optional).
- Safety charging interlock.
- ac input system.

**11-16** Jim Harris' prototype "magic box" for 1987 Ford Ranger.

In Jim's 1987 Ranger conversion, he mounted the onboard K & W BC-20 charger on the driver's side of the aluminum heatsink plate, directly above the motor in the engine compartment, as shown in FIG. 11-17 (top). The K & W LB-20 line booster, required by the charger for 120-volt operation, is mounted next to the throttle potentiometer on the driver's side fender well (FIG. 11-17, bottom).

The preferred location for most ac input charging connections in conversion vehicles is usually the location vacated by the gas tank filler neck opening. Figure 11-18 (top) shows that Jim's choice follows this pattern—it's behind the original gas cap door. Jim also chose to implement a male "twist-lock" three-prong ac charging connector, enabling him to use a standard extension cable (one end male, other end female) with the male plug on the extension cord end able to mate conveniently with the standard 120-volt ac female service outlets. (Be sure to use a three-conductor extension cable with at least AWG 12 wire in it.) Figure 11-18 (middle) shows an onboard female twist-lock three-prong ac connector. Figure 11-18 (bottom) shows that behind the gas tank filler door is not the only location; in this case, a male twist-lock three-prong ac connector has been recessed in the front bumper using a conventional outdoor ac junction wiring box.

For the charging system, you should use AWG 10 (30-amp rating) stranded insulated copper wire for both the charger-to-battery and the charger-to-ac-input receptacle connections. In order to prevent you (or anyone else) from casually driving away with the extension cord attached while charging, it's a good idea to implement a *charger interlock* system. Jim's approach is to use a relay whose coil is energized by

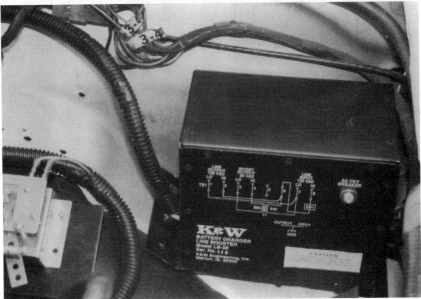

**11-17**  1987 Ford Ranger mounted K & W battery charger (top) and mounted K & W line booster (bottom).

the presence of 120 volts ac, and whose contacts are in series with the 12-volt key-switch line. When the 120-volt ac line cord is plugged in, this relay latches open and keeps the main battery pack disconnected from the controller and motor—the vehicle is immobilized.

**11-18** ac charging connector options—Jim Harris' male (top), optional female (middle), optional male bumper mount (bottom).

Other EV converters have also used the charger interlock feature to energize battery compartment fans (forced ventilation of the batteries) while charging. Additional interlock possibilities include: sensors that inhibit drive-away when the fault conditions of engine compartment hood open, battery compartment hood open, or ac charging connector access door open are detected; sensors that inhibit the charging function during fault conditions such as engine or battery compartment hood open (because you don't want outsiders prying when charging currents and battery gases are present). You might consider any of these options.

In order to prevent you (or anyone else) from getting shocked when touching your EV's body while it's charging, the body and frame should be grounded to the ac neutral line (the third prong of the connector with the green wire leading to it). This neutral wire is connected between the ac input connector and the body or frame, and is utilized only when charging. The batteries should be floating—no terminals touch the frame—and might even be further isolated by locating them inside their own compartment or battery box. This is particularly appropriate for not recommended (but done anyway) inside-the-passenger-compartment battery installations. To guarantee no shocks, transformerless chargers should always have a ground fault interrupter installed, and transformer-based chargers should be of the isolation-type.

If you prefer not to use an onboard charger, this changes your wiring and interlock design plans. In this case, you are going to be providing a 120-volt dc input at currents up to 30 amps from a stationary charger. You will also need to charge the onboard 12-volt accessory battery (unless you utilize a dc-to-dc converter). This means your charging receptacle input needs at least two connectors: one for the high-current floating 120-volt plus and minus input leads, and the other for the low-voltage grounded-to-frame 12-volt plus and minus input leads. The charger interlock design for the offboard charger is identical to the onboard case, except that you use a relay whose coil is energized by the presence of 120 volts dc and whose contacts are in series with the 12-volt key-switch line. When the 120-volt dc extension cord from the charger is plugged in, this relay latches open, disconnects the main battery pack, and immobilizes the vehicle.

## BATTERY

Buying and installing your main propulsion battery pack batteries is the last step in the EV conversion process. You buy them last because you don't want to be tripping over them, nor do you want to keep charging them during the weeks (or months) of assembly. When your EV conversion is nearly complete, after all the battery mounting frames are built and the wiring is done, you bring in the batteries. Let's take a closer look at the steps.

### Battery installation

The most important consideration in battery installation is to mount them in a location that will be easily accessible for servicing later. You've previously decided on your battery type and quantity, obtained their dimensions, allocated your front and rear mounting space, and designed and built your battery-mounting frames slightly oversized to accommodate battery expansion with use. All that remains is to install them.

In Jim's 1987 Ranger conversion, he mounted a bank of four 6-volt batteries in the front of the engine compartment space (FIG. 11-19, top). Jim chose U.S. Battery Model 2200 batteries. These provide 220-AH capacity—slightly less than that of the Trojan T-125 (235-AH) batteries recommended in chapter 9. Size (and volume) is approximately the same. Figure 11-19 (middle) shows how a wooden spacer strip is used to wedge the batteries in place. You could also use a frame around the top of the batteries to hold them in place, but unless you are going Baja racing, the weight of the batteries does a very good job by itself.

**11-19** 1987 Ford Ranger battery mounting—front batteries (top), wood strip wedges (middle), rear batteries (bottom).

Jim's rear pickup box four-by-four of 16 6-volt batteries is shown (FIG. 11-19, bottom) along with the now wired-up main circuit breaker. Notice the plywood has been painted black to match the steel frame and chassis rails. The batteries are wedged in place with wood strips along the base as in mounting the front batteries. The top battery pack tie-down frame is again optional unless you plan on doing some rugged driving.

### Battery wiring

The most important consideration in battery wiring is to make the connections clean and tight. Figure 11-19 also shows both front and rear battery packs in their wired-up condition. Check that you haven't accidentally reversed the wiring to any battery in the string as you go. Double-check your work when you finish, and use a voltmeter to measure across the completed battery pack to see that it produces the nominal 120 volts you expect. If not, measure across each battery separately to determine the problem. If reversed wiring was the culprit, the correctly installed and wired battery should fix it. If a badly discharged or defective battery is the culprit, check to see that it comes up on charging and/or replace it with a good battery from your dealer.

Important: *Make sure that the main circuit breaker is off before you connect the last power cable in the battery circuit.* Better still, switch the main circuit breaker off and wait until the system checkout phase before final battery connection.

### Accessory battery

This is also the time to mount your 12-volt accessory battery. Jim initially used the 12-volt starter battery that came with his conversion chassis (FIG. 11-20) for two reasons: it allowed him to do component testing during the wiring phase, and it was already mounted and wired in place (saving him a few steps). Figure 11-20 (top) shows that the inside of Jim's "magic box" is now down to two power cables (the leftmost cable's battery destination is now clearly visible) and two instrumentation cables. Figure 11-20 (bottom) shows the outside of Jim's magic box with the cover in place installed in its initial location.

## AFTER CONVERSION

This is the system checkout, trial run, and finishing touches stage. First make sure everything works, then find out how well it works, then try to make it work even better. When you're satisfied, you paint, polish, and sign your work. Let's look at the individual areas.

### System checkout on blocks

Jack up the drive wheels of your conversion vehicle (or raise them up on work stands) for this phase. The objective is to see that everything works right before you drive it out on the street. With your vehicle's drive wheels off the ground and the transmission in first gear, do the following:

1. Before connecting the last battery cable, verify that the proper battery polarity connections have been made to the controller's B+ and B– terminals.
2. Obtain a 100- to 200-ohm 5- or 10-watt resistor, and wire it in place across the main contactor's terminals. With the key switch off but the last battery cable connected and the main circuit breaker on, measure the voltage across the

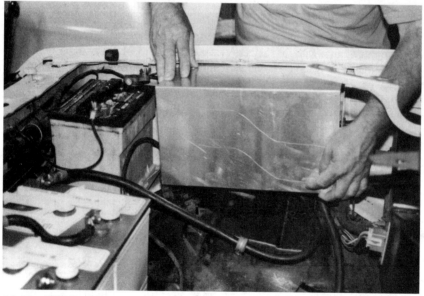

**11-20** 1987 Ford Ranger mounted Jim Harris' "magic box" (top) and with cover held in place (bottom).

controller's B+ and B– terminals. It should measure approximately 90 percent of the main battery pack voltage (in the neighborhood of 108 volts) with the correct polarity to match the terminals. If this does not happen, troubleshoot the wiring connections. If it does, you're ready to turn the key switch.

3. Turn on the key switch with your foot off the accelerator pedal. If the motor runs without the accelerator pedal depressed, turn off the key switch and troubleshoot your wiring connections. If nothing happens when you turn on the key switch, go to the next step.
4. With the transmission in first gear, slowly press the accelerator pedal and see if the wheels turn. If the wheels turn, good. Now look to see which way the wheels are turning. If the wheels are turning in the right direction, this is doubly good. If not, turn off the key switch and main breaker and interchange the dc series motor's field connections. If you are moving in the right direction, go to the next step.
5. If you have the high pedal disable option on the Curtis controller, turn off the key switch, depress the accelerator pedal and turn on the key switch. The motor should not run. Now completely take your foot off the accelerator and slowly reapply it. The motor should run as before. If this does not work correctly, troubleshoot your wiring connections. If it works, you are ready for a road test. Turn off the key switch.

## Neighborhood trial run

Check the state of charge of your main battery pack. If it's fully charged or nearly full, you can proceed. If it's not charged, recharge it before taking the next step. It's just too embarrassing to run out of juice on your first neighborhood cruise—and once you drive it a little, you'll want to drive it a lot more.

After the batteries are fully charged, remove the jack and/or wheel stands from under your EV, open the garage door (believe me, it's a necessary step), check to see that all tools, parts and electrical cords are out of the vehicle's path, and turn on the key switch. Put it in gear, disengage the parking brake, step on the accelerator and cruise off into the neighborhood. Neat, eh?

The vehicle should have smooth acceleration, good top speed, and should brake and handle normally. The overwhelming silence should enable you to hear anything out of the ordinary with the drivetrain, motor, or brake linings, etc. Now, you're ready to get fancy.

## First visitor does a second take

Before Jim had gotten very far on his first test drive he was already drawing attention. Jim had just parked his newly-completed EV conversion against the scenic backdrop of the Chula Vista marina and gotten ready to take some photographs when a city employee drove by and said, "Hey, you're not allowed to park there."

Jim replied, "I'm just taking some photographs of my electric vehicle. I'll only be a minute."

"Your what?"

"Electric vehicle."

"Where's the engine?"

One quick spin around the park loop and the city employee, now grinning ear to ear, emerged and said, "Wow, that was neat, take as long as you like."

You too can have fun and make new friends.

## Improved cooling

Jim noticed after a few trial runs that he was getting an unusual thermal cutout indication on the controller. Before finally isolating the culprit—it was a marginal controller that Curtis quickly replaced under warranty—Jim did a thermal and mechanical redesign on the entire control electronics area. Figure 11-22 shows the result.

Figure 11-21A shows Jim's "magic box" redesign. In this incarnation, everything is accessible from the outside, the two power cables bolt on, the two instrumentation cables plug in, and the two fuses are accessible from outside the case. Figure 11-21A shows a front view of the thermally redesigned controller mounting. In this design the controller

**11-21A** Jim Harris's redesigned "magic box" and controller mounting for 1987 Ford Ranger.

**11-21B** "Magic box" and controller, driver-side view.

is flipped over and a heat sink is mounted to its bottom (silicone thermal lubricant was spread over the area for the highest heat dissipation efficiency). The combined package was then centrally mounted in the location of maximum engine compartment airflow. Figure 11-21B shows a driver-side view: charger (in the original location), newly inverted controller with heat sink on top, and revision two of Jim's magic box.

When the replacement controller from Curtis was plugged into the new layout, everything worked fine with no more thermal cutout.

### Further improved cooling

The eternal mechanic in Jim was still not satisfied—he wanted to have controller cooling independent of the vagaries of engine compartment airflow. So he added the fan and directional cooling shroud to the top of the heat sink (FIG. 11-22). He also widened the mounting space between his magic box and the controller (everything is now mounted on the common heat sink aluminum plate), and relocated the battery charger to the passenger side fender well—all to maximize cooling and airflow.

### Paint, polish, and sign

After everything is running the way you like, it's time to embellish the outside of your EV conversion. Why? The paint and the polish goes without saying—you are proud of your work and want to show it off in its best light. Jim's original paint was pretty good, so he opted to leave it alone and apply only a basic wax job to it. But without an outside *ELECTRIC* sign, the only way you can tell Jim's vehicle is electric-powered is to look closely through its front grille, where the dc motor is visible just above the license plate (although the silence would definitely make you suspicious).

Because few people (if any) are likely to examine your EV conversion closely, you have to advertise instead. A couple of well-placed large letters is all it takes—or an

**11-22** Jim Harris' improved "magic box," fan-cooled controller and battery charger mounting for 1987 Ford Ranger.

electric vehicle license tag. Jim opted to place his letters on both front fenders and on the tail gate as shown in FIG. 11-23. It makes a world of difference. Be sure to carry plenty of literature because people will talk to you whether moving (neighborhood or freeway) or stopped (with the hood up or down). While you can draw a crowd whenever you visit the local convenience store—and certainly will at a gas station—you can really impress your date while waving to the folks on the freeway as you pass them at 75 mph and watch their mouths drop open.

## Onward and upward

It's easier the second time around with EV conversions. Contrast the visual, practical, and functional appeal of Jim's second EV conversion—the 1993 Ford Ranger

**11-23** 1987 Ford Ranger with "Electric Vehicle" lettering.

pickup shown in FIG. 11-24 with that of his first. He learned a lot and put it all to use. You will too.

**11-24** 1993 Ford Ranger "magic box."

Notice the evolution of Jim's "magic box" for the 1993 Ford Ranger. Everything—controller, junction box, heatsinks, and fans—is packaged in a protective metal enclosure that is now centrally mounted above the motor in the engine compartment. With everything in one box, you just have to connect a few wires—the conversion is even easier to do and is much more reliable an operation.

## Put yourself in the picture

Now that you've seen how it's done, you can do it yourself. It's a very simple project that virtually anyone can accomplish today—just by asking for help in a few appropriate places and taking advantage of today's kit components and pre-built conversion packages. Imagine yourself in FIG. 11-25, and take the steps to make it happen.

**11-25** Put yourself in this picture—do an EV conversion of your own.

# 12

# Maximize your electric vehicle enjoyment

*Now that you are driving for only pennies a day, it takes little more to make your pleasure complete.*

Once you've driven your EV conversion around the block for the first time, it's time to start planning for the future. You need to license and insure it so you can drive it farther than just around the block. You also need to learn how to drive and care for it to maximize your driving pleasure and its economy and longevity.

## LICENSING & INSURANCE OVERVIEW

No one anywhere would seriously question licensing and insuring Jim Harris' 1993 Ford Ranger pickup EV conversion you met in chapter 11. It goes 75 mph and Ford has already guaranteed its chassis compliance with FMVSS and NHTSA safety standards. On the other hand, many jurisdictions balked at licensing and insuring the home-built and/or commercially produced golf-cart style EVs of the 1970s that could barely reach 45 mph and did raise issues of safety.

Which end of the spectrum your EV conversion of today resembles directly determines its ability to clear any licensing and insuring hurdles. Let's take a closer look at each area.

### Getting licensed

Any vehicle's license falls under the jurisdiction of the state in which it resides. All state motor vehicle codes, although based on common federal standards, are just a little bit different. While your internal combustion engine vehicle conversion chassis might be fully compliant with federal FMVSS and NHTSA safety standards, you need to check out what your state's motor vehicle code says. If you're doing a from-the-ground-up EV, you'd be well-advised to check out these rules and regulations in advance.

In general, few states have specific EV regulations. You'll find that states with larger vehicle populations, such as California, New York, and Florida, are on the leading edge in terms of establishing guidelines for EVs. Check with your own state's motor vehicle department to be sure.

As for the licensing process, most people are far more involved with the smog-certification or DEQ (Department of Environmental Quality) check than with vehicle compliance of the rules and regulations. Owning an EV conversion short-circuits this process. In fact, you can offer to buy the whole local DEQ inspection team lunch if their meters find any emissions coming out of your EV at all (hybrid EV owners please don't make this offer!)

In Los Angeles, California, you are entitled to a reduction in your electric power rate (from the Los Angeles Department of Water and Power) and tax credit (from the California Energy Commission). Check with your local utility and city and state governments to see if you and your EV are entitled to something similar in your area.

### Getting insured

Insurance is roughly the same as licensing—you're not likely to have any trouble with your EV if it's been converted from an internal combustion engine chassis. Most large national insurance companies write EV insurance coverage without blinking an eye. Verify that your design meets insurance requirements in advance.

### Safety footnote

My basic assumption in this section is that you would put safety high on your list of desirable characteristics for your converted EV. This line of reasoning assumes you leave intact the safety systems of the original internal combustion engine chassis: lights, horn, steering, brakes, parking brake, seat belts, windshield wipers, etc. It also assumes you are thinking "safety first" when installing your new EV components.

In particular, your electric motor control system should be failsafe and have safety interlocks, and your batteries should be mounted for minimal danger to people, other objects, or themselves—both in normal operation and in the event of an accident. Don't even think about going to a vehicle inspection appointment if your from-the-ground-up EV design or your standard EV conversion has left any of these out, or your control and/or battery systems don't emphasize safety first.

## DRIVING & MAINTENANCE OVERVIEW

An EV is easier to drive and requires less maintenance than its internal combustion counterpart. But because its driving and caring requirements are different, you'll need to adjust your acquired internal combustion engine vehicle habit patterns. The driving part is very similar to the experience of a lifelong stick-shift driver who drives an automatic transmission vehicle for the first time. As for the caring part, there's a whole lot less to do, but it has to be done conscientiously. Let's take a closer look at each area.

### Driving your electric vehicle

Your EV conversion may still look like its internal combustion engine ancestor, but it drives very differently. Here's a short list of reminders:

**Starting**  When starting, put it in gear (1st, 2nd etc.), then press the accelerator pedal. No need to use the clutch on startup because the motor's not turning when your foot is off the pedal. On the other hand, a very definite need to have it in gear because you always want to start your series dc motor with a load on it so it doesn't runaway to high RPM and hurt itself. If you forget and accidentally leave the clutch in (or the transmission in neutral), back off immediately when you hear the electric motor winding up.

**Shifting**  Don't feel you need to shift gears. If you do city driving, you'll wind up mostly using the first two gears. The lower the gear, the better your range, so use the lowest gear possible at any given speed.

**Economical driving**  If you keep an eye on your ammeter while driving, you'll soon learn the most economical way to drive, shift gears, and brake. For maximum range, the objective is to use the least current at all times. You'll immediately notice the difference in drag racing and going up hills—either alter your driving habits or plan on recharging more frequently.

**Coasting**  If you don't have regeneration, coasting in an EV is unlike anything you've ever encountered in your internal combustion engine vehicle—there's no engine compression to slow you down. By definition, an EV is designed to be as frictionless as possible, so take advantage of this characteristic. Learn to pulse your accelerator and coast to the next light or to the vehicle ahead of you in traffic.

**Regeneration**  If you do have regeneration and can adjust it, use it at highway speeds (and on long downgrades, etc.) but disengage it in stop and go traffic in favor of the pulse and coast method.

**Determining range**  Use the voltmeter as a fuel gauge in conjunction with the odometer to tell you how far your batteries can take you. Tape a note to your dashboard of the elapsed mileage between full charges and voltmeter reading (or use a notebook)—the voltmeter reading was $x$ after you drove $y$ miles—and you'll quickly get an idea of the pattern. Keep in mind that your battery pack will not reach its peak range until you've deep-cycled it a few times.

**Running out of power**  If you totally run out of power and cannot find an electric outlet, turn off your key switch and allow your batteries to rest for 20 to 30 minutes (shut down everything else electrical at this time also). Amazingly, you'll find extra energy in the batteries that may just be enough to take you to the power outlet you need. The convenience of an onboard charger is welcomed most of all in this particular circumstance.

**Regular driving**  Drive your EV regularly—several times a week. Remember, the chemical clock inside your lead-acid batteries is ticking whether you use them or not, so use them. Better yet, think of how much money you are saving by using your EV. The 20 6-volt 220-AH capacity batteries gave Jim Harris' 1987 Ford Ranger conversion of chapter 11 an on-board capacity of

$$Capacity = 220 \text{ AH} \times 120 \text{ volts} = 26.6 \text{ kWh}$$

If Jim got a 60-mile range out of one charge, his average energy use would be

$$Average \text{ energy} = (26.6 \text{ kWh})/(60 \text{ mi}) = 0.44 \text{ kWh/mi}$$

At an average of $0.05 per kWh, that works out to be

$$Cost \text{ per mile} = 0.44 \text{ kWh/mi} \times \$0.05 \text{ per kWh} = 2.2 \text{ cents per mile}$$

If you compare that with gasoline at $1.25 per gallon and a typical 20 mpg for the internal combustion engine pickup before conversion, that works out to

$$Cost \text{ per mile} = 1/20 \text{ gal/mi} \times \$1.25 \text{ per gal} = 6.25 \text{ cents per mile}$$

That's a three-to-one savings—take advantage of it.

## Caring for your electric vehicle

Now that you are driving for only pennies a day, it takes little more to make your pleasure complete. Actually, a properly designed and built EV conversion requires surprisingly little attention compared to an internal combustion engine vehicle. It comes down to the care and feeding of your batteries, minimizing friction, and preventative maintenance.

**Battery care** Of course you are going to be charging your batteries on a regular basis—using the guidelines of chapter 9—so battery maintenance really comes down to periodically checking to see that your batteries are properly watered. Jim Harris prefers to use the U.S. Batteries "quick disconnect" caps. All three caps on the 6-volt battery are ganged together, and a quick flip releases them all—it makes watering a breeze. Use steam distilled water only—not the water flowing from your faucet tap that could be heavily mineralized. Observe the condition of the battery tops when watering; any acid overspray will cling to the battery tops and attract dirt. Clean this off immediately after replacing all the battery caps with a solution of water and baking soda as described in chapter 9. Also check the battery terminals for corrosion, and correct any deficiencies using the baking soda solution.

**Tire care** Tire rolling resistance is a main contributor to friction so switch to a low rolling resistance tire (if possible) and frequently check that your tires are properly inflated using a more accurate meter-style tire gauge. Proper inflation means 32 psi and up—EV tires should be inflated hard. Talk to your local tire specialist about inflation limits versus loading. Also listen to be sure that no brake shoes or pads are dragging, etc.

**Lubricants** The weight of viscosity of your drivetrain fluids—transmission and rear axle lubricant—also contributes to friction losses on an ongoing basis, so experiment with light weight lubricant in both these areas. The EV conversion puts a much smaller design load on the mechanical drivetrain so you ought to be able to drop down to a 50 weight lubricant for the rear axle and a lower-loss transmission fluid grade.

**Checking connections** Preventative maintenance mostly involves periodically checking the high-current wiring connections for tightness. Use your hands here. Warmth is bad—it means a loose connection—and anything that moves when you pull it is also bad. A few open end and box wrenches ought to make quick work of your retightening preventive maintenance routine.

**Emergency kit** While carrying extra onboard weight is a no-no, carrying a small highway kit ought to be just enough to give you on-the-road peace of mind, knowing you have planned for most contingencies. As a minimum, your kit should have: a small fire extinguisher, a small bottle of baking soda solution, a small toolkit (wrenches, flat and wire cutting pliers, screwdrivers, wire and tape), and a heavy-duty charger extension cord with multiple adapter plugs (male and female).

# References

## Preface

1. Aubrey Pilgrim's *Build Your Own 80486 PC and Save a Bundle*, *Upgrade Your IBM Compatible and Save a Bundle* and earlier books in the series cover the DOS PC area; my *Build Your Own Macintosh and Save a Bundle—2nd Edition* and *Upgrade Your Macintosh and Save a Bundle* books cover the Macintosh side.

## Chapter 1

1. Michael Parrish, "Quest for a Local Electric-Car Industry to Begin," *Los Angeles Times*, 8 June 1992, p. D2.

2. David H. Freedman, "Batteries Included," *Discover*, March 1992, p. 90.

3. Quanlu Wang and Mark A. DeLuchi, "Impacts of Electric Vehicles on Primary Energy Consumption and Petroleum Displacement," Institute of Transportation Studies, UC Davis, reprinted in *Energy*, 1991, vol. 17-4, p. 355.

4. Kim Reynolds, "AC Propulsion CRX: Harbinger of Things Electric," *Road & Track*, October 1992, p. 126.

5. U.S. Dept of Transportation, *1990 National Personal Transportation Survey*, p. 24.

6. H.J. Schwartz, "Computer Simulation of Automobile Use Patterns," 4th International EV Symposium, 1976.

7. John R. Dabels, "Is The Electric Vehicle For You?," *GM Electric Vehicles*, Winter 1992, p. 12-14.

## Chapter 2

1. Darwin Gross, "Your Right To Choose," Chapter 11—A Look At The Sun, SOS Publishing, 1986.

2. Vice President Al Gore, *Earth in the Balance—Ecology and the Human Spirit*, Houghton Mifflin, 1992, p. 40.

3. Energy Information Administration, *Annual Energy Review 1991*, Department of Energy. This is the source for all the data in this section unless otherwise noted.

4. Energy Information Administration, *1992 Annual Energy Outlook, with Projections to 2010*, Department of Energy.

5. Ross Perot, "United We Stand," Hyperion, 1992, p.50 (and material from TV ads).

6. Philip DiLavore, *Energy: Insights from Physics*, John Wiley & Sons, 1984.

7. Quanlu Wang and Mark A. DeLuchi, "Impacts of Electric Vehicles on Primary Energy Consumption and Petroleum Displacement," Institute of Transportation Studies, UC Davis, 1991, reprinted in *Energy*, vol. 17-4, p. 362.

8. United Nations Environmental Summit, Nairobi, Africa, 1991.

9. Gary E. Maciel, Daniel D. Traficante, and David Lavallee, *Chemistry*, D.C. Heath and Company, 1978.

10. Mark A. DeLuchi, "Emissions of Greenhouse Gases from the Use of Transportation Fuels and Electricity," Argonne National Laboratory, 1991, amended by 4/22/92 memo.

11. *Facts & Figures '91*, Motor Vehicle Manufacturer's Association, 1991, p. 21.

12. "Draft 1988 Air Quality Management Plan," South Coast Air Quality Management District, 1989.

13. "National Air Pollutant Emission Estimates, 1940-1987," Environmental Protection Agency, 1989.

14. File number 88-0829, May 6, 1988, City Clerk, Room 395, City of Los Angeles.

15. *Facts & Figures '91*, Motor Vehicle Manufacturer's Association, 1991.

16. *Facts & Figures '91.*

17. "Draft 1988 Air Quality Management Plan."

18. *Facts & Figures '91.*

19. Quanlu Wang, Mark A. DeLuchi, and Daniel Sperling, Institute of Transportation Studies, UC Davis, "Emission Impacts of Electric Vehicle," *Air Waste Management Association*, September 1990, vol. 40-9, p. 1275.

20. Clean Air Act amendments of 1990, Public Law 101-549, November 15, 1990.

21. "Annual Energy Review 1991," Energy Information Administration, Department of Energy, p. 211.

## Chapter 3

1. A. Nevins, *Ford: The Times, The Man, The Company*, Charles Scribner's Sons, 1954.

2. Robert Lacy, *Ford, The Men and the Machine*, Little, Brown and Company, 1986.

3. Philip Van Doren Stern, *Tin Lizzie—Story of the Fabulous Model T Ford*, Simon and Schuster, 1955.

4. Daniel Yergin, *The Prize: The Epic Quest for Oil, Money and Power*, Simon and Schuster, 1991.

5. T. R. Nicholson, *Passenger Cars 1863-1904*, The Macmillan Company, 1970.

6. L. Scott Bailey, "GM: The First 75 Years of Transportation Products," General Motors, 1983.

7. George H. Dammann, *Illustrated History of Ford, 1903-1970*, Crestline Publishing, 1970.

8. Ray Miller, *Chevrolet: The Coming of Age 1911-1942*, Evergreen Press, 1976.

9. Jonathan Kwitny, "The Great Transportation Conspiracy," *Harper's*, February 1981.

10. Len Frank and Dan McCosh, "Electric Vehicles Only," *Popular Science*, May 1991.

11. Don Sherman, "Speed Picnic," *Motor Trend*, December 1992.

12. Dan McCosh, Automotive Newsfront, "Electric Record," *Popular Science*, July 1992.

13. "H.R. 5470 + Hearings = H.R. 8800," *Electric Vehicle News*, August 1975.

14. Ron Cogan, "Electric Cars: The Silence of the Cams," *Motor Trend*, August 1991.

15. Hans Rudiger Etzold, *The Beetle: The Chronicles of the People's Car*, Haynes Publishing Group, 1990.

16. Marco Ruiz, *The Complete History of the Japanese Car: 1907 to Present*, Crown Publishers, 1986.

17. David H. Freedman, "Batteries Included," *Discover*, March 1992.

18. Kathy Jackson, "Ford upgrades its electric vehicle project," *Automotive News*, July 20, 1992.

19. Dennis Simanaitis, "Ford Ecostar Electric," *Automobile*, February, 1993.

20. Kim Reynolds, "AC Propulsion CRX: Harbinger of Things Electric," *Road & Track*, October, 1992.

21. Luca Ciferri, "Ample Performance," (Cover Story), *AutoWeek*, September 7, 1992.

## Chapter 4

1. William R. Diem, "Ford's First—Michigan outlet named to service Ecostar," *Automotive News*, December 21, 1992, p. 6.

2. Jack Keebler, "Ford VP Believes Big 3 Should Share Electric Technology," *Automotive News*, January 18, 1993, p. 20.

3. Raymond Stein, "GM learning how to plug electric cars," *Automotive News*, January 18, 1993, p. 20.

4. Kristine Stiven Breese, "Entrepreneur wrestles with electric market," *Automotive News*, December 14, 1992, p. 26.

## Chapter 6

1. U. Adler, *Automotive Handbook—2nd Edition*, Robert Bosch GmbH, 1986.

2. Thomas D. Gillespie, *Fundamentals of Vehicle Dynamics*, Society of Automotive Engineers, 1992.

3. D.W. Kurtz, "Aerodynamic Design of Electric and Hybrid Vehicles," JPL Publication 80-69, 1980.

4. *1991 Yearbook*, Tire and Rim Association, Copley, Ohio, 1991.

5. Leslie R. Sachs and James R. Bennett, *Cheap Wheels*, Pocket Books, 1989.

## Chapter 8

1. Irving M. Gottlieb, *Electric Motors and Control Techniques*, Tab Books Inc., 1982.

2. S. K. Datta, *Power Electronics and Controls*, Reston Publishing Inc., 1985.

3. Gopal. K. Dubey, *Power Semiconductor Controlled Drives*, Prentice-Hall Inc., 1989.

4. Richard Valentine et al, "Semiconductor Technologies for Electric Vehicles," SAE 92C038, presented at Convergence 92, Detroit, October 1992.

5. *General Electric SCR Manual*, 4th Edition, 1967, p. 237-243.

6. *Motorola Circuits Manual*, 1965, p. 6-8.

7. Richard Valentine, *DC Motor Control For Electric Vehicles*, Motorola, 1993.

8. "Using the MC68332 Microcontroller for AC Induction Motor Control," Motorola Application Note AN 1310.

9. Kim Reynolds, "AC Propulsion CRX - Harbinger of Things Electric," *Road and Track*, October 1992, p. 126.

10. Lawrence A. Berardinis, "Clear Thinking On Fuzzy Logic," *Machine Design*, 4/23/92, p. 46.

11. "DSP card give golf simulator power of virtual reality," *Personal Instrumentation and Engineering News*, March 1993, p. 62.


## Chapter 9

1. Richard A. Perez, *The Complete Battery Book*, Tab Books Inc., 1985.

2. David Linden, *Handbook of Batteries and Fuel Cells*, McGraw-Hill Publishing Co., 1984.

3. Matthew A. Dziechiuch, "International Battery Technology Overview," Ford Motor Co., Convergence 1992 Paper 92C036, October, 1992.


## Chapter 10

1. Shoji Tange and Masato Fukino (Nissan Motor Co.), Takahiko Yamamoto (Tokyo Electric Power Co.), "Feasibility Study of Super Quick Charging System," Convergence 1992 Paper 92C039, October, 1992.

2. "A Quicker Charge for Electric Cars," *New York Times*, July 21, 1992, v 141, C3.

3. James Graham, "Electric Cars? Cut the Cord," *New York Times*, September 26, 1991, v 141, A19.

4. Michael Parrish, "Edison, Air Agency Will Plug Solar Carport," *Los Angeles Times*, v 110, D2; Emily T. Smith, "How to Fill 'er Up with California Sunshine," in "Developments to Watch," *Business Week*, November 25, 1991, p. 232.

5. Robert K. Arnold, "How Urban America Can Drive Itself Out of Commuter Gridlock—Imagine Electric Cars Tying Every Neighborhood and Job to Rapid Transit," *Washington Post*, February 2, 92, v 115, C2; "Electric Cars Draw Power from Roadway," *Design News*, October 1, 1990, v 46, p. 41; and Art Spinella, "A Zap from the Road Propels Electric Cars," *Ward's Auto World*, May 1990, p. 87.


## Chapter 11

1. Peter Coxhead and Martin Foster, *The Kitcar Builder's Manual*, Haynes Publications Inc., 1990.

# Index